Fundamentals of Bayesian Epistemology 2

Fundamentals of Bayesian Epistemology 2

Arguments, Challenges, Alternatives

MICHAEL G. TITELBAUM

OXFORD
UNIVERSITY PRESS

Great Clarendon Street, Oxford, OX2 6DP,
United Kingdom

Oxford University Press is a department of the University of Oxford.
It furthers the University's objective of excellence in research, scholarship,
and education by publishing worldwide. Oxford is a registered trade mark of
Oxford University Press in the UK and in certain other countries

First Edition published in 2022

Published in the United States of America by Oxford University Press
198 Madison Avenue, New York, NY 10016, United States of America

British Library Cataloguing in Publication Data
Data available

Library of Congress Control Number: 2021949533

ISBN 978–0–19–286314–0 (hbk.)
ISBN 978–0–19–286315–7 (pbk.)

DOI: 10.1093/oso/9780192863140.001.0001

Printed and bound by
CPI Group (UK) Ltd, Croydon, CR0 4YY

Contents

VOLUME 1

I. OUR SUBJECT

II. THE BAYESIAN FORMALISM

VOLUME 2

III. APPLICATIONS

IV. ARGUMENTS FOR BAYESIANISM

V. CHALLENGES AND OBJECTIONS

Quick Reference

Non-Negativity: For any P in $\mathcal{L}$, $cr(P) \geq 0$.

Normality: For any tautology T in $\mathcal{L}$, $cr(\mathsf{T}) = 1$.

Finite Additivity: For any mutually exclusive P and Q in $\mathcal{L}$, $cr(P \lor Q) = cr(P) + cr(Q)$.

Ratio Formula: For any P and Q in $\mathcal{L}$, if $cr(Q) > 0$ then $cr(P \mid Q) = \frac{cr(P \& Q)}{cr(Q)}$.

Conditionalization: For any time t_i and later time t_j, if E in $\mathcal{L}$ represents everything the agent learns between t_i and t_j, and $cr_i(E) > 0$, then for any H in $\mathcal{L}$, $cr_j(H) = cr_i(H \mid E)$.

CONSEQUENCES OF THESE RULES

Negation: For any P in $\mathcal{L}$, $cr(\sim P) = 1 - cr(P)$.

Maximality: For any P in $\mathcal{L}$, $cr(P) \leq 1$.

Contradiction: For any contradiction F in $\mathcal{L}$, $cr(\mathsf{F}) = 0$.

Entailment: For any P and Q in $\mathcal{L}$, if $P \vDash Q$ then $cr(P) \leq cr(Q)$.

Equivalence: For any P and Q in $\mathcal{L}$, if $P =\!\!\vDash Q$ then $cr(P) = cr(Q)$.

General Additivity: For any P and Q in $\mathcal{L}$, $cr(P \lor Q) = cr(P) + cr(Q) - cr(P \& Q)$.

Finite Additivity (Extended): For any finite set of mutually exclusive propositions $\{P_1, P_2, \ldots, P_n\}$, $cr(P_1 \lor P_2 \lor \ldots \lor P_n) = cr(P_1) + cr(P_2) + \ldots + cr(P_n)$.

Decomposition: For any P and Q in $\mathcal{L}$, $cr(P) = cr(P \& Q) + cr(P \& \sim Q)$.

Partition: For any finite partition of propositions in $\mathcal{L}$, the sum of their unconditional cr-values is 1.

Law of Total Probability: For any proposition P and finite partition $\{Q_1, Q_2, \ldots, Q_n\}$ in $\mathcal{L}$,

$$cr(P) = cr(P \mid Q_1) \cdot cr(Q_1) + cr(P \mid Q_2) \cdot cr(Q_2) +$$
$$\ldots + cr(P \mid Q_n) \cdot cr(Q_n).$$

Bayes's Theorem: For any H and E in $\mathcal{L}$,

$$cr(H \mid E) = \frac{cr(E \mid H) \cdot cr(H)}{cr(E)}.$$

Multiplication: P and Q with nonextreme cr-values are independent relative to cr if and only if $cr(P \& Q) = cr(P) \cdot cr(Q)$.

PART III

APPLICATIONS

Volume 1 of this book explained why we might want to theorize about degrees of belief, then laid out the five core normative rules of Bayesian epistemology: Kolmogorov's three probability axioms, the Ratio Formula for conditional credences, and updating by Conditionalization. We also (in Chapter 5) investigated some further norms that might be adopted in addition to—or in place of—the core rules.

I'm sure as you worked through that material, a number of objections to and problems for those norms occurred to you. Hopefully many of those will be addressed when we get to Part V of the book. First, though, we will ask why Bayesian epistemologists believe in the norms to begin with. Part IV considers explicit, premise-conclusion style philosophical arguments for the Bayesian rules. But as I see it, what actually convinced most practioners to adopt Bayesian epistemology—to accept that agents can be usefully represented as assigning numerical degrees of belief, and that rationality requires those degrees of belief to satisfy certain mathematical constraints—were the applications in which this approach found success.

Our discussion has already covered some minor successes of Bayesian epistemology. For example, while a purely binary doxastic view has trouble furnishing agents with a rational, plausible set of attitudes to adopt in the Lottery Paradox (Section 1.1.2), Bayesian epistemology has no trouble sketching a set of lottery credences that are intuitively appropriate and entirely consistent with Bayesian norms (Section 2.2.2).

Now we are after bigger targets. At one time or another, Bayesianism has been applied to offer positive theories of such central philosophical concepts as explanation, coherence, causation, and information. Yet the two applications most central to the historical development of Bayesian epistemology were confirmation theory and decision theory. As these two subjects grew over the course of the twentieth century and cemented their significance in philosophy (as well as economics and other nearby disciplines), Bayesian epistemology came to be viewed more and more as an indispensible philosophical tool.

Each chapter in this part of the book takes up one of those two applications. Confirmation is tied to a number of central notions in theoretical rationality, such as induction, justification, evidential support, and epistemic reasons. Bayesian epistemology provides the most detailed, substantive, and plausible account of confirmation philosophers have available, not only accounting for the broad contours of the concept but also yielding particular results concerning specific evidential situations. Decision theory, meanwhile, concerns rational action under uncertainty, and so is a central plank of practical rationality and the theory of rational choice. Degrees of belief have been indispensible to decision theory since its inception.

Volumes have been written on each of these subjects, so these two chapters aim merely to introduce you to their historical development, identify some successes achieved, and point to some lingering controversies. More information can be found through the Further Reading sections in each chapter. As for the applications of Bayesian epistemology not covered here, you might start with the book cited below.

Further Reading

Luc Bovens and Stephan Hartmann (2003). *Bayesian Epistemology*. Oxford: Oxford University Press

Discusses applications of Bayesian epistemology to information, coherence, reliability, confirmation, and testimony.

6

Confirmation

When evidence supports a hypothesis, philosophers of science say that the evidence "confirms" that hypothesis. Bayesians place this confirmation relation at the center of their theory of induction. But confirmation is also closely tied to such epistemological notions as justification and reasons. Bayesian epistemology offers a systematic theory of confirmation (and its opposite, disconfirmation) that not only deepens our understanding of this relation but also provides specific answers about which hypotheses are supported (and to what degree) in particular evidential situations.

Since its early days, the analysis of confirmation has been driven by a perceived analogy to deductive entailment. In Chapter 4 we discussed epistemic standards that relate a body of evidence (represented as a proposition) to the doxastic *attitudes* it supports. But confirmation—though intimately linked with epistemic standards in ways we'll presently see—is a different kind of relation: instead of relating a proposition and an attitude, it relates two propositions (evidence and hypothesis). Confirmation shares this feature with deductive entailment. In fact, Rudolf Carnap thought of confirmation as a generalization of traditional logical relations, with deductive entailment and refutation as two extremes of a continuous confirmational scale.

In the late nineteenth and early twentieth centuries, logicians produced ever-more-powerful syntactical theories capable of answering specific questions about which propositions deductively entailed which. Impressed by this progress, theorists such as Carl Hempel and Carnap envisioned a syntactical theory that would do the same for confirmation. As Hempel put it:

> The theoretical problem remains the same: to characterize, in precise and general terms, the conditions under which a body of evidence can be said to confirm, or to disconfirm, a hypothesis of empirical character. (1945a, p. 7)

Hempel identified various formal properties that the confirmation relation might or might not possess. Carnap then argued that we get a confirmation relation with exactly the right formal properties by identifying confirmation with positive probabilistic relevance.

Fundamentals of Bayesian Epistemology 2: Arguments, Challenges, Alternatives. Michael G. Titelbaum,
Oxford University Press. © Michael G. Titelbaum 2022. DOI: 10.1093/oso/9780192863140.003.0006

This chapter begins with Hempel's formal conditions on the confirmation relation. Identifying the right formal conditions for confirmation will not only help us assess various theories of confirmation; it will also help us understand exactly what relation philosophers of science have in mind when they talk about "confirmation".[1] We then move on to Carnap's Objective Bayesian theory of confirmation, which grounds confirmation in probability theory. While Carnap's theory has a number of attractive features, we will also identify two drawbacks: its failure to capture particular patterns of inductive inference that Carnap found appealing, and the language-dependence suggested by Goodman's "grue" problem. We'll respond to these problems with a confirmation theory based on Subjective Bayesianism (in the normative sense).

Confirmation is fairly undemanding, in one sense: we say that evidence confirms a hypothesis when it provides *any* amount of support for that hypothesis, no matter how small. But we might want to distinguish bodies of evidence that *strongly* support hypotheses from those that provide only weak support. Probabilistic theories of confirmation offer a number of different ways to measure the strength of confirmation in any particular case. In Section 6.4.1 we'll survey these different measures of confirmational strength, assessing the pros and cons of each. Finally, we'll apply probabilistic confirmation theory to provide a Bayesian solution to Hempel's Paradox of the Ravens.

6.1 Formal features of the confirmation relation

6.1.1 Confirmation is weird! The Paradox of the Ravens

One way to begin thinking about confirmation is to consider the simplest possible cases in which a piece of evidence confirms a general hypothesis. For example, the proposition that a particular frog is green seems to confirm the hypothesis that all frogs are green. On the other hand, the proposition that a particular frog is not green disconfirms the hypothesis that all frogs are green. (In fact, it *refutes* that hypothesis!) If we think this pattern always holds, we will maintain that confirmation satisfies the following constraint:

Nicod's Criterion: For any predicates F and G and constant a of $\mathcal{L}$, $(\forall x)$ $(Fx \supset Gx)$ is confirmed by $Fa \,\&\, Ga$ and disconfirmed by $Fa \,\&\, {\sim}Ga$.

Hempel (1945a,b) named this condition after Jean Nicod (1930), who built his theory of induction around the criterion. We sometimes summarize the

Nicod Criterion by saying that a universal generalization is confirmed by its **positive instances** and disconfirmed by its **negative instances**. Notice that one can endorse the Nicod Criterion as a sufficient condition for confirmation without taking it to be necessary; we need not think *all* cases of confirmation follow this pattern.

Yet Hempel worries about the Nicod Criterion even as a sufficient condition for confirmation, because of how it interacts with another principle he endorses:

Equivalence Condition (for hypotheses):   Suppose H and H' in $\mathcal{L}$ are logically equivalent ($H \dashv\vDash H'$). Then any E in $\mathcal{L}$ that confirms H also confirms H'.

Hempel endorses the Equivalence Condition because he doesn't want confirmation to depend on the particular way a hypothesis is formulated; logically equivalent hypotheses say the same thing, so they should enter equally into confirmation relations. Hempel is also concerned with how working scientists *use* confirmed hypotheses; for instance, practitioners will often deduce predictions and explanations from confirmed hypotheses. Equivalent hypotheses have identical deductive consequences, and scientists don't hesitate to substitute logical equivalents for each other.

But combining Nicod's Criterion with the Equivalence Condition yields counterintuitive consequences, which Hempel calls the "paradoxes of confirmation". The most famous of these is the **Paradox of the Ravens**. Consider the hypothesis that all ravens are black, representable as $(\forall x)(Rx \supset Bx)$. By Nicod's Criterion this hypothesis is confirmed by the evidence that a particular raven is black, $Ra \,\&\, Ba$. But now consider the evidence that a particular non-raven is non-black, $\sim Ba \,\&\sim Ra$. This is a positive instance of the hypothesis $(\forall x)(\sim Bx \supset \sim Rx)$, so by Nicod's Criterion it confirms that hypothesis. By contraposition, that hypothesis is equivalent to the hypothesis that all ravens are black. So by the Equivalence Condition, $\sim Ba \,\&\, \sim Ra$ confirms $(\forall x)(Rx \supset Bx)$ as well. The hypothesis that all ravens are black is confirmed by the observation of a red herring, or a white shoe. This result seems counterintuitive, to say the least.

Nevertheless, Hempel writes that "the impression of a paradoxical situation . . . is a psychological illusion" (1945a, p. 18); on his view, we reject the confirmational result because we misunderstand what it says. Hempel notes that in everyday life people make confirmation judgments relative to an extensive corpus of background knowledge. For example, a candidate's performance in an interview may confirm that she'd be good for the job, but only relative to a great deal of background information about how the questions asked relate to the job requirements, how interview performance predicts job performance, etc. In assessing confirmation, then, we should always be explicit about the background we're assuming. This is especially important because background knowledge can dramatically alter confirmation relations. For example, in Section 4.3 we discussed a poker game in which you receive the cards that will make up your hand one at a time. At the beginning of the game, your background knowledge contains facts about how a deck is constructed and about which poker hands are winners. At that point the proposition that your last card will be the two of clubs does not confirm the proposition that you will win the hand. But as the game goes along and you're dealt some other twos, your total background knowledge changes such that the proposition that you'll receive the two of clubs now strongly confirms that you'll win.

Nicod's Criterion does not mention background corpora, so it's tempting to apply the criterion with full generality—that is, regardless of which corpus which use. Yet there are clearly some evidential situations in which a positive instance does *not* confirm a universal generalization. For example, suppose I know I'm in the Hall of Atypically Colored Birds. A bird is placed in the Hall only if the majority of his species-mates are one color but he happens to be another color. If my background corpus includes that I'm in the Hall of Atypically Colored Birds, then observing a black raven *disconfirms* the hypothesis that all ravens are black.[2] Because of such examples, Hempel thinks the Nicod Criterion should be applied only against a **tautological background**. A tautological background corpus contains no contingent propositions; it is logically equivalent to a tautology T.

When we intuitively reject the Nicod Criterion's consequence that a red herring confirms the ravens hypothesis, we are sneaking non-tautological information into the background of our judgment. Hempel thinks we're imagining a situation in which we already know in advance (as part of the background) that we will be observing a herring and checking its color. Relative to that background—which includes the information $\sim Ra$—we know that whatever we're about to observe will have no evidential import for the hypothesis that ravens are black. So when we then get the evidence that $\sim Ba$, that evidence is confirmationally inert with respect to the hypothesis $(\forall x)(Rx \supset Bx)$.

But the original question was whether $\sim Ba \mathbin{\&} \sim Ra$ (taken all together, at once) confirmed $(\forall x)(Rx \supset Bx)$. On Hempel's view, this is a fair test of the Nicod Criterion only against an empty background corpus (since that's the background against which he thinks the Criterion applies). And against that corpus, Hempel thinks the confirmational result is correct. Here's a way of understanding why: Imagine you've decided to test the hypothesis that all ravens are black. You will do this by selecting objects from the universe one at a time and checking them for ravenhood and blackness. It's the beginning of the experiment, you haven't checked any objects yet, and you have no background information about the tendency of objects to be ravens and/or black. Moreover, you've found a way to select objects from the entire universe at random, so you have no background information about what kind of object you'll be getting. Nevertheless, you start thinking about what sorts of objects might be selected, and whether they would be good or bad news for the hypothesis. Particularly important would be any ravens that weren't black, since any such negative instance would immediately refute the hypothesis. (Here it helps to realize that the ravens hypothesis is logically equivalent to $\sim(\exists x)(Rx \ \& \ \sim Bx)$.) So when the first object arrives and you see it's a red herring—$\sim Ba \& \sim Ra$—this is good news for the hypothesis (at least, *some* good news). After all, the first object could've been a non-black raven, in which case the hypothesis would've been sunk.

This kind of reasoning defuses the seeming paradoxicality of a red herring's confirming that all ravens are black, and the objection to the Nicod Criterion that results. As long as we're careful not to smuggle in illicit background information, observing a red herring confirms the ravens hypothesis to at least a small degree. Hempel concludes that on his reading of the Nicod Criterion, it is consistent with the Equivalence Condition.

Nevertheless, I.J. Good worries about the Nicod Criterion, even against a tautological background:

> [T]he closest I can get to giving [confirmation relative to a tautological background] a practical significance is to imagine an infinitely intelligent newborn baby having built-in neural circuits enabling him to deal with formal logic, English syntax, and subjective probability. He might now argue, after defining a crow in detail, that it is initially extremely unlikely that there are any crows, and therefore that it is extremely likely that all crows are black. "On the other hand," he goes on to argue, "if there are crows, then there is a reasonable chance that they are of a variety of colors. Therefore, if I were to discover that even a black crow exists I would consider [the hypothesis that all crows are black] to be less probable than it was initially." (1968, p. 157)[3]

Here Good takes advantage of the fact that $(\forall x)(Rx \supset Bx)$ is true if there are no ravens (or crows, in his example).[4] Before taking any samples from the universe, the intelligent newborn might consider four possibilities: there are no ravens; there are ravens but they come in many colors; there are ravens and they're all black; there are ravens and they all share some other color. The first and third of these possibilities would make $(\forall x)(Rx \supset Bx)$ true. When the baby sees a black raven, the first possibility is eliminated; this might be a large enough blow to the ravens hypothesis that the simultaneous elimination of the fourth possibility would fail to compensate. In other words, the observation of a black raven might not confirm that all ravens are black, violating the Nicod Criterion even against a tautological background.

6.1.2 Further adequacy conditions

We have already seen two general conditions (Nicod's Criterion and the Equivalence Condition) that one might take the confirmation relation to satisfy. We will now consider a number of other such conditions, most of them discussed (and given the names we will use) by Hempel. Sorting out which of these are genuine properties of confirmation has a number of purposes. First, Hempel thought the correct list provided a set of adequacy conditions for any positive theory of confirmation. Second, sorting through these conditions will help us understand the abstract features of evidential support. These are features about which epistemologists, philosophers of science, and others (including working scientists and ordinary folk!) often make strong assumptions—many of them incorrect. Finally, we are going to use the word "confirmation" in subsequent sections as a somewhat technical term, distinct from some of the ways "confirm" is used in everyday speech. Working through the properties of the confirmation relation will help illustrate how we're using the term.

The controversy between Hempel and Good leaves it unclear whether the Nicod Criterion should be endorsed as a constraint on confirmation, even when it's restricted to the tautological background. On the other hand, the Equivalence Condition can be embraced in a highly general form:

Equivalence Condition (full version): For any H, H', E, E', K and K' in $\mathcal{L}$, suppose $H \dashv\vDash H'$, $E \dashv\vDash E'$, and $K \dashv\vDash K'$. Then E confirms (/disconfirms) H relative to background K just in case E' confirms (/disconfirms) H' relative to background K'.

Here we can think of K as a conjunction of all the propositions in an agent's background corpus, just as E is often a conjunction of multiple pieces of evidence. This full version of the Equivalence Condition captures the idea that logically equivalent propositions should enter into all the same confirmation relations.

Our next candidate constraint is the

Entailment Condition: For any consistent E, H, and K in $\mathcal{L}$, if $E \,\&\, K \vDash H$ but $K \nvDash H$, then E confirms H relative to K.

This condition captures the idea that entailing a hypothesis is one way to support, or provide evidence for, that hypothesis. If E entails H in light of background corpus K (in other words, if E and K together entail H), then E confirms H relative to K. The only exception to this rule is when K already entails H, in which case the fact that E and K together entail H does not indicate any particular relation between E and H.[5] Notice that a tautological H will be entailed by every K, so the restriction on the Entailment Condition prevents the condition from saying anything about the confirmation of tautologies. Hempel thinks of his adequacy conditions as applying only to empirical hypotheses and bodies of evidence, so he generally restricts them to logically contingent Es and Hs.

Hempel considers a number of adequacy conditions motivated by the following intuition:

Confirmation Transitivity: For any A, B, C, and K in $\mathcal{L}$, if A confirms B relative to K and B confirms C relative to K, then A confirms C relative to K.

It's tempting to believe confirmation is transitive, as well as other nearby notions such as justification or evidential support. This temptation is buttressed by the fact that logical entailment is transitive. Confirmation, however, is not in general transitive. Here's an example of Confirmation Transitivity failure: Suppose our background is the fact that a card has just been selected at random from a standard fifty-two-card deck. Consider these three propositions:

A: The card is a jack.
B: The card is the jack of spades.
C: The card is a spade.

Relative to our background, A would confirm B, at least to some extent. And relative to our background, B would clearly confirm C (in accordance with the

Entailment Condition). But relative to the background that a card was picked from a fair deck, *A* does nothing to support conclusion *C*.

The failure of Confirmation Transitivity has a number of important consequences. First, it explains why in the study of confirmation we take evidence to be propositional rather than objectual. In everyday language we often use "evidence" to refer to objects rather than propositions; police don't store propositions in their Evidence Room. But as possible entrants into confirmation relations, objects have an ambiguity akin to the reference class problem (Section 5.1.1). Should I consider this bird evidence that all ravens are black (against tautological background)? If we describe the bird as a black raven, the answer might be yes. But if we describe it as a black raven found in the Hall of Atypically Colored Birds, the answer seems to be no. Yet a black raven in the Hall of Atypically Colored birds is still a black raven. If confirmation were transitive, knowing that a particular description of an object confirmed a hypothesis would guarantee that more precise descriptions confirmed the hypothesis as well. Logically stronger descriptions (it's a black raven in the Hall of Atypically Colored Birds) entail logically weaker descriptions (it's a black raven) of the same object; by the Entailment Condition, the logically stronger description confirms the logically weaker; so if confirmation were transitive anything confirmed by the weaker description would be confirmed by the stronger as well.

But confirmation isn't transitive, so putting more or less information in our description *of the very same object* can alter what's confirmed. (Black raven? Might confirm ravens hypothesis. Black raven in Hall of Atypically Colored Birds? Disconfirms. Black raven mistakenly placed in the Hall of Atypically Colored Birds when it shouldn't have been? Perhaps confirms again.) We solve this problem by letting propositions rather than objects enter into the confirmation relation. If we state our evidence as a proposition—such as the proposition *that a black raven is in the Hall of Atypically Colored Birds*—there's no question how the objects involved are being described.[6]

Confirmation's lack of transitivity also impacts epistemology more broadly. For instance, it may cause trouble for Richard Feldman's (2007) principle that "evidence of evidence is evidence". Suppose I read in a magazine that anthropologists have reported evidence that Neanderthals cohabitated with *homo sapiens*. I don't actually have the anthropologists' evidence for that hypothesis—the body of information that they think supports it. But the magazine article constitutes evidence that they have such evidence; one might think that the magazine article therefore also constitutes evidence that Neanderthals and *homo sapiens* cohabitated. (After all, reading the article seems to provide me with some justification for that hypothesis.) Yet we cannot adopt this "evidence of evidence is evidence" principle with full generality. Suppose I've randomly picked a card from a standard deck and examined it carefully. If I tell you my card is a jack, you have evidence for the proposition that I know my card is the jack of spades. If I know my card is the jack of spades, I have (very strong) evidence for the proposition that my card is a spade. So your evidence (jack) is evidence that I have evidence (jack of spades) for the conclusion spade. Yet your evidence is no evidence at all that my card is a spade.[7]

Finally, the failure of Confirmation Transitivity shows what's wrong with two confirmation constraints Hempel embraces:

Consequence Condition: If E in $\mathcal{L}$ confirms every member of a set of propositions relative to K and that set jointly entails H' relative to K, then E confirms H' relative to K.

Special Consequence Condition: For any E, H, H', and K in $\mathcal{L}$, if E confirms H relative to K and $H \,\&\, K \vDash H'$, then E confirms H' relative to K.

The Consequence Condition states if a set of propositions together entails a hypothesis (relative to some background), then any evidence that confirms every member of the set also confirms that hypothesis (relative to that background). The Special Consequence Condition says that if a *single* proposition entails some hypothesis, then anything that confirms the proposition also confirms the hypothesis (again, all relative to some background corpus).

The Special Consequence Condition is so-named because it can be derived from the Consequence Condition (by considering singleton sets). Yet each of these conditions is a bad idea, as can be demonstrated by our earlier jack of spades example. Proposition A that the card is a jack confirms proposition B that it's the jack of spades; B entails proposition C that the card is a spade; yet A does not confirm C. We can even create examples in which H entails H' relative to K, but evidence E that confirms H *dis*confirms H' relative to K. Relative to the background corpus K most of us have concerning the kinds of animals people keep as pets, the evidence E that Bob's pet is hairless confirms (at least slightly) the hypothesis H that Bob's pet is a Peruvian Hairless Dog. Yet relative to K that same evidence E disconfirms the hypothesis H' that Bob's pet is a dog.[8]

Why might the Special Consequence Condition seem plausible? It certainly looks tempting if one reads "confirmation" in a particular way. In everyday language it's a fairly strong claim that a hypothesis has been "confirmed"; this suggests our evidence is sufficient for us to accept the hypothesis. (Consider the sentences "That confirmed my suspicion" and "Your reservation has been confirmed.") If we combine that reading of "confirmed" with Glymour's position that "when we accept a hypothesis we commit ourselves to accepting all of its logical consequences" (1980, p. 31), we get that evidence confirming a hypothesis also confirms its logical consequences, as the Special Consequence Condition requires. But hopefully the discussion to this point has indicated that we are not using "confirms" in this fashion. On our use, evidence confirms a hypothesis if it provides *any* amount of support for that hypothesis; the support need not be decisive. For us, evidence can confirm a hypothesis without requiring or even permitting acceptance of the hypothesis. If your only evidence about a card is that it's a jack, this evidence *confirms* in our sense that the card is the jack of spades. But this evidence doesn't authorize you to *accept* or *believe* that the card is the jack of spades.

Another motivation for the Special Consequence Condition—perhaps this was Hempel's motivation—comes from the way we often treat hypotheses in science. Suppose we make a set of atmospheric observations confirming a particular global warming hypothesis. Suppose further that in combination with our background knowledge, the hypothesis entails that average global temperatures will increase by five degrees in the next fifty years. It's very tempting to report that the atmospheric observations support the conclusion that temperatures will rise five degrees in fifty years. Yet that's to unthinkingly apply the Special Consequence Condition.

I hope you're getting the impression that denying Confirmation Transitivity can have serious consequences for the ways we think about everyday and scientific reasoning. Yet it's important to realize that denying the Special Consequence Condition as a *general* principle does not mean that the transitivity it posits *never* holds. It simply means that we need to be careful about assuming confirmation will transmit across an entailment, and perhaps also that we need a precise, positive theory of confirmation to help us understand when it will and when it won't.

Rejecting the Special Consequence Condition does open up some intriguing possibilities in epistemology. Consider these three propositions:

E: I am having a perceptual experience as of a hand before me.
H: I have a hand.
H′: There is a material world.

This kind of evidence figures prominently in G.E. Moore's (1939) proof of the existence of an external world. Yet for some time it was argued that *E* could not possibly be evidence for H. The reasoning was, first, that E could not discriminate between H' and various skeptical hypotheses (such as Descartes's evil demon), and therefore could not provide evidence for H'. Next, H entails H', so if E were evidence for H it would be evidence for H' as well. But this step assumes the Special Consequence Condition. Epistemologists have recently explored positions that allow E to support H without supporting H', by denying Special Consequence.[9]

Hempel's unfortunate endorsement of the Consequence Condition also pushes him toward the

Consistency Condition: For any E and K in $\mathcal{L}$, the set of all hypotheses confirmed by E relative to K is logically consistent with $E \& K$.

In order for the set of all hypotheses confirmed by E to be consistent with $E \& K$, it first has to be a logically consistent set in its own right. So among other things, the Consistency Condition bans a single piece of evidence from confirming two hypotheses that are mutually exclusive with each other.

It seems easy to generate confirmational examples that violate this requirement: evidence that a randomly drawn card is red confirms both the hypothesis that it's a heart and the hypothesis that it's a diamond, but these two confirmed hypotheses are mutually exclusive. Hempel also notes that scientists often find themselves entertaining theoretical alternatives that are mutually exclusive with each other; experimental data eliminates some of those theories while confirming all of the ones that remain. Yet Hempel is trapped into the Consistency Condition by his allegiance to the Consequence Condition. Taken together, the propositions in an inconsistent set entail a contradiction; so any piece of evidence that confirmed all the members of an inconsistent set would also (by the Consequence Condition) confirm a contradiction. Hempel refuses to grant that anything could confirm a contradiction! So he tries to make the Consistency Condition work.[10]

Hempel rightly rejects the

Converse Consequence Condition: For any E, H, H', and K in $\mathcal{L}$ (with H' consistent with K), if E confirms H relative to K and $H' \& K \vDash H$, then E confirms H' relative to K.

The Converse Consequence Condition says that relative to a given background, evidence that confirms a hypothesis also confirms anything that entails that hypothesis.

Here's a counterexample. Suppose our background knowledge is that a fair six-sided die has been rolled, and our propositions are:

E: The roll outcome is prime.
H: The roll outcome is odd.
H′: The roll outcome is 1.

In this case E confirms H relative to our background, H′ entails H, yet E refutes H′. (Recall that 1 is not a prime number!)

Still, there's a good idea in the vicinity of Converse Consequence. Suppose our background consists of the fact that we are going to run a certain experiment. A particular scientific theory, in combination with that background, entails that the experiment will produce a particular result. If this result does in fact occur when the experiment is run, we take that to support the theory. This is an example of the

Converse Entailment Condition: For any consistent E, H, and K in $\mathcal{L}$, if $H \& K \vDash E$ but $K \nvDash E$, then E confirms H relative to K.

Converse Entailment says that if, relative to a given background, a hypothesis entails some evidence, then that evidence confirms that hypothesis relative to that background. (Again, this condition omits cases in which the background K entails the experimental result E all on its own, because such cases need not reveal any connection between H and E.)

Converse Entailment doesn't give rise to examples like the die roll case above (because in that case E is not *entailed* by either H or H′ in combination with K). But because deductive entailment is transitive, Converse Entailment does generate the **problem of irrelevant conjunction**. Consider the following propositions:

E: My pet is a flightless bird.
H: My pet is an ostrich.
H′: My pet is an ostrich and beryllium is a good conductor.

Here H entails E, so by the Converse Entailment Condition E confirms H, which seems reasonable.[11] Yet despite the fact that H′ also entails E (because H′ entails H), it seems worrisome that E would confirm H′. What does my choice in pets indicate about the conductivity of beryllium?

Nothing—and that's completely consistent with the Converse Entailment Condition. Just because E confirms a conjunction one of whose conjuncts concerns beryllium doesn't mean E confirms that beryllium-conjunct all on its own. To assume that it does would be to assume the Special Consequence Condition, which we've rejected. So facts about my pet don't confirm any conclusions that are about beryllium but not about birds. On the other hand, it's reasonable that E would confirm H' to at least some extent, by virtue of eliminating such rival hypotheses as "beryllium is a good conductor and my pet is an iguana."

Rejecting the Special Consequence Condition therefore allows us to accept Converse Entailment. But again, this should make us very careful about how we reason in our everyday lives. A scientific theory, for instance, will often have wide-ranging consequences, and might be thought of as a massive conjunction. When the theory makes a prediction that is borne out by experiment, that experimental result confirms the theory. But it need not confirm the rest of the theory's conjuncts, taken in isolation. In other words, experimental evidence that confirms a theory may not confirm that theory's further predictions.[12]

Finally, we should say something about disconfirmation. Hempel takes the following position:

Disconfirmation Duality: For any E, H, and K in $\mathcal{L}$, E confirms H relative to K just in case E disconfirms $\sim H$ relative to K.

Disconfirmation Duality pairs confirmation of a hypothesis with disconfirmation of its negation. It allows us to immediately convert many of our constraints on confirmation into constraints on disconfirmation. For example, the Entailment Condition now tells us that if E & K deductively refutes H (yet K doesn't refute H all by itself), then E disconfirms H relative to K. (See Exercise 6.2.) We should be careful, though, not to think of confirmation and disconfirmation as exhaustive categories: for many propositions E, H, and K, E will neither confirm nor disconfirm H relative to K.

Figure 6.1 summarizes the formal conditions on confirmation we have accepted and rejected. The task now is to find a positive theory of which bodies of evidence confirm which hypotheses relative to which backgrounds that satisfies the right conditions and avoids the wrong ones.

Name	Brief, somewhat imprecise description	Verdict
Equivalence Condition	equivalent hypotheses, evidence, backgrounds behave same confirmationally	accepted
Entailment Condition	evidence confirms what it entails	accepted
Converse Entailment Condition	a hypothesis is confirmed by what it entails	accepted
Disconfirmation Duality	a hypothesis is confirmed just when its negation is disconfirmed	accepted
Confirmation Transitivity	anything confirmed by a confirmed hypothesis is also confirmed	rejected
Consequence Condition	anything entailed by a set of confirmed hypotheses is also confirmed	rejected
Special Consequence Condition	anything entailed by a confirmed hypothesis is also confirmed	rejected
Consistency Condition	all confirmed hypotheses are consistent	rejected
Converse Consequence Condition	anything that entails a confirmed hypothesis is also confirmed	rejected
Nicod's Criterion	Fa & Ga confirms $(\forall x)(Fx \supset Gx)$	???

Figure 6.1 Accepted and rejected conditions on confirmation

6.2 Carnap's theory of confirmation

6.2.1 Confirmation as relevance

Carnap saw that we could get a confirmation theory with exactly the properties we want by basing it on probability. Begin by taking *any* probabilistic distribution Pr over $\mathcal{L}$. (I've named it "Pr" because we aren't committed at this stage to its being any *kind* of probability in particular—much less a credence distribution. All we know is that it's a distribution over the propositions in $\mathcal{L}$ satisfying the Kolmogorov axioms.) Define Pr's background corpus K as the conjunction of all propositions X in $\mathcal{L}$ such that $\Pr(X) = 1$.[13] Given an E and H in $\mathcal{L}$, we apply the Ratio Formula to calculate $\Pr(H \mid E)$. Two distinct theories of confirmation now suggest themselves: (1) E confirms H relative to K just in case $\Pr(H \mid E)$ is high; (2) E confirms H relative to K just in case $\Pr(H \mid E) > \Pr(H)$. In the preface to the second edition of his *Logical Foundations of Probability*, Carnap calls the first of these options a "**firmness**" concept of confirmation and the second an "**increase in firmness**" concept (1962a, p. xvff).[14]

The firmness concept of confirmation has a number of problems. First, there are questions about where exactly the threshold for a "high" value of $\Pr(H \mid E)$ falls, what determines that threshold, how we discover it, etc. Second, there will be cases in which E is irrelevant to H—or even negatively relevant to H!—yet $\Pr(H \,|\, E)$ is high because $\Pr(H)$ was already high. For example, take the background K that a fair lottery with a million tickets has been held, the hypothesis H that ticket 942 did not win, and the evidence E that elephants have trunks. In this example $\Pr(H \,|\, E)$ may very well be high, but that need not be due to any confirmation of lottery results by the endowments of elephants. Finally, the firmness concept doesn't match the confirmation conditions we approved in the previous section. Wherever the threshold for "high" is set, whenever E confirms H relative to K it will also confirm any H' entailed by H. As a probability distribution, Pr must satisfy the Entailment rule and its extension to conditional probabilities (see Section 3.1.2), so if $H \vDash H'$ then $\Pr(H' \,|\, E) \geq \Pr(H \,|\, E)$. If $\Pr(H \,|\, E)$ surpasses the threshold, $\Pr(H' \,|\, E)$ will as well. But that means the firmness concept of confirmation satisfies the Special Consequence Condition, to which we've already seen counterexamples.

Warning

Conflating firmness and increase in firmness, or just blithely assuming the firmness concept is correct, is one of the most frequent mistakes made in the confirmation literature and more generally in discussions of evidential support.[15] For example, it is often claimed that an agent's evidence supports or justifies a conclusion just in case the conclusion is probable on that evidence. But for conclusions with a high prior, the conclusion may be probable on the evidence not because of anything the evidence is doing, but instead because the conclusion was probable all along. Then it's not *the evidence* that's justifying anything!

Increase in firmness has none of these disadvantages; it is the concept of confirmation we'll work with going forward. Given a probability distribution Pr with background K (as defined above), E confirms H relative to K just in case $\Pr(H \,|\, E) > \Pr(H)$. In other words, given Pr, evidence E confirms H relative to K just in case E is *positively relevant* to H. We identify disconfirmation with *negative relevance*: Given Pr, E disconfirms H relative to K just in case $\Pr(H \,|\, E) < \Pr(H)$. If $\Pr(H \,|\, E) = \Pr(H)$, then E is irrelevant to H and neither confirms nor disconfirms it relative to K.

This account of confirmation meets exactly those conditions we endorsed in the previous section: Disconfirmation Duality and the Equivalence,

Entailment, and Converse Entailment Conditions. Disconfirmation Duality follows immediately from our definitions of positive and negative relevance. The Equivalence Condition follows from the Equivalence rule for probability distributions; logically equivalent propositions will always receive identical Pr-values. We get the Entailment Condition because if E, H, and K are consistent, $E \& K \models H$, but $K \not\models H$, then $\Pr(H \mid E) = 1$ while $\Pr(H) < 1$. (You'll prove this in Exercise 6.4.) The key result for Converse Entailment was established in Exercise 4.4. Identifying confirmation with positive relevance yields an account of confirmation with the general contours we want, prior to our committing to anything about the specific numerical values of Pr.

6.2.2 Finding the right function

Yet Carnap wants more than the general contours of confirmation—he wants a substantive theory that says in every case which bodies of evidence support which hypotheses relative to which backgrounds. A theory like that seems obtainable to Carnap because he sees confirmation as a *logical* relation. As with other logical relations, whether E confirms H relative to K is independent of the truth-values of those propositions and of particular attitudes individuals adopt toward them. Like Hempel, Carnap thinks confirmation relations emerge from the logical form of propositions, and therefore can be captured by a syntactical theory working with strings of symbols representing logical forms. (Nicod's Criterion is a good example of a confirmation principle that works with logical form.) Enormous progress in formal deductive logic in the decades just before *Logical Foundations* makes Carnap confident that a formalism for inductive logic is within reach.

To construct the formalism Carnap wants, we begin with a formal language $\mathcal{L}$.[16] We then take each consistent corpus in that language (represented as a non-contradictory, conjunctive proposition K) and associate it with a particular Pr distribution over $\mathcal{L}$. That done, we can test whether evidence E confirms hypothesis H relative to a particular K by seeing whether E is positively relevant to H on the Pr associated with that K.

The crucial step for Carnap is the association of each K with a unique distribution Pr. A full Pr distribution must be specified for each K so that for any E, H, and K we might select in $\mathcal{L}$, there will be a definite answer to the question of whether E confirms, disconfirms, or is irrelevant to H on K. (Just as there's always a definite answer as to whether a given P deductively entails a given Q, refutes it, or neither.) The Pr associated with a given K must assign

1 to each conjunct of that K, but this leaves a lot of latitude with respect to the Pr-values of other propositions. And it's important to get the *right* Pr for each K; the wrong Pr distribution could make evidential support counterinductive, or could have everyday evidence confirming skeptical hypotheses.

Even in a language $\mathcal{L}$ with finitely many atomic propositions, there will typically be many, many possible non-contradictory background corpora K. Specifying a Pr-distribution for each such K could be a great deal of trouble. Carnap simplifies the process by constructing every such Pr from a single, regular probability distribution over $\mathcal{L}$ that he calls $\mathbf{m}$. As a regular probability distribution, $\mathbf{m}$ reflects no contingent truths. (In other words, $\mathbf{m}$ has a tautological background corpus.) To obtain the $\text{Pr}(\cdot)$ distribution relative to some substantive, non-contradictory K, we calculate $\mathbf{m}(\cdot \mid K)$. (This guarantees that $\text{Pr}(K) = 1$.) Evidence E confirms hypothesis H relative to K just in case $\text{Pr}(H \mid E) > \text{Pr}(H)$, which is equivalent to $\mathbf{m}(H \mid E \& K) > \mathbf{m}(H \mid K)$. So instead of working with particular Pr-distributions, we can focus our attention on $\mathbf{m}$.[17]

$\mathbf{m}$ also fulfills a number of other roles for Carnap. Carnap thinks of an agent's background corpus at a given time as her total evidence at that time. If an agent's total evidence is K, Carnap thinks $\mathbf{m}(H \mid K)$ provides the logical probability of H on her total evidence. Moreover, Carnap thinks that logical probabilities dictate rational credences. A rational agent with total evidence K will assign credence $\text{cr}(H) = \mathbf{m}(H \mid K)$ for any H in $\mathcal{L}$. Since $\mathbf{m}$ is a particular, unique distribution, this means there is a unique credence any agent is required to assign each proposition H given body of total evidence K. So Carnap endorses the Uniqueness Thesis (Section 5.1.2), with $\mathbf{m}$ playing the role of the uniquely rational hypothetical prior distribution. On Carnap's view, logic provides the unique correct epistemic standards that all rational agents should apply, represented numerically by the distribution $\mathbf{m}$. Carnap is thus an Objective Bayesian in both senses of the term: in the normative sense, because he thinks there's a unique rational hypothetical prior; and in the semantic sense, because he defines "probability" as an objective concept independent of agents' particular attitudes.[18]

$\mathbf{m}$ allows us to separate out two questions that are sometimes run together in the confirmation literature. Up until now we have been asking whether evidence E confirms hypothesis H relative to background corpus K. For Carnap this question can be read: For a rational agent with total evidence K, would some *further* piece of evidence E be positively relevant to H? Carnap answers this question by checking whether $\mathbf{m}(H \mid E \& K) > \mathbf{m}(H \mid K)$. But we might also ask about confirmational relations involving K itself. Bayesians sometimes ask how an agent's total evidence bears on a hypothesis—does the sum total of information in the agent's possession tell in favor of or against *H*? From a Carnapian perspective this question is usually read as comparing the probability of *H* on *K* to *H*'s probability relative to a tautological background. So we say that the agent's total evidence *K* confirms *H* just in case $m(H \mid K) > m(H)$.

Carnap doesn't just assert that this rational prior m must exist; he provides a recipe for calculating its numerical values. To understand Carnap's calculations, let's begin with a very simple language, containing only one predicate *F* and two constants *a* and *b*. This language has only two atomic propositions (*Fa* and *Fb*), so we can specify distribution m over the language using a probability table with four state-descriptions. Carnap runs through a few candidates for distribution m; he calls the first one $m^\dagger$:

Fa	Fb	$m^\dagger$
T	T	1/4
T	F	1/4
F	T	1/4
F	F	1/4

In trying out various candidates for m, Carnap is attempting to determine the logical probabilities of particular propositions relative to a tautological background. $m^\dagger$ captures the natural thought that a tautological background should treat each of the available possibilities symmetrically. $m^\dagger$ applies a principle of indifference and assigns each state-description the same value.[19]

Yet $m^\dagger$ has a serious drawback:

$$m^\dagger(Fb \mid Fa) = m^\dagger(Fb) = 1/2 \qquad (6.1)$$

On $m^\dagger$, *Fa* is irrelevant to *Fb*; so according to $m^\dagger$, *Fa* does not confirm *Fb* relative to the empty background. Carnap thinks the fact that one object has property *F* should confirm that the next object will have *F*, even against a tautological background. Yet $m^\dagger$ does not yield this result. Even worse, this flaw remains as $m^\dagger$ is extended to larger languages. $m^\dagger$ makes each proposition *Fa*, *Fb*, *Fc*, etc. independent not only of each of the others but also of logical combinations of the others; even the observation that ninety-nine objects all have property *F* will not confirm that the 100th object is an *F*. (See Exercise 6.5.) This is especially bad because $m^\dagger$ is supposed to represent the unique hypothetical prior for all rational agents. According to $m^\dagger$, if a rational agent's total evidence consists of the fact that ninety-nine objects all have property *F*, this total evidence does not confirm in the slightest that the next object will have F. $\mathbf{m}^\dagger$ does not allow "learning from experience"; as Carnap puts it:

> The choice of $[\mathbf{m}^\dagger]$ as the degree of confirmation would be tantamount to the principle never to let our past experiences influence our expectations for the future. This would obviously be in striking contradiction to the basic principle of all inductive reasoning.   (1950, p. 565)

Carnap wants a theory of confirmation that squares with commonsense notions of rational inductive reasoning; $\mathbf{m}^\dagger$ clearly fails in that role.

To address this problem, Carnap offers distribution $\mathbf{m}^*$. According to $\mathbf{m}^*$, logical probability is indifferent not among the state-descriptions in a language, but instead among its **structure-descriptions**. To understand structure-descriptions, start by thinking about property profiles. A property profile specifies exactly which of the language's predicates an object does or does not satisfy. In a language with the single predicate F, the two available property profiles would be "this object has property F" and "this object lacks property F." In a language with two predicates F and G, the property profiles would be "this object has both F and G," "this object lacks property F but has property G"; etc. Given language $\mathcal{L}$, a structure-description describes *how many* objects in the universe of discourse possess each of the available property profiles, but doesn't say which *particular* objects possess which profiles.[20] For example, the language containing one property F and two constants a and b has the two property profiles just mentioned. Since there are two objects, this language allows three structure-descriptions: "both objects have F", "one object has F and one object lacks F", and "both objects lack F". Written in disjunctive normal form, the three structure-descriptions are:

$$\begin{aligned} &\text{i. } Fa \,\&\, Fb \\ &\text{ii. } (Fa \,\&\, {\sim}Fb) \lor ({\sim}Fa \,\&\, Fb) \\ &\text{iii. } {\sim}Fa \,\&\, {\sim}Fb \end{aligned} \qquad (6.2)$$

Note that one of these structure-descriptions is a disjunction of multiple state-descriptions.[21]

$\mathbf{m}^*$ assigns equal value to each structure-description in a language. If a structure-description contains multiple state-description disjuncts, $\mathbf{m}^*$ divides the value of that structure-description equally among its state-descriptions. For our simple language, the result is:

Fa	Fb	m^*
T	T	1/3
T	F	1/6
F	T	1/6
F	F	1/3

Each structure-description from (6.2) above receives an m^*-value of 1/3; the structure-description containing the middle two lines of the table divides its m^*-value between them.

m^* allows learning from experience. From the table above, we can calculate

$$m^*(Fb \mid Fa) = 2/3 > 1/2 = m^*(Fb) \tag{6.3}$$

On m^*, the fact that a possesses property F confirms that b will have F relative to the tautological background.

Nevertheless, m^* falls short in a different way. Suppose our language contains two predicates F and G and two constants a and b. Carnap thinks that on the correct, logical m distribution we should have

$$m(Fb \mid Fa \& Ga \& Gb) > m(Fb \mid Fa) > m(Fb \mid Fa \& Ga \& {\sim} Gb) > m(Fb) \tag{6.4}$$

While evidence that a has F should increase a rational agent's confidence that b has F, that rational confidence should increase even higher if we throw in the evidence that a and b share property G. If a and b both have G, in some sense they're the same kind of object, so one should expect them to be alike with respect to F as well. When I tell you that one object in my possession has a beak, this might make you more confident that the other object in my possession is beaked as well. But if you already know that both objects are animals of the same species, beak information about one is much more relevant to your beak beliefs about the other. On the other hand, information that a and b are *un*alike with respect to G should make F-facts about a less relevant to F-beliefs about b. Telling you that my two objects are *not* animals of the same species reduces the relevance of beak information about one object to beak conclusions about the other. Carnap thinks a successful m-distribution would capture these **analogical effects**, expressed in Equation (6.4).

To see if Equation (6.4) holds for m^*, one would need to identify the structure-descriptions in this language. The available property profiles are: object has both F and G, object has F but not G, object has G but not F, object has neither. Some examples of structure-descriptions are: both objects have

F and G, one object has both F and G while the other has neither, one object has F but not G while the other object has G but not F, etc. I'll leave the details to the reader (see Exercise 6.6), but suffice it to say that $\mathbf{m}^*$ is unable to capture the analogical effects of Equation (6.4).

Carnap responded to this problem (and others) by introducing a continuum of $\mathbf{m}$-distributions with properties set by two adjustable parameters.[22] (1971, 1980) The parameter λ was an "index of caution", controlling how reluctant $\mathbf{m}$ made an agent to learn from experience. Given the details of how Carnap defined λ, $\mathbf{m}^\dagger$ was the $\mathbf{m}$-distribution with λ-value ∞ (because it made the agent infinitely cautious and forbade learning from experience), while $\mathbf{m}^*$ had λ-value 2. Adjusting the other parameter, γ, made analogical effects possible. Carnap suggested we set the values of these parameters using pragmatic considerations, which threatened the Objective Bayesian aspects of his project. But even then, Mary Hesse (1963, p. 121) and Peter Achinstein (1963) uncovered more subtle learning effects that Carnap's parameterized $\mathbf{m}$-distributions were unable to capture. In the end, Carnap never constructed an $\mathbf{m}$-distribution (or set of $\mathbf{m}$-distributions) with which he was entirely satisfied.

6.3 Grue

Nelson Goodman (1946, 1955) offered another kind of challenge to Hempel and Carnap's theories of confirmation. Here is the famous passage:

> Suppose that all emeralds examined before a certain time t are green. At time t, then, our observations support the hypothesis that all emeralds are green; and this is in accord with our definition of confirmation. Our evidence statements assert that emerald a is green, that emerald b is green, and so on; and each confirms the general hypothesis that all emeralds are green. So far, so good.

> Now let me introduce another predicate less familiar than "green". It is the predicate "grue" and it applies to all things examined before t just in case they are green but to other things just in case they are blue. Then at time t we have, for each evidence statement asserting that a given emerald is green, a parallel evidence statement asserting that that emerald is grue. And the statements that emerald a is grue, that emerald b is grue, and so on, will each confirm the general hypothesis that all emeralds are grue. Thus according to our definition, the prediction that all emeralds subsequently examined will be green and the prediction that all will be grue are alike confirmed by evidence

statements describing the same observations. But if an emerald subsequently examined is grue, it is blue and hence not green. Thus although we are well aware which of the two incompatible predictions is genuinely confirmed, they are equally well confirmed according to our definition. Moreover, it is clear that if we simply choose an appropriate predicate, then on the basis of these same observations we shall have equal confirmation, by our definition, for any prediction whatever about other emeralds.   (1955, pp. 73–4)

The target here is any theory of confirmation on which the observation that multiple objects all have property F confirms that the next object will have F as well. As we saw, Carnap built this "learning from experience" feature into his theory of confirmation. It was also a feature of Hempel's positive theory of confirmation, so Goodman is objecting to both Carnap's and Hempel's theories. We will focus on the consequences for Carnap, since I did not present the details of Hempel's positive approach.

Goodman's concern is as follows: Suppose we have observed ninety-nine emeralds before time t, and they have all been green. On Carnap's theory, this total evidence confirms the hypothesis that the next emerald observed will be green. So far, so good. But Goodman says this evidence can be re-expressed as the proposition that the first ninety-nine emeralds are grue. On Carnap's theory, this evidence confirms the hypothesis that the next emerald observed will be grue. But for the next emerald to be grue it must be blue. Thus it seems that on Carnap's theory our evidence confirms both the prediction that the next emerald will be green and the prediction that the next emerald will be blue. Goodman thinks it's intuitively obvious that the former prediction is confirmed by our evidence while the latter is not, so Carnap's theory is getting things wrong.

Let's look more carefully at the details. Begin with a language $\mathcal{L}$ containing constants a_1 through a_{100} representing objects, and predicates G and O representing the following properties:

> Gx: x is green
> Ox: x is observed by time t

We then define "grue" as follows in language $\mathcal{L}$:

> $Gx \equiv Ox$: x is grue; it is either green and observed by time t or non-green and
> not observed by t

The grue predicate says that the facts about whether an emerald is green match the facts about whether it was observed by t.[23] Goodman claims that according to Carnap's theory, our total evidence in the example confirms $(\forall x)Gx$ and Ga_{100} (which is good), but also $(\forall x)(Gx \equiv Ox)$ and $Ga_{100} \equiv Oa_{100}$ (which are supposed to be bad).

But what exactly *is* our evidence in the example? Goodman agrees with Hempel that in assessing confirmation relations we must explicitly and precisely state the contents of our total evidence. Evidence that the first ninety-nine emeralds are green would be:

E: Ga_1 & Ga_2 & ... & Ga_{99}

But E neither entails nor is equivalent to the statement that the first ninety-nine emeralds are grue (because it doesn't say anything about whether those emeralds' G-ness matches their O-ness), nor does E confirm $(\forall x)(Gx \equiv Ox)$ on Carnap's theory.

A better statement of the evidence would be:

E': $(Ga_1$ & $Oa_1)$ & $(Ga_2$ & $Oa_2)$ & ... & $(Ga_{99}$ & $Oa_{99})$

Here we've added an important fact included in the example: that emeralds a_1 through a_{99} are observed by t. This evidence statement entails both that all those emeralds were green and that they all were grue. A bit of technical work with Carnap's theory[24] will also show that according to that theory, E' confirms $(\forall x)Gx$, Ga_{100}, $(\forall x)(Gx \equiv Ox)$, and $Ga_{100} \equiv Oa_{100}$.

It looks like Carnap is in trouble. As long as his theory is willing to "project" past observations of any property onto future predictions that that property will appear, it will confirm grue predictions alongside green predictions. The theory seems to need a way of preferring greenness over grueness for projection purposes; it seems to need a way to play favorites among properties.

Might this need be met by a technical fix? One obvious difference between green and grue is the more complex logical form of the grue predicate in $\mathcal{L}$. There's also the fact that the definition of "grue" involves a predicate O that mentions time; perhaps induction on predicates referring to times is suspicious. So maybe we could build a new version of Carnap's theory that only projects logically simple predicates, or predicates that don't mention times. Yet Goodman turns these proposals against themselves by re-expressing the problem in an alternate language $\mathcal{L}'$, built on the following two predicates:

GRx: x is grue

Ox: x is observed by time t

Defined in language $\mathcal{L}'$, the predicate "green" looks like this:

$GRx \equiv Ox$:   x is green; it is either grue and observed by time t or non-grue and not observed by t

Expressed in $\mathcal{L}'$, the evidence E' is:

$$E': \quad (GRa_1 \ \& \ Oa_1) \ \& \ (GRa_2 \ \& \ Oa_2) \ \& \ \ldots \ \& \ (GRa_{99} \ \& \ Oa_{99})$$

This expression of E' in $\mathcal{L}'$ is true in exactly the same possible worlds as the expression of E' we gave in $\mathcal{L}$. And once more, when applied to $\mathcal{L}'$ Carnap's theory has E' confirming both that all emeralds are grue and that they are green, and that a_{100} will be grue and that it will be green.

But in $\mathcal{L}'$ all the features that discriminated against grue now work against green—it's the definition of greenness that is logically complex and involves the predicate O referring to time. If you believe that logical complexity or reference to times makes the difference between green and grue, you now need a reason to prefer the expression of the problem in language $\mathcal{L}$ over its expression in $\mathcal{L}'$. This is why Goodman's grue problem is sometimes described as a problem of **language dependence**: We could build a formal confirmation theory that projected logically simple predicates but not logically complex, yet such a theory would yield different projections when the *very same situation* was expressed in different languages (such as $\mathcal{L}$ and $\mathcal{L}'$).

Why is language dependence such a concern? Recall that Hempel endorsed the Equivalence Condition in part because he didn't want confirmation to depend on the particular way hypotheses and evidence were presented. For theorists like Hempel and Carnap who take confirmational relations to be objective, it shouldn't make a difference how particular subjects choose to represent certain propositions linguistically. Two scientists shouldn't draw different conclusions from the same data just because one speaks English and the other speaks Japanese![25]

Hempel and Carnap sought a theory of confirmation that worked exclusively with the syntactical forms of propositions represented in language. Goodman charges that such theories can yield consistent verdicts only when applied within a carefully selected subset of languages. And since a syntactical theory may operate only once a language is provided, such a theory cannot select appropriate languages for us. Goodman concludes, "Confirmation of a hypothesis by an instance depends rather heavily upon features of the hypothesis other than its syntactical form" (1955, pp. 72–3).

Warning

It is sometimes suggested that—although this is certainly not a *syntactical* distinction—the grue hypothesis may be dismissed on the grounds that it is "metaphysically weird". This usually involves reading "All emeralds are grue" as saying that all the emeralds in the universe are green before time *t* then *switch* to being blue after *t*. But that reading is neither required to get the problem going nor demanded by anything in Goodman (1946) or Goodman (1955). Suppose, for instance, that each emerald in the universe is either green or blue, and no emerald ever changes color. By an unfortunate coincidence, it just so happens that the emeralds you observe before *t* are all and only the green emeralds. In that case it will be true that all emeralds are grue, with no metaphysical sleight-of-hand required.

As the previous warning suggests, the metaphysical details of Goodman's grue example have sometimes obscured its philosophical point. "Grue" indicates a correlation between two properties: being green and being observed by time *t*. It happens to be a perfect correlation, expressed by a biconditional. Some such correlations are legitimately projectible in science: If you observe that certain types of fish are born with a fin on the left side just when they are born with a fin on the right, this bilateral symmetry is a useful, projectible biconditional correlation. The trouble is that any sizable body of data will contain many correlations, and we need to figure out which ones to project as regularities that will extend into the future. (The women in this meeting room all have non-red hair, all belong to a particular organization, and all are under 6 feet tall. Which of those properties will also be exhibited by the next woman to enter the room?) Grue is a particularly odd, particularly striking example of a spurious correlation, but is emblematic of the problem of sorting projectible from unprojectible hypotheses.[26] It is not essential to the example that one of the properties involved refers to times, or that one of the properties is a relatively simple physical property (color). Sorting spurious from significant correlations is a general problem, for all kinds of variables.[27]

Goodman offers his own proposal for detecting projectible hypotheses, and many authors have made further proposals since then. Instead of investigating those, I'd like to examine exactly what the grue problem establishes about Carnap's theory (and others). The first thing to note is that although evidence E' confirms on Carnap's theory that emerald a_{100} is grue, it does *not* confirm that emerald a_{100} is blue. Recall that Carnap offers a hypothetical prior distribution $\mathbf{m}^*$ that is supposed to capture the unique, logically-mandated ultimate epistemic standards. A hypothesis H is confirmed by total evidence E' (the evidence we settled on for the grue example) just in case $\mathbf{m}^*(H \mid E') > \mathbf{m}^*(H)$. For example, it turns out that

$$\mathbf{m}^*(Ga_{100} \mid E') > \mathbf{m}^*(Ga_{100}) \tag{6.5}$$

So on Carnap's theory, E' confirms Ga_{100}. But if that's true, then E' must be *negatively* relevant to $\sim Ga_{100}$, the proposition that emerald a_{100} is blue. So while E' confirms that a_{100} is green and confirms that a_{100} is grue, it does *not* confirm that a_{100} is blue.

This seems impossible—given that a_{100} is not observed by t (that is, given that $\sim Oa_{100}$), a_{100} is grue just in case a_{100} is blue! But $\sim Oa_{100}$ is *not* given in the total evidence E'. E' says that every emerald a_1 through a_{99} was observed by t and is green. If that's all we put into the evidence, that evidence is going to confirm that a_{100} both is green *and was observed by t*. After all, if every object described in the evidence has the property $Gx \& Ox$, Carnapian "learning from experience" will confirm that other objects have this property as well. Once we understand that Carnap's theory is predicting from E' that a_{100} bears both Ox and Gx, the prediction that a_{100} will have $Gx \equiv Ox$ is no longer so startling.

In fact, the assessment of E' one gets from Carnap's theory is intuitively plausible. If *all you knew* about the world was that there existed ninety-nine objects and all of them were green and observed before t, you would expect that if there were a 100th object it would be green and observed before t as well. In other words, you'd expect the 100th object to be grue—by virtue of being green (and observed), not blue![28] We can read the prediction that the 100th object is grue as a prediction that it's not green only if we smuggle extra background knowledge into the case—namely, the assumption that a_{100} is an unobserved emerald. (This is similar to what happened in Hempel's analysis of the Paradox of the Ravens.) So what happens if we explicitly state this extra fact, by adding to the evidence that a_{100} is not observed by t?

E'': $(Ga_1 \& Oa_1) \& (Ga_2 \& Oa_2) \& \ldots \& (Ga_{99} \& Oa_{99}) \& \sim Oa_{100}$

Skipping the calculations (see Exercise 6.7), it turns out that

$$m^*(Ga_{100} \equiv Oa_{100} \mid E'') = m^*(\sim Ga_{100} \mid E'')$$
$$= m^*(Ga_{100} \mid E'') = m^*(Ga_{100}) \qquad (6.6)$$
$$= 1/2$$

On Carnap's probabilistic distribution m^*, E'' confirms neither that a_{100} will be grue, nor that a_{100} will be green, nor—for that matter—that all emeralds are grue or that all emeralds are green.

Perhaps it's a problem for Carnap's theory that none of these hypotheses is confirmed by E'', when intuitively some of them should be. Or perhaps it's a problem that on m^*, E' confirms that all emeralds are grue—even if that doesn't have the consequence of confirming that the next emerald will be blue. Suffice it to say that while language-dependence problems can be found for Carnap's theory as well as various other positive theories of confirmation,[29] it's very subtle to determine exactly where those problems lie and what their significance is.

6.4 Subjective Bayesian confirmation

I began my discussion of Carnap's confirmation theory by pointing out its central insight: We can get a confirmation relation with exactly the right formal features by equating confirmation with probabilistic relevance. In 1980, Clark Glymour reported the influence of this insight on philosophical theories of confirmation:

Almost everyone interested in confirmation theory today believes that confirmation relations ought to be analysed in terms of *probability* relations. Confirmation theory is the theory of probability plus introductions and appendices. Moreover, almost everyone believes that confirmation proceeds through the formation of conditional probabilities of hypotheses on evidence. The basic tasks facing confirmation theory are thus just those of explicating and showing how to determine the probabilities that confirmation involves, developing explications of such metascientific notions as "confirmation," "explanatory power," "simplicity," and so on in terms of functions of probabilities and conditional probabilities, and showing that the canons and patterns of scientific inference result. It was not always so. Probabilistic accounts of confirmation really became dominant only after the publication of Carnap's *Logical Foundations of Probability*, although of course many probabilistic accounts had preceded Carnap's. An eminent contemporary philosopher has compared Carnap's achievement in inductive logic with Frege's in deductive logic: just as before Frege there was only a small and theoretically uninteresting collection of principles of deductive inference, but after him the foundation of a systematic and profound theory of demonstrative reasoning, so with Carnap and inductive reasoning.

(1980, pp. 64-5)

Carnap holds that if a rational agent's credence distribution is cr, evidence E confirms hypothesis H relative to that agent's background corpus just in case $cr(H \mid E) > cr(H)$. The distinctive feature of Carnap's positive theory is that he thinks only one credence distribution is rationally permissible for each agent given her total evidence: the distribution obtained by conditionalizing the "logical probability" distribution m on her background corpus. So if we wanted, we could give Carnap's entire account of confirmation without mentioning agents at all: E confirms H relative to K just in case $m(H \mid E \& K) > m(H \mid K)$. In the end, confirmation facts are logical and objective, existing "out there" among the propositions.

Carnap's commitment to Uniqueness makes him an Objective Bayesian in the normative sense. Subjective Bayesians appreciate Carnap's central insight about probabilistic relevance, and agree with him that a piece of evidence E confirms hypothesis H relative to an agent's credence distribution cr just in case $cr(H \mid E) > cr(H)$. But these points of agreement are separable from the commitment to a unique distribution m determining the correct cr-distribution relative to total evidence for all rational agents. Subjective Bayesians think that specifying an agent's background corpus/total evidence K is insufficient to fully determine her rational credences. They are willing to let different rational agents construct their credences using different hypothetical priors, encoding those agents' differing epistemic standards. So two rational agents with the same total evidence may assign different credences to the same propositions.

This makes agents' particular credence distributions much more significant to the Subjective Bayesian account of confirmation than they are to Carnap's approach. For Subjective Bayesians, whether E confirms H cannot be relative simply to a background corpus, because such a corpus is insufficient to determine an entire probability distribution. Without a unique function m to rely on, Subjective Bayesians need something else to fill out the details around K and generate a full distribution. For this, they usually rely on the opinions of a particular agent. A Subjective Bayesian will say that E confirms

H for a specific agent just in case $cr(H \mid E) > cr(H)$ on that agent's current credence distribution cr. A piece of evidence confirms a hypothesis for an agent when the evidence is positively relevant to the hypothesis relative to that agent's current credences. Put another way, evidence confirms a hypothesis for an agent just in case conditionalizing on that evidence would increase her confidence in the hypothesis.

Since Subjective Bayesians permit rational agents to assign different credence distributions (even against the same background corpus!), this means that the same evidence will sometimes confirm different hypotheses for different rational agents. Two agents with, say, different levels of trust in authority might draw differing conclusions about whether a particular report confirms that Oswald acted alone or is further evidence of a government conspiracy. For a Subjective Bayesian, it may turn out that due to differences in the agents' credence distributions, the fact in question confirms one hypothesis for one agent and a different hypothesis for the other. There need be no independent, absolute truth about what's *really* confirmed. And this confirmatory difference need not be traceable to any difference in the agents' background corpora; the agents may possess different credence distributions because of differences in their epistemic standards, even when their bodies of total evidence are the same.

The Subjective Bayesian approach to confirmation is still a probabilistic relevance account, so it still satisfies all the adequacy conditions we endorsed in Section 6.1.2. Take any credence distribution cr a rational agent might assign that satisfies the Kolmogorov axioms and Ratio Formula. Let K be the conjunction of all propositions X in $\mathcal{L}$ such that $cr(X) = 1$. Now specify that E confirms H relative to K *and that credence distribution* just in case $cr(H \mid E) > cr(H)$. Confirmation relative to cr will now display exactly the features we accepted in Figure 6.1: Disconfirmation Duality and the Equivalence, Entailment, and Converse Entailment Conditions. So, for instance, one will get the desirable result that relative to any rational credence distribution, a hypothesis is confirmed by evidence it entails.[30]

While Subjective Bayesians usually talk about confirmation relative to a particular *agent's* credence distribution, they are not committed to doing so. The central claim of the Subjective Bayesian account of confirmation is that confirmation is always relative to *some* probability distribution, which is underdetermined by a corpus of background evidence. The distribution that does the work is often—but need not *always* be—an agent's credence distribution.[31] For example, a scientist may assess her experimental data relative to a commonly accepted probabilistic model of the phenomenon under investigation (such as a statistical model of gases), even if that model doesn't match her personal credences about the events in question. Similarly, a group may agree to assess evidence relative to a probability distribution distinct from the credence distributions of each of its members. Whatever probability distribution we consider, the Kolmogorov axioms and Ratio Formula ensure that the confirmation relation relative to that distribution will display the formal features we desire.

The most common objection to the Subjective Bayesian view of confirmation is that for confirmation to play the objective role we require in areas like scientific inquiry, it should *never* be relative to something so subjective as an agent's opinions about the world. (A Congressional panel's findings about the evidence bearing on the Kennedy assassination shouldn't depend on the committee members' levels of trust in authority!) We will return to this objection—and to some theories of confirmation that try to avoid it—in Chapter 13. For now I want to consider another objection to the Subjective Bayesian view, namely that it is so empty as to be near-useless. There are so many probability distributions available that for any E and H we will be able to find *some* distribution on which they are positively correlated (except in extreme cases when $E \vDash \sim H$). It looks, then, like the Subjective Bayesian view tells us almost nothing substantive about which particular hypotheses are confirmed by which bodies of evidence.

While the Subjective Bayesian denies the existence of a unique probability distribution to which all confirmation relations are relative, the view need not be anything-goes.[32] Chapter 5 proposed a number of plausible constraints beyond the Kolmogorov axioms and Ratio Formula on the credences of a rational agent. If these constraints are accepted, they will impose some shape upon any confirmation relation defined relative to a rational agent's credences. For example, David Lewis argues at his (1980, p. 285ff.) that if a credence distribution satisfies the Principal Principle, then relative to that distribution the evidence that a coin has come up heads on x percent of its tosses will confirm that the objective chance of heads on a single toss is close to x. By confirming such hypotheses about the chances, this evidence will also confirm predictions about the outcomes of future flips, achieving some of the learning from experience that Carnap sought in a theory of confirmation.

So one way to get substantive results from a Subjective Bayesian theory of confirmation is to impose more substantive requirements on rational credences than just the probability axioms. But there's another way to get substance out of the theory, without adding additional constraints. Notice that Lewis's Principal Principle result has the form: if your credences have features such-and-such, then confirmation relative to those credences will have features so-and-so. We can derive results like this even when the credal features such-and-such are not mandated by rationality. For example, if you assign equal credence to each possible outcome of a six-sided-die roll, then relative to your credence distribution the evidence that the roll came up odd will confirm that it came up prime. This will be true regardless of whether your total evidence rationally required equanimity over the possible outcomes. Subjective Bayesianism can yield interesting, informative results about which bodies of evidence confirm which hypotheses relative to a particular probability distribution, regardless of whether that distribution was rationally required.

The theory can also work in the opposite direction: given a target confirmational relation, it can tell us features of a probability distribution that will yield that relation. But before I describe some of Subjective Bayesianism's more interesting results on that front, I need to explain how Bayesians measure the strength of evidential support.

6.4.1 Confirmation measures

We have been considering a *classificatory* question: Under what conditions does a body of evidence E confirm a hypothesis H? Related to that classificatory question are various *comparative* confirmational questions: Which of E or E' confirms H more strongly? Is E better evidence for H or H'? etc. These comparative questions could obviously be answered if we had the answer to an underlying *quantitative* question: To what degree does E confirm H? (If we knew the degree to which E confirms H and the degree to which E' confirms the same H, we could say whether E or E' confirms H more strongly.) Popper (1935/1959) introduced the notion of *degrees* of confirmation. Since then various Bayesian **confirmation measures** have been proposed to quantify degree of confirmation: they take propositions E and H and some probability distribution Pr (perhaps an agent's credence distribution) and try to measure how much E confirms H relative to Pr.

Warning

When we set out to understand degree of confirmation in terms of Pr, it's important not to conflate firmness with increase in firmness (Section 6.2.1).

It's also important to get clear on how degree of confirmation relates to various notions involving justification.

Compare the following:

- the degree to which E confirms H relative to Pr
- $\Pr(H \mid E)$
- the degree to which an agent with total evidence E would be justified in believing (or accepting) H

The degree to which E confirms H relative to Pr cannot be measured by $\Pr(H \mid E)$. $\Pr(H \mid E)$ tells us how probable H is given E (relative to Pr). If Pr represents an agent's hypothetical prior distribution, $\Pr(H \mid E)$ tells us the degree of confidence rationality requires that agent to assign H when her total evidence is E. The confirmation of H by E is a *relation* between E and H, while the value of $\Pr(H \mid E)$ may be affected as much by the value of $\Pr(H)$ as it is by E. So $\Pr(H \mid E)$ is not solely reflective of the confirmational relationship between H and E (relative to Pr).

Some authors discuss the degree to which E "justifies" H. This may or may not be meant as synonymous with the degree to which E confirms H. Even so, it cannot be identical to $\Pr(H \mid E)$, for the reasons just explained. But other authors think it's a category mistake to speak of one proposition's justifying another; evidence may only justify particular *attitudes* toward H. When Pr is an agent's hypothetical prior, we might speak of $\Pr(H \mid E)$ as the credence in H an agent is justified in possessing when her total evidence is E.

Yet even this is distinct from the degree to which such an agent is justified in *believing* H. Belief is a binary doxastic attitude. We might propose a theory that quantifies how much an agent is justified in possessing this binary attitude. But there's no particular reason to think that the resulting measure should satisfy the Kolmogorov axioms, much less be precisely equal to $\Pr(H \mid E)$ for any Pr with independent significance. (See Shogenji 2012.)

Finally, there is the view that an agent is justified in believing or accepting H only if $\Pr(H \mid E)$ is high (where E represents total evidence and Pr her hypothetical prior). Here $\Pr(H \mid E)$ is not supposed to measure how justified such an acceptance would be; it's simply part of a necessary condition for such acceptance to be justified. Whether one accepts this proposal depends on one's views about the rational relations between credences and binary acceptances/beliefs.

So if the degree to which E confirms H relative to Pr cannot be measured by $\Pr(H \mid E)$, how *should* it be measured? There is a now a sizable literature that attempts to answer this question. Almost all of the measures that have been seriously defended are **relevance measures**: They agree with our earlier analysis that E confirms H relative to Pr just in case $\Pr(H \mid E) > \Pr(H)$. In other words, the relevance measures all concur that confirmation goes along with positive probabilistic relevance (and disconfirmation goes with negative probabilistic relevance). Yet there turn out to be a wide variety of confirmation measures satisfying this basic constraint. The following measures have all been extensively discussed in the historical literature:[33]

$$d(H, E) = \Pr(H \mid E) - \Pr(H)$$

$$s(H, E) = \Pr(H \mid E) - \Pr(H \mid {\sim}E)$$

$$r(H, E) = \log\left[\frac{\Pr(H \mid E)}{\Pr(H)}\right]$$

$$l(H, E) = \log\left[\frac{\Pr(E \mid H)}{\Pr(E \mid {\sim}H)}\right]$$

These measures are to be read such that, for instance, $d(H, E)$ is the degree to which E confirms H relative to Pr on the d-measure. For the measures containing fractions, we stipulate that the value of the fraction is infinite when the denominator but not the numerator is 0. (Just think of it as taking the limit of the fraction as the denominator goes to 0.)

Each of the measures has been defined such that if H and E are positively relevant on Pr, then the value of the measure is positive; if H and E are negatively relevant, the value is negative; and if H is independent of E then the value is 0. In other words: positive values represent confirmation, negative values represent disconfirmation, and 0 represents irrelevance.[34] For example, if Pr assigns each of the six faces on a die equal probability of coming up on a given roll, then

$$
\begin{aligned}
d(2, \text{prime}) &= \Pr(2 \mid \text{prime}) - \Pr(2) \\
&= 1/3 - 1/6 \quad\quad\quad\quad\quad (6.7) \\
&= 1/6
\end{aligned}
$$

This value is positive because evidence that the die roll came up prime would confirm the hypothesis that it came up two. Beyond the fact that it's positive, the particular number returned by the d-measure has no intrinsic significance.

(It's not as if a d-value of, say, 10 has a particular meaning.) But we can make meaningful comparisons *among* values assigned by the same confirmation measure. For example, $d(3 \vee 5, \text{prime}) = 1/3$, which is greater than the $1/6$ calculated above. So according to the d-measure (sometimes called the "difference measure"), on this Pr-distribution evidence that the die came up prime more strongly supports the disjunctive conclusion that it came up three or five than the conclusion that it came up two.

Since they are all relevance measures, the confirmation measures I listed will agree on *classificatory* facts about whether a particular E supports a particular H relative to a particular Pr. Nevertheless, they are distinct measures because they disagree about various *comparative* facts. A bit of calculation will reveal that $r(2, \text{prime}) = \log(2)$. Again, that particular number has no special significance, nor is there really much to say about how an r-score of $\log(2)$ compares to a d-score of $1/6$. (r and d measure confirmation on different scales, so to speak.) But it *is* significant that $r(3 \vee 5, \text{prime}) = \log(2)$ as well. According to the r-measure (sometimes called the "log ratio measure"), evidence that the roll came up prime confirms the hypothesis that it came up two to the *exact same degree* as the hypothesis that it came up either three or five. That is a substantive difference with the d-measure about a comparative confirmation claim.

Since the various confirmation measures can disagree about comparative confirmation claims, to the extent that we are interested in making such comparisons we will need to select among the measures available. Arguing for some measures over others occupies much of the literature in this field. What kinds of arguments can be made? Well, we might test our intuitions on individual cases. For instance, it might just seem intuitively obvious to you that the primeness evidence favors the $3 \vee 5$ hypothesis more strongly than the 2 hypothesis, in which case you may prefer the d-measure over the r-measure. Another approach parallels Hempel's approach to the qualitative confirmation relation: We first identify formal features we want a confirmation measure to display, then we test positive proposals for each of those features.

For example, suppose E confirms H strongly while E' confirms H only weakly. If we let c represent the "true" confirmation measure (whichever measure that turns out to be), $c(H, E)$ and $c(H, E')$ will both be positive numbers (because E and E' both confirm H), but $c(H, E)$ will be the larger of the two. Intuitively, since E is such good news for H it should also be very *bad* news for $\sim H$; since E' is only weakly good news for H it should be only weakly bad news for $\sim H$. This means that while $c(\sim H, E)$ and $c(\sim H, E')$ are both negative, $c(\sim H, E)$ is the lower (farther from zero) of the two. That relationship is guaranteed by the following formal condition:

Hypothesis Symmetry: For all H and E in $\mathcal{L}$ and every probabilistic Pr,
$$c(H, E) = -c(\sim H, E).$$

Hypothesis Symmetry says that evidence which favors a hypothesis will disfavor the negation of that hypothesis just as strongly. It guarantees that if $c(H, E) > c(H, E')$ then $c(\sim H, E) < c(\sim H, E')$.[35]

Hypothesis Symmetry won't do all that much work in narrowing our field; of the confirmation measures under consideration, only r is ruled out by this condition. A considerably stronger condition can be obtained by following Carnap's thought that entailment and refutation are the two extremes of confirmation.[36] If that's right, then confirmation measures must satisfy the following adequacy condition:

Logicality: All entailments receive the same degree of confirmation, and have a higher degree of confirmation than any non-entailing confirmations.[37]

If we combine Logicality with Hypothesis Symmetry, we get the further result that refutations are the strongest form of disconfirmation, and all refutations are equally strong.

Logicality is violated by, for instance, confirmation measure d. It's easy to see why. d subtracts the prior of H from its posterior. Since the posterior can never be more than 1, the prior will therefore put a cap on how high d can get. For example, if $\Pr(H) = 9/10$, then no E will be able to generate a d-value greater than $1/10$, which is the value one will get when $E \vDash H$. On the other hand, we saw in Equation (6.7) that d-values greater than $1/10$ are possible even for evidence that doesn't entail the hypothesis (e.g., $d(2, \text{prime}) = 1/6$), simply because the prior of the hypothesis in question is so much lower. As with the firmness concept of confirmation, the prior of H interferes with the d-score's assessment of the relation *between* E and H. This interference makes the d-measure violate Logicality.

Out of all the confirmation measures prominently defended in the historical literature (including all the measures described above), only measure l satifies Logicality. This constitutes a strong argument in favor of measure l (sometimes called the "log likelihood-ratio measure" of confirmation). If l looks familiar to you, that may be because it simply applies a logarithm to the Bayes factor, which we studied in Section 4.1.2. There we saw that the Bayes factor equals the ratio of posterior odds to prior odds, and is a good way of measuring the impact a piece of evidence has on an agent's opinion about a hypothesis. Moreover, the log-likelihood ratio has a convenient mathematical feature often cited approvingly by statisticians: When pieces of evidence E_1 and E_2 are screened off by H on Pr, $l(H, E_1 \& E_2) = l(H, E_1) + l(H, E_2)$. (See Exercise 6.9.) We often have cases in which independent pieces of evidence stack up in favor of a hypothesis. Measure l makes confirmation by independent evidence additive; the strength of a stack of independent pieces of evidence equals the sum of the individual pieces' strengths.

However, a newer confirmation measure[38] has been proposed by Crupi, Tentori, and Gonzalez (2007) that also satisfies both Hypothesis Symmetry and Logicality:

$$z(H, E) = \begin{cases} \frac{\Pr(H \mid E) - \Pr(H)}{1 - \Pr(H)} & \text{if } \Pr(H \mid E) \geq \Pr(H) \\[2ex] \frac{\Pr(H \mid E) - \Pr(H)}{\Pr(H)} & \text{if } \Pr(H \mid E) < \Pr(H) \end{cases}$$

This measure is particulary interesting because it measures confirmation differently from disconfirmation (hence the piecewise definition). That means confirmation and disconfirmation may satisfy different general conditions under the z-measure. For example, the following condition is satisfied for cases of disconfirmation but not for cases of confirmation:

$$z(H, E) = z(E, H) \tag{6.8}$$

Interestingly, Crupi, Tentori, and Gonzalez have conducted empirical studies in which subjects' comparative judgments seem to match z-scores more than they match other confirmation measures. In particular, subjects seem to intuitively treat disconfirmation cases differently from confirmation cases. (See Exercise 6.10.)

6.4.2 Subjective Bayesian solutions to the Paradox of the Ravens

Earlier (Section 6.1.1) we saw Hempel endorsing conditions on confirmation according to which the hypothesis that all ravens are black would be confirmed not only by the observation of a black raven but also by the observation of a red herring. Hempel explained this result—the so-called Paradox of the Ravens—by arguing that its seeming paradoxicality results from background assumptions we illicitly smuggle into the question. Our confirmation intuitions are driven by contingent facts we typically know about the world, but for Hempel the only fair test of ravens confirmation was against an empirically empty background. Hempel would ultimately defend a positive theory of confirmation on which black raven and red herring observations stand in exactly the same relations to the ravens hypothesis, as we long as we stick to a tautological background corpus.

Subjective Bayesians take the paradox in exactly the opposite direction. They examine our contingent background assumptions about what the world is like, and try to explain the intuitive confirmation judgments that result. As Charles Chihara puts it (in a slightly different context), the problem is "that of trying to see why we, who always come to our experiences with an encompassing complex web of beliefs," assess the paradox the way we do. (1981, p. 437)

Take the current knowledge you actually have of what the world is like. Now suppose that against the background of that knowledge, you are told that you will soon be given an object a to observe. You will record whether it is a raven and whether it is black; you are not told in advance whether a will have either of these properties. Recall that on the Subjective Bayesian view of confirmation, evidence E confirms hypothesis H relative to probability distribution Pr just in case E is positively relevant to H on Pr. In this situation it's plausible that, when you gain evidence E about whether a is a raven and whether it is black, you will judge the confirmation of various hypotheses by this evidence relative to your personal credence distribution. So we will let your cr play the role of Pr.

The key judgment we hope to explain is that the ravens hypothesis (all ravens are black) is *more strongly confirmed* by the observation of a black raven than by the observation of a non-black non-raven (a red herring, say). One might go further and suggest that observing a red herring shouldn't confirm the ravens hypothesis at all. But if we look to our considered judgments (rather than just our first reactions) here, we should probably grant that insofar as a non-black raven would be absolutely disastrous news for the ravens hypothesis, any observation of an object that doesn't turn out to be a non-black raven should be at least *some* good news for the hypothesis.[39]

Expressing our key judgment formally requires us to measure degrees of confirmation, a topic we discussed in the previous section. If $c(H, E)$ measures the degree to which E confirms H relative to cr, the Bayesian claims that

$$c(H, Ba \& Ra) > c(H, \sim Ba \& \sim Ra) \qquad (6.9)$$

where H is the ravens hypothesis $(\forall x)(Rx \supset Bx)$. Again, the idea is that relative to the credence distribution cr you assign before observing a, observing a to be a black raven would confirm H more strongly than observing a to be a non-black non-raven.

Fitelson and Hawthorne (2010b) show that Equation (6.9) will hold relative to cr if both the following conditions are met:

$$cr(\sim Ba) > cr(Ra) \qquad (6.10)$$

$$\frac{cr(\sim Ba \mid H)}{cr(Ra \mid H)} \leqslant \frac{cr(\sim Ba)}{cr(Ra)} \qquad (6.11)$$

These conditions are jointly sufficient for the confirmational result in Equation (6.9). They are not necessary; in fact, Bayesians have proposed a number of different sufficient sets over the years.[40] But these have the advantage of being simple and compact; they also work for every construal of c canvassed in the previous section except for confirmation measure s.

What do these conditions *say*? You satisfy Equation (6.10) if you are more confident prior to observing the object a that it will be non-black than you are that a will be a raven. This would make sense if, for example, you thought a was going to be randomly selected for you from a universe that contained more non-black things than ravens.[41] Equation (6.11) then considers the *ratio* of your confidence that a will be non-black to your confidence that it will be a raven. Meeting condition (6.10) makes this ratio greater than 1; now we want to know how the ratio would *change* were you to suppose all ravens are black. Equation (6.11) says that when you make this supposition the ratio doesn't go up—supposing all ravens are black wouldn't, say, dramatically increase how many non-black things you thought were in the pool or dramatically decrease your count of ravens. (It turns out from the math that for the confirmational judgment in Equation (6.9) to go false, the left-hand ratio in (6.11) would have to be *much* larger than the right-hand ratio; hence my talk of *dramatic* changes.) This constraint seems sensible. Under normal circumstances, for instance, supposing that all ravens are black should if anything increase the number of black things you think there are, not increase your count of *non-black* items.

Subjective Bayesians suggest that relative to our real-life knowledge of the world, were we to confront a selection situation like the one proposed in the ravens scenario, our credence distribution would satisfy Equations (6.10) and (6.11). Relative to such a credence distribution, the observation of a black raven confirms the ravens hypothesis more strongly than the observation of a red herring. This is how a Subjective Bayesian explains the key intuitive judgment that the ravens hypothesis is better confirmed by a black raven observation than a red herring observation: by showing how that judgment follows from more general assumptions we make about the composition of the world. Given that people's outlook on the world typically satisfies Equations (6.10) and (6.11), it follows from the Subjective Bayesian's quantitative theory of confirmation that if they are rational, most people will take the black raven observation to confirm more strongly.[42]

Now one might object that people who endorse the key ravens judgment have credence distributions that don't actually satisfy the conditions specified (or other sets of sufficient conditions Bayesians have proposed). Or an Objective Bayesian might argue that a confirmation judgment can be vindicated only by grounding it in something firmer than personal credences. I am not going to take up those arguments here. But I hope to have at least fought back the charge that Subjective Bayesianism about confirmation is empty. The Subjective Bayesian account of confirmation tells us when evidence E confirms hypothesis H relative to credence distribution cr. You might think that because it does very little to constrain the values of cr, this account can tell us nothing interesting about when evidence confirms a hypothesis. But we have just seen a substantive, unexpected result. It was not at all obvious at the start of our inquiry that any rational credence distribution satisfying Equations (6.10) and (6.11) would endorse the key ravens judgment. Subjective Bayesian results about confirmation often take the form, "If your credences are such-and-such, then these confirmation relations follow," but such conditionals can be highly informative.

For instance, the result we've just seen not only reveals what confirmational judgments agents will make in typical circumstances but also what kinds of atypical circumstances may legitimately undermine those judgments. Return to the Hall of Atypically Colored Birds, where a bird is displayed only if the majority of his species-mates are one color but his color is different. Suppose it is part of an agent's background knowledge (before she observes object a) that a is to be selected from the Hall of Atypically Colored Birds. If at that point—before observing a—the agent were to suppose that all ravens are black, that would dramatically decrease her confidence that a will be a raven. If all ravens are black, there are no atypically colored ravens, so there should be no ravens in the Hall.[43] Thus given the agent's background knowledge about the Hall of Atypically Colored Birds, supposing the ravens hypothesis H decreases her confidence that a will be a raven (that is, Ra). This makes the left-hand side of Equation (6.11) greater than the right-hand side, and renders Equation (6.11) false. So one of the sufficient conditions in our ravens result fails, and its route to Equation (6.9) is blocked. This provides a tidy explanation of why, if you know you're in the Hall of Atypically Colored Birds, observing a black raven should not be better news for the ravens hypothesis than observing a non-black non-raven.

Besides this account of the Paradox of the Ravens, Subjective Bayesians have offered solutions to various other confirmational puzzles. For example, Hawthorne and Fitelson (2004) approach the problem of irrelevant conjunction (Section 6.1.2) by specifying conditions under which adding an irrelevant conjunct to a confirmed hypothesis yields a new hypothesis that—while still confirmed—is less strongly confirmed than the original. Similarly, Chihara (1981) and Eells (1982, Ch. 2) respond to Goodman's grue example (Section 6.3) by specifying credal conditions under which a run of observed green emeralds more strongly confirms the hypothesis that all emeralds are green than the hypothesis that all emeralds are grue.[44]

Even more intriguingly, the Subjective Bayesian account of confirmation has recently been used to provide rationalizing explanations for what otherwise look like irrational judgments on the part of agents. The idea is that sometimes when subjects are asked questions about probability, they respond with answers about confirmation instead. In Tversky and Kahneman's Conjunction Fallacy experiment (Section 2.2.4), the hypothesis that Linda is a bank teller is entailed by the hypothesis that Linda is a bank teller and active in the feminist movement. This entailment means that an agent satisfying the probability axioms must be at least as confident in the former hypothesis as the latter. But it does *not* mean that evidence must confirm the former as strongly as the latter. Crupi, Fitelson, and Tentori (2008) outline credal conditions under which the evidence presented to subjects in Tversky and Kahneman's experiment would confirm the feminist-bank-teller hypothesis more strongly than the bank-teller hypothesis. It may be that subjects who rank the feminist-bank-teller hypothesis more highly in light of that evidence are reporting confirmational judgments rather than credences.

Similarly, in analyzing the Base Rate Fallacy (Section 4.1.2) we noted the strong Bayes factor of the evidence one gets from a highly reliable disease test. Since the Bayes factor tracks the log likelihood-ratio measure of confirmation, this tells us that a positive result from a reliable test strongly confirms that the patient has the disease (as it should!). When doctors are asked for their confidence that the patient has the disease in light of such a positive test result, the high values they report may reflect their confirmational judgments.

The Subjective Bayesian account of confirmation may therefore provide an explanation of what subjects are doing when they seem to make irrational credence reports. Nevertheless, having an explanation for subjects' behavior does not change the fact that these subjects may be making serious *mistakes*. It's one thing when a doctor is asked in a study to report a credence value and reports a confirmation value instead. But if that doctor goes on to make treatment decisions based on the confirmation value rather than the posterior probability, this can have significant consequences. Confusing how probable a hypothesis is on some evidence with how strongly that hypothesis is confirmed by that evidence is a version of the firmness/increase-in-firmness conflation. If the doctor recommends a drastic treatment for a patient on the basis that the test applied was highly reliable (even though, with the base rates taken into account, the posterior probability that a disease is present remains quite low), her confusion about probability and confirmation may prove highly dangerous for her patient.

6.5 Exercises

Unless otherwise noted, you should assume when completing these exercises that the distributions under discussion satisfy the probability axioms and Ratio Formula. You may also assume that whenever a conditional probability expression occurs, the needed proposition has nonzero unconditional probability so that conditional probabilities are well defined.

Problem 6.1. *♪* Salmon (1975) notes that each of the following confirmational situations can arise:
  (a) Two pieces of evidence each confirm some hypothesis, but their conjunction disconfirms it.
  (b) Two pieces of evidence each confirm some hypothesis, but their disjunction disconfirms it.
  (c) A piece of evidence confirms each of two hypotheses, but it disconfirms their conjunction.
  (d) A piece of evidence confirms each of two hypotheses, but it disconfirms their disjunction.
Provide a real-world example of each of these four situations. (And don't make it easy on yourself—none of propositions (dis)confirming or being (dis)confirmed should be a tautology or a contradiction!)

Problem 6.2. 🌙 For purposes of this problem, assume that the Equivalence Condition, the Entailment Condition, and Disconfirmation Duality are all true of the confirmation relation.

    (a) Show that if E & K deductively refutes H but K does not refute H on its own, then E disconfirms H relative to K.

    (b) Show that if H & K deductively refutes E but K does not refute H on its own, then E disconfirms H relative to K.

Problem 6.3. 🌙🌙 Suppose the Special Consequence Condition and Converse Consequence Condition were both true. Show that under those assumptions, if evidence E confirms some proposition H relative to K, then relative to K evidence E will also confirm any other proposition X we might choose.[45] (<u>Hint</u>: Start with the problem of irrelevant conjunction.)

Problem 6.4. 🌙🌙 Suppose we have propositions E, H, and K in $\mathcal{L}$ meeting the following conditions: (1) the set containing E, H, and K is logically consistent; (2) E & $K \vDash H$; and (3) $K \nvDash H$. Suppose also probabilistic distribution Pr over $\mathcal{L}$ is such that for any proposition X, $\Pr(X) = 1$ just in case $K \vDash X$.

    (a) Prove that $\Pr(E) > 0$.

    (b) Prove that $\Pr(H \mid E) = 1$.

    (c) Prove that $\Pr(H) < 1$.

Problem 6.5. 🌙🌙 Suppose we have a language whose only atomic propositions are $Fa_1, Fa_2, \ldots, Fa_n$ for some integer $n > 1$. In that case, $\mathbf{m}^\dagger(Fa_n) = 1/2$.

    (a) Show that for any non-contradictory proposition K expressible solely in terms of Fa_1 through Fa_{n-1}, $\mathbf{m}^\dagger(Fa_n \mid K) = 1/2$.

    (b) What does the result you demonstrated in part (a) have to do with Carnap's point that $\mathbf{m}^\dagger$ does not allow "learning from experience"?

Problem 6.6. 🌙🌙

    (a) Make a probability table over state-descriptions for the four atomic propositions Fa, Fb, Ga, Gb. In the right-hand column, enter the values Carnap's $\mathbf{m}^*$ assigns to each state-description. (<u>Hint</u>: Keep in mind that Fa & $\sim Fb$ & Ga & $\sim Gb$ belongs to a different structure-description than Fa & $\sim Fb$ & $\sim Ga$ & Gb.)

    (b) Use your table to show that $\mathbf{m}^*(Fb \mid Fa$ & Ga & $Gb) > \mathbf{m}^*(Fb \mid Fa)$.

    (c) Use your table to show that $\mathbf{m}^*(Fb \mid Fa$ & Ga & $\sim Gb) = \mathbf{m}^*(Fb)$.

(d) For each of problem (b) and (c) above, explain how your answer relates to m^*'s handling of "analogical effects".[46]

Problem 6.7. 🌙🌙🌙 Suppose E'' is the proposition

$$(Ga_1 \& Oa_1) \& (Ga_2 \& Oa_2) \& \ldots \& (Ga_{99} \& Oa_{99}) \& \sim Oa_{100}$$

Without actually making a probability table, prove that on Carnap's confirmation theory:
(a) $m^*(Ga_{100} \equiv Oa_{100} \mid E'') = m^*(\sim Ga_{100} \mid E'')$
(b) $m^*(Ga_{100} \equiv Oa_{100} \mid E'') = m^*(Ga_{100} \mid E'')$
(c) $m^*(Ga_{100} \mid E'') = 1/2$
(d) $m^*(Ga_{100}) = 1/2$
(e) $m^*(Ga_{100} \equiv Oa_{100} \mid E'') = m^*(Ga_{100} \mid E'') = m^*(Ga_{100})$

Problem 6.8. 🌙 Provide examples showing that the r-measure of confirmation violates each of the following constraints:
(a) Hypothesis Symmetry
(b) Logicality

Problem 6.9. 🌙🌙 Prove that on the l-measure of degree of confirmation, if E_1 is screened off from E_2 by H on Pr, then the degree to which $E_1 \& E_2$ confirms H can be found by summing the degrees to which E_1 and E_2 each confirm H individually. (<u>Hint</u>: Remember that $\log(x \cdot y) = \log x + \log y$.)

Problem 6.10. 🌙 Crupi, Tentori, and Gonzalez think it's intuitive that on whatever measure c correctly gauges confirmation, the following constraint will be satisfied for cases of disconfirmation but not confirmation:

$$c(H, E) = c(E, H)$$

(a) Provide a real-world example of two propositions A and B such that, intuitively, A confirms B but B does not confirm A to the same degree. (Don't forget to specify the Pr distribution to which your confirmation judgments are relative!)
(b) Provide a real-world example of two propositions C and D such that, intuitively, C disconfirms D and D disconfirms C to the same degree. (Don't make it too easy on yourself—pick a C and D that aren't mutually exclusive!)

(c) Does it seem to you intuitively that for *any* propositions C and D and probability distribution Pr, if C disconfirms D then D will disconfirm C to the same degree? Explain why or why not.

Problem 6.11.

(a) 🎵🎵 Provide an example in which the *l*- and *z*-measures disagree on a comparative confirmational claim. That is, provide an example in which the *l*-measure says that E_1 confirms H_1 more strongly than E_2 confirms H_2, but the *z*-measure says E_2 confirms H_2 more strongly than E_1 confirms H_1.

(b) 🎵🎵🎵 Prove that the *l*- and *z*-measures never disagree on how strongly two pieces of evidence confirm the same hypothesis. That is, prove that there do not exist H, E_1, E_2, and Pr such that $l(H, E_1) > l(H, E_2)$ but $z(H, E_1) < z(H, E_2)$.

Problem 6.12. 🎵🎵🎵 The solution to the Paradox of the Ravens presented in Section 6.4.2 is not the only Subjective Bayesian solution that has been defended. An earlier solution[47] invoked the following four conditions (where H abbreviates $(\forall x)(Rx \supset Bx)$):

(i) $Pr(Ra \ \& \sim Ba) > 0$

(ii) $Pr(\sim Ba) > Pr(Ra)$

(iii) $Pr(Ra \mid H) = Pr(Ra)$

(iv) $Pr(\sim Ba \mid H) = Pr(\sim Ba)$

Assuming Pr satisfies these conditions, complete each of the following. (<u>Hint</u>: Feel free to write H instead of the full, quantified proposition it represents, but don't forget what H entails about Ra and Ba.)

(a) Prove that $Pr(\sim Ra \ \& \sim Ba) > Pr(Ra \ \& \ Ba)$.

(b) Prove that $Pr(Ra \ \& \ Ba \ \& \ H) = Pr(H) \cdot Pr(Ra)$.

(c) Prove that $Pr(\sim Ra \ \& \sim Ba \ \& \ H) = Pr(H) \cdot Pr(\sim Ba)$.

(d) Show that on confirmation measure *d*, if Pr satisfies conditions (i) through (iv) then Ra & Ba confirms H more strongly than $\sim Ra$ & $\sim Ba$ does.

(e) Where in your proofs did you use condition (i)?

(f) Suppose Pr is your credence distribution when you know you are about to observe an object *a* drawn from the Hall of Atypically Colored Birds. Which of the conditions (i) through (iv) will Pr probably not satisfy? Explain.

6.6 Further reading

INTRODUCTIONS AND OVERVIEWS

Ellery Eells (1982). *Rational Decision and Causality*. Cambridge Studies in Philosophy. Cambridge: Cambridge University Press

The latter part of Chapter 2 (pp. 52–64) offers an excellent discussion of Hempel's adequacy conditions for confirmation, how the correct conditions are satisfied by a probabilistic relevance approach, and Subjective Bayesian solutions to the Paradox of the Ravens and Goodman's grue puzzle.

Rudolf Carnap (1955/1989). Statistical and Inductive Probability. In: *Readings in the Philosophy of Science*. Ed. by Baruch A. Brody and Richard E. Grandy. 2nd edition. Hoboken: Prentice-Hall

A brief, accessible overview by Carnap of his position on the meaning of "probability" and the development of his various confirmation functions. (Here he uses "individual distribution" to refer to state-descriptions and "statistical distribution" to refer to structure-descriptions.) Includes a probability table with diagrams!

Alan Hájek and James M. Joyce (2008). Confirmation. In: *The Routledge Companion to Philosophy of Science*. Ed. by Stathis Psillos and Martin Curd. New York: Routledge, pp. 115–28

Besides providing an overview of much of the material in this chapter, suggests that there may not be one single correct function for measuring degree of confirmation.

CLASSIC TEXTS

Janina Hosiasson-Lindenbaum (1940). On Confirmation. *Journal of Symbolic Logic* 5, pp. 133–48

Early suggestion that the Paradox of the Ravens might be resolved by first admitting that both a black raven and a red herring confirm that all ravens are black, but then second arguing that the former confirms more strongly than the latter. (Though she uses a different example than the color of ravens.) Also anticipates some of Carnap's later conclusions about which adequacy conditions could be met by a confirmation theory based on probability.

> Carl G. Hempel (1945a). Studies in the Logic of Confirmation (I). *Mind* 54, pp. 1–26
> Carl G. Hempel (1945b). Studies in the Logic of Confirmation (II). *Mind* 54, pp. 97–121

Hempel's classic papers discussing his adequacy conditions on the confirmation relation and offering his own positive, syntactical account of confirmation.

> Rudolf Carnap (1950). *Logical Foundations of Probability*. Chicago: University of Chicago Press

While much of the material earlier in this book is crucial for motivating Carnap's probabilistic theory of confirmation, his discussion of distributions $m^\dagger$ and m^* occurs in the Appendix. (Note that the preface distinguishing "firmness" from "increase in firmness" concepts of confirmation does not appear until the second edition of this text, in 1962.)

> Nelson Goodman (1955). *Fact, Fiction, and Forecast*. Cambridge, MA: Harvard University Press

Chapter III contains Goodman's "grue" discussion.

EXTENDED DISCUSSION

> Michael G. Titelbaum (2010). Not Enough There There: Evidence, Reasons, and Language Independence. *Philosophical Perspectives* 24, pp. 477–528

Proves a general language-dependence result for all objective accounts of confirmation (including accounts that are Objective Bayesian in the normative sense), then evaluates the result's philosophical significance.

> Katya Tentori, Vincenzo Crupi, and Selena Russo (2013). On the Determinants of the Conjunction Fallacy: Probability versus Inductive Confirmation. *Journal of Experimental Psychology: General* 142, pp. 235–55

Assessment of various explanations of the Conjunction Fallacy in the psychology literature, including the explanation that subjects are reporting confirmation judgments rather than posterior credences.

Notes

1. Scientists—and philosophers of science—are interested in a number of other properties and relations of evidence and hypotheses in addition to confirmation. These include predictive power, informativeness, simplicity, unification of disparate phenomena, etc. An interesting ongoing Bayesian line of research asks whether and how these various other notions relate to confirmation.

2. Good (1967) offers a more detailed example in the same vein. Good describes the population distributions of two worlds constructed so that observing a black raven confirms that we are in the world in which not all ravens are black.

3. As I pointed out in Chapter 4's note 13, this passage may have been the inspiration for David Lewis's referring to hypothetical priors (numerical distributions satisfying the probability axioms yet containing no contingent evidence) as "superbaby" credences.

4. In discussing the Paradox of the Ravens, one might worry whether $(\forall x)(Rx \supset Bx)$—with its material conditional, and lack of existential import—is a faithful translation of "All ravens are black." Strictly speaking, Hempel is examining what confirms the proposition expressed by the sentence in logical notation, rather than the proposition expressed by the sentence in English. But if the two come apart, intuitions about "All ravens are black" may become less relevant to Hempel's discussion.

5. Despite his attention to background corpora, Hempel isn't careful about backgrounds in the adequacy conditions he proposes. So I have added the relevant background restrictions to Hempel's official definitions of his conditions, and will explain the motivations for those restrictions as we go along.

   Also, in case you're wondering why E, H, and K are required to be consistent in the Entailment Condition, consider a case in which K refutes E and E is entirely irrelevant to H. $E \& K$ will be a contradiction, and so will entail H, but we don't want to say E confirms H relative to K. I will include similar consistency requirements as needed going forward.

6. Similar considerations tell against citing events as evidence. Was it the discovery of Troy that supported the historicity of Homer, the discovery of Troy in northwestern Turkey, or the discovery of Troy in northwestern Turkey by Heinrich Schliemann?

7. An interesting literature has sprung up among Bayesian epistemologists about the precise conditions under which evidence of evidence constitutes evidence for a hypothesis. See, for example, Tal and Comesaña (2017), Roche (2014), and Fitelson (2012), building off foundational work in Shogenji (2003).

8. I learned of this example from Bradley (2015, §1.3); as far as I know it first appeared at Pryor (2004, pp. 350–1).

   Note, by the way, that while one might want a restriction to keep the Special Consequence Condition from applying when $K \vDash H'$, in the stated counterexamples H' is not entailed by K. Out of desperation we could try to save Special Consequence by claiming it holds only relative to tautological backgrounds (as Hempel did with Nicod's Criterion). But we can recreate our cards counterexample to Special Consequence by emptying out the background and adding facts about how the card was drawn as conjuncts to each of A, B, and C. Similar remarks apply to the counterexamples we'll soon produce for other putative confirmation constraints.

9. For one of many articles on confirmational intransitivity and skepticism, see White (2006). The broader epistemology community's recent acknolwedgment that reasons for a conclusion need not transmit to propositions entailed by that conclusion often feels to me like a rediscovery of Bayesian epistemology's rejection of Hempel's Special Consequence Condition decades earlier.

10. We could provide another argument for the Consistency Condition from the premises that (1) if a hypothesis is confirmed by evidence we possess then we should accept that hypothesis; and (2) one should never accept inconsistent propositions. But we've already rejected (1) for our notion of confirmation.

11. I'm assuming the definition of an ostrich includes its being a flightless bird, and whatever K is involved doesn't entail E, H, or H' on its own.

12. **Hypothetico-deductivism** is a positive view of confirmation that takes the condition in Converse Entailment to be not only sufficient but also necessary for confirmation: E confirms H relative to K just in case $H \& K \vDash E$ and $K \nvDash E$. This is implausible for a number of reasons (see Hempel 1945b). Here's one: Evidence that a coin of unknown bias has come up heads on exactly half of a huge batch of flips supports the hypothesis that the coin is fair; yet that evidence isn't *entailed* by that hypothesis.

13. Strictly speaking there will be infinitely many X in $\mathcal{L}$ such that $\Pr(X) = 1$, so we will take K to be a proposition in $\mathcal{L}$ logically equivalent to the conjunction of all such X. (I'll ignore this detail in what follows.) Notice that if Pr represents the credence distribution of an agent at a given time t_i, this definition makes the agent's background corpus equivalent to her total evidence E_i at that time, as that notion was discussed in Section 4.3.

14. Carnap's preface to the second edition distinguishes the firmness and increase in firmness concepts because he had equivocated between them in the first edition. Carnap was roundly criticized for this by Popper (1954).

15. I even made this mistake once in an article, despite my intense awareness of the issue! Luckily the error was caught before the offending piece was published.

16. Here we assume that, as pointed out in Chapter 2's note 5, the atomic propositions of $\mathcal{L}$ are logically independent.

17. A word about Carnap's notation in his (1950). Carnap actually introduces two confirmation functions, $\mathfrak{m}(\cdot)$ and $\mathfrak{c}(\cdot, \cdot)$. For any non-contradictory proposition K in $\mathcal{L}$, $\mathfrak{c}(\cdot, K)$ is just the function I've been describing as $\Pr(\cdot)$ relative to K; in other words, $\mathfrak{c}(\cdot, K) = \mathfrak{m}(\cdot \mid K) = \mathfrak{m}(\cdot \& K)/\mathfrak{m}(K)$. As I've just mentioned in the main text, this makes $\mathfrak{c}$ somewhat redundant in the theory of confirmation, so I won't bring it up again.

18. As I mentioned in Chapter 5, note 8, Carnap actually thinks "probability" is ambiguous between two meanings. What he calls "probability$_1$" is the logical notion of probability we've been discussing. Carnap's "probability$_2$" is based on frequencies, and is therefore objective as well.

19. Among other things, $m^\dagger$ represents the technique for determining probabilities that Ludwig Wittgenstein proposed in his *Tractatus Logico-Philosophicus* (1921/1961, Proposition 5.15ff.)

20. Jonathan Kvanvig suggested to me the following helpful illustration: Suppose I roll two dice, a red one and a blue one. A structure-description would say "One die came up six, the other came up three", while a state-description would say "The red die came up six, the blue die came up three". The structure-description tells us how many objects display particular combinations of features, but—unlike the state-description—it doesn't tell us which specific objects display which combinations.

21. Formally, two state-descriptions are disjuncts of the same structure-description just in case one state-description can be obtained from the other by permuting its constants.

22. Interestingly, Carnap's continuum proposal shared a number of features with a much earlier proposal by Johnson (1932).

23. To simplify matters, I'm going to assume going forward that (1) each object under discussion in the grue example is observed exactly once (so that "not observed by t" is equivalent to "observed after t"); (2) each object is either green or blue (so "not green" is equivalent to "blue"); and (3) each object is an emerald. Strictly speaking these assumptions should be made explicit as part of the agent's total evidence, but since doing so would make no difference to the forthcoming calculations, I won't bother. This approach is backed up by Goodman's position in his (1955, p. 73, n. 9) that the grue problem is "substantially the same" as the problem he offered in Goodman (1946). The earlier version of the problem was both more clearly laid-out and cleaner from a logical point of view. For instance, instead of green and blue, he used red and not-red. The earlier paper also made clearer exactly whose positive theories of confirmation Goodman took the problem to target.

24. I'm going to assume Goodman is criticizing the version of Carnap's theory committed to m^*; Carnap's subsequent changes to handle analogical effects make little difference here.

25. Compare the difficulties with partition selection we encountered for indifference principles in Section 5.3.

26. To emphasize that not every pattern observed in the past should be expected to hold in the future, John Venn once provided the following example: "I have given a false alarm of fire on three different occasions and found the people came to help me each time" (1866, p. 180). One wonders if his false alarms were intentional experiments in induction (quoted in Galavotti 2005, pp. 77–8.)

27. Hume's (1739–40/1978) problem of induction asked what justifies us in projecting *any* past correlations into the future. Goodman's "new riddle of induction" asks, given that we are justified in projecting some correlations, *which* ones we ought to project.

28. Hempel's theory of confirmation displays a similar effect. And really, any close reader of Hempel should've known that some of Goodman's claims against Hempel were overstated. I mentioned that Hempel endorses the Consistency Condition (Section 6.1.2); he goes on to prove that it is satisfied by his positive theory of confirmation. On Hempel's theory, the hypotheses confirmed by any piece of evidence must be consistent with both that evidence and each other. So a single body of evidence cannot confirm

both Ga_{100} and $\sim Ga_{100}$. *Contra* Goodman, it just can't be that on Hempel's theory we get the "devastating result that any statement will confirm any statement" (1955, p. 81).

29. For more on this topic, see Hooker (1968), Fine (1973, Ch. VII), Maher (2010), and Titelbaum (2010).

30. Also, the Subjective Bayesian account of confirmation does not suffer from language-dependence problems. Suppose credence distribution cr, defined over language $\mathcal{L}$, makes it the case that $cr(H \mid E) > cr(H)$. We might define a different language $\mathcal{L}'$ that expresses all the same propositions as $\mathcal{L}$, and a distribution cr' over $\mathcal{L}'$. Intuitively, cr' expresses the same credences as cr just in case $cr'(X') = cr(X)$ whenever $X' \in \mathcal{L}'$ expresses the same proposition as $X \in \mathcal{L}$. If that condition is met, then cr' will satisfy the Kolmogorov axioms just in case cr does. And if $cr(H \mid E) > cr(H)$, we will have $cr'(H' \mid E') > cr'(H')$ for the H' and E' in $\mathcal{L}'$ that express the same propositions as H and E (respectively). So confirmation relations are unaffected by translations across languages. (The same will be true of the confirmation *measures* we discuss in Section 6.4.1.)

31. Keep in mind that we're discussing Subjective Bayesians *in the normative sense* (Section 5.1.2). Their Subjective Bayesianism is an *epistemological* position, about the variety of rational hypothetical priors available. Subjective Bayesians in the normative sense need not be Subjective Bayesians in the semantic sense; they need not read every assertion containing the word "probability" as attributing a credence to an individual.

32. As we put it in Chapter 5, the Subjective Bayesian need not be an extreme Subjective Bayesian, who denies any constraints on rational hypothetical priors beyond the probability axioms.

33. For citations to various historical authors who defended each measure, see Eells and Fitelson (2002).

34. The logarithms have been added to the r- and l-measures to achieve this centering on 0. Removing the logarithms would yield measures ordinally equivalent to their logged versions, but whose values ran from 0 to infinity (with a value of 1 indicating probabilistic independence). Notice also that the base of the logarithms is irrelevant for our purposes.

35. Hypothesis Symmetry was defended as a constraint on degree of confirmation by Kemeny and Oppenheim (1952); see also Eells and Fitelson (2002), who gave it that particular name.

36. Carnap thought of confirmation as a "generalization of entailment" in a number of senses. Many Subjective Bayesians are happy to accept Carnap's idea that deductive cases are limiting cases of confirmation, but aren't willing to follow Carnap in taking those limiting cases as a model for the whole domain. Whether E entails H relative to K depends just on the content of those propositions, and Carnap thought matters should be the same for all confirmatory relations. To a Subjective Bayesian, though, whether E confirms H relative to K depends on something more—a full probability distribution Pr.

37. See Fitelson (2006) for Logicality. Note that in assessing whether confirmation measures satisfy Logicality, we restrict our attention to "contingent" cases, in which neither E nor H is entailed or refuted by the K associated with Pr.

38. Glass and McCartney (2015) notes that Crupi, Tentori, and Gonzalez's z-measure adapts to the problem of confirmation the so-called "certainty factor" that has been used in the field of expert systems since Shortliffe and Buchanan (1975).
39. Another thought one might have is that while a red herring confirms that all ravens are black, its degree of confirmation of that hypothesis is exceedingly weak in absolute terms. While some Bayesian analyses of the paradox also try to establish this result, we won't consider it here. (See Vranas 2004 for discussion and citations on the proposal that a red herring confirms the ravens hypothesis to a degree that is "positive but minute.")
40. For citations to many historical proposals, see Fitelson and Hawthorne (2010a, esp. n. 10). Fitelson and Hawthorne (2010b) goes beyond these historical sources by also proposing necessary conditions for Equation (6.9), which unfortunately are too complex to detail here.
41. The result assumes you assign non-extreme unconditional credences to the proposition that a is black and to the proposition that it's a raven. This keeps various denominators in the Ratio Formula positive. We also assume you have a non-extreme prior in H.
42. Why "if they are rational"? The mathematical result assumes not only that the credence distribution in question satisfies Equations (6.10) and (6.11) but also that it satisfies the probability axioms and Ratio Formula. (This allows us to draw out conclusions about values in the credence distribution beyond those directly specified by Equations (6.10) and (6.11).) Subjective Bayesians assume a rational credence distribution satisfies the probability axioms and Ratio Formula.
43. Perhaps even with the supposition that all ravens are black, the agent's confidence that a will be a raven is slightly above zero because once in a long while the Hall's curators make a mistake.
44. At the 2020 Central Division Meeting of the American Philosophical Assocation, Sven Moritz Silvester Neth presented a paper arguing that a sufficient condition for favoring the green hypothesis over the grue hypothesis after observing a run of green emeralds is to have a prior on which an emerald's being green is probabilistically independent of whether it has been observed by time t. This fits my diagnosis that what's at stake in the grue problem is which correlations we should take seriously and which we should treat as spurious. If one antecedently assumes that emerald color is independent of time of observation, then any correlations between those two variables one notices in a sample will be discarded as unimportant, rather than projected forward through a grue hypothesis.
45. For purposes of this problem you may assume that E, H, X, and K stand in no special logical relationships. (Being inconsistent counts as a "special logical relationship"; being consistent does not.)
46. I owe this entire problem to Branden Fitelson.
47. See Fitelson and Hawthorne (2010a, Sect. 7) for discussion.

7

Decision Theory

Up to this point most of our discussion has been about epistemology. But probability theory originated in attempts to understand games of chance, and historically its most extensive application has been to practical decision-making. The Bayesian theory of probabilistic credence is a central element of decision theory, which developed throughout the twentieth century in philosophy, psychology, and economics. **Decision theory** searches for rational principles to evaluate the acts available to an agent at any given moment. Given what she values (her utilities) and how she sees the world (her credences), decision theory recommends the act that is most efficacious for achieving those values from her point of view.

Decision theory has always been a crucial application of Bayesian theory. In his *The Foundations of Statistics*, Leonard J. Savage wrote:

> Much as I hope that the notion of probability defined here is consistent with ordinary usage, it should be judged by the contribution it makes to the theory of decision.   (1954, p. 27)

Decision theory has also been extensively studied, and a number of excellent book-length introductions are now available. (I recommend one in the Further Reading section of this chapter.) As a result, I haven't packed as much information into this chapter as into the preceding chapter on confirmation. I hope only to equip the reader with the terminology and ideas we will need later in this book, and that she would need to delve further into the philosophy of decision theory.

We will begin with the general mathematical notion of an expectation, followed by the philosophical notion of utility. We will then see how Savage calculates expected utilities to determine rational preferences among acts, and the formal properties of rational preference that result. Next comes Richard Jeffrey's Evidential Decision Theory, which improves on Savage's by applying to probabilistically dependent states and acts. We will then discuss Jeffrey's troubles with certain kinds of risk-aversion (especially the Allais Paradox), and with Newcomb's Problem. Causal Decision Theory will be proposed as

Fundamentals of Bayesian Epistemology 2: Arguments, Challenges, Alternatives. Michael G. Titelbaum,
Oxford University Press. © Michael G. Titelbaum 2022. DOI: 10.1093/oso/9780192863140.003.0007

a better response to Newcomb. I will close by briefly tracing some of the historical back-and-forth about which decision theory handles Newcomb's problem best.

7.1 Calculating expectations

Suppose there's a numerical quantity—say, the number of hits a particular batter will have in tonight's baseball game—and you have opinions about what value that quantity will take. We can then calculate your **expectation** for the quantity. While there are subtleties we will return to later, the basic idea of an expectation is to multiply each value the quantity might take by your credence that it'll take that value, then add up the results. So if you're 30% confident the batter will have one hit, 20% confident she'll have two hits, and 50% confident she'll have three, your expectation for the number of hits is

$$1 \cdot 0.30 + 2 \cdot 0.20 + 3 \cdot 0.50 = 2.2 \tag{7.1}$$

Your expectation of a quantity is *not* the value you anticipate the quantity will actually take, or even the value you think it's most probable the quantity will take—in the baseball example, you're certain the batter won't have 2.2 hits in tonight's game! Your expectation of a quantity is a kind of *estimate* of the value the quantity will take. When you're uncertain about the value of a quantity, a good estimate may straddle the line between multiple options.

While your expectation for a quantity isn't necessarily the exact value you think it will take on a given occasion, it should equal the *average* value you expect that quantity to take in the long run. Suppose you're certain that our batter will play in many, many games. The **law of large numbers** says that if you satisfy the probability axioms, you'll have credence 1 that as the number of games increases, her average number of hits per game will tend toward your expectation for that quantity. In other words, you're highly confident that as the number of games approaches the limit, the batter's average hits per game will approach 2.2.[1]

We've already calculated expectations for a few different quantities in this book. For example, when you lack inadmissible evidence the Principal Principle requires your credence in a proposition to equal your expectation of its chance. (See especially our calculation in Equation (5.7).) But by far the most commonly calculated expectations in life are monetary values. For example, suppose you have the opportunity to buy stock in a company just before it announces quarterly earnings. If the announcement is good you'll be able to sell shares at $100 each, but if the announcement is bad you'll be forced to sell at $10 apiece. The value you place in these shares depends on your confidence in a good report. If you're 40% confident in a good earnings report, your expected value for each share is

$$\$100 \cdot 0.40 + \$10 \cdot 0.60 = \$46 \tag{7.2}$$

As a convention, we let positive monetary values stand for money accrued to the agent; negative monetary values are amounts the agent pays out. So your expectation of how much money you will receive for each share is $46.

An agent's **fair price** for an investment is what she takes to be that investment's break-even point—she'd pay anything *up to* that amount of money in exchange for the investment. If you use expected values to make your investment decisions, your fair price for each share of the stock just described will be $46. If you buy shares for less than $46 each, your expectation for that transaction will be positive (you'll expect to make money on the deal). If you buy shares for more than $46, you'll expect to lose money.

The idea that your fair price for an investment should equal your expectation of its monetary return dates to Blaise Pascal, in a famous seventeenth-century correspondence with Pierre Fermat (Fermat and Pascal 1654/1929). There are a couple of reasons why this is a sensible idea. First, suppose you know you're going to be confronted with this exact investment situation many, many times. The law of large numbers says that you should anticipate a long-run average return of $46 per share. So if you're going to adopt a standing policy for buying and selling such investments, you are highly confident that any price higher than $46 will lose you money and any price lower than $46 will make you money in the long term. Second, expectations vary in intuitive ways when conditions change. If you become more confident in a good earnings report, each share becomes more valuable to you, and you should be willing to pay a higher price. This is exactly what the expected value calculation predicts. If you learn that a good earnings report will send the share value to only $50, this decreases the expected value of the investment and also decreases the price you should be willing to pay.

An investment is a type of bet, and fair betting prices play a significant role in Bayesian lore. (We'll see one reason why in Chapter 9.) A bet that pays $1 if proposition P is true and nothing otherwise has an expected value of

$$\$1 \cdot cr(P) + \$0 \cdot cr(\sim P) = \$cr(P) \tag{7.3}$$

If you use expectations to calculate fair betting prices, your price for a gamble that pays $1 on *P* equals your unconditional credence in *P*.

We can also think about fair betting prices using odds. We saw in Section 2.3.4 that an agent's odds against *P* equal $cr(\sim P) : cr(P)$. So if the agent's credence in *P* is 0.25, her odds against *P* are 3 : 1. What will she consider a fair bet on *P*? Consider what the casinos would call a bet on *P* at 3 : 1 odds. If you place such a bet and win, you get back the original amount you bet plus three times that amount. If you lose your bet, you're out however much you bet. In terms of net returns, a bet at 3 : 1 odds offers you a possible net gain that's three times your possible net loss.

So suppose an agent with 0.25 credence in *P* places a $20 bet on *P* at 3 : 1 odds. Her expected net return is

$$\text{(net return on winning bet)} \cdot cr(P) + \text{(net return on losing bet)} \cdot cr(\sim P)$$
$$= \$60 \cdot 0.25 + -\$20 \cdot 0.75 = \$0 \qquad (7.4)$$

This agent expects a bet on *P* at 3 : 1 odds to be a break-even gamble—from her perspective, it's a fair bet. She will be willing to bet on *P* at those odds or anything higher. In general, an agent who bets according to her expectations will accept a bet on a proposition at odds equal to her odds against it, or anything higher. Remember that an agent's odds against a proposition *increase* as her credence in the proposition *decreases*. So if an agent becomes less confident in *P*, you need to offer her higher odds on *P* before she'll be willing to gamble.

A lottery ticket is a type of bet, and in the right situation calculating its expected value can be highly lucrative. Ellenberg (2014, Ch. 11) relates the story of Massachusetts's Cash WinFall state lottery game, which was structured in such a way that if the jackpot got high enough, the expected payout for a single ticket grew larger than the price the state charged for that ticket. For example, on February 7, 2005 the expected value of a $2 lottery ticket was $5.53. The implications of this arrangement were understood by three groups of individuals—led respectively by an MIT student, a medical researcher in Boston, and a retiree in Michigan who had played a short-lived similar game in his home state. Of course, the expected value of a ticket isn't necessarily what you will win if you buy a single ticket, but because of the long-run behavior of expectations your confidence in a net profit goes up the more tickets you buy. So these groups bought a *lot* of tickets. For instance, on August 13, 2010 the MIT group bought around 700,000 tickets, almost 90% of the Cash WinFall tickets purchased that day. Their $1.4 million investment netted about $2.1

million in payouts, for a 50% profit in one day. Expected value theory can be *extremely* effective.

7.1.1 The move to utility

Yet sometimes we value something other than money. For example, suppose it's late at night, it's cold out, you're trying to catch a bus that costs exactly $1 to ride, and you've got no money on you. A stranger offers either to give you $1 straight up, or to flip a fair coin and give you $2.02 if it comes up heads. It might be highly rational for you to prefer the guaranteed dollar even though its expected monetary value is less than that of the coin bet.

Decision theorists and economists explain this preference with the notion of **utility**. Introduced by Daniel Bernoulli and Gabriel Cramer in the eighteenth century,[2] utility is a numerical quantity meant to directly measure how much an agent values an arrangement of the world. Just as we suppose that each agent has her own credence distribution, we will suppose that each agent has a real-valued utility distribution over the propositions in language $\mathcal{L}$. The utility an agent assigns to a proposition represents how much she values that proposition's being true (or if you like, how happy that proposition's being true would make her). If an agent would be just as happy for one proposition to be true as another, she assigns them equal utility. But if it would make her happier for one of those propositions to be true, she assigns it the higher utility of the two.

Utilities provide a uniform value-measurement scale. In the bus example above, you don't value each dollar equally. Going from zero dollars to one dollar would mean a lot to you; it would get you out of the cold and on your way home. Going from one dollar to two dollars would not mean nearly as much in your present context. Not every dollar represents the same amount of value in your hands, so counting the number of dollars in your possession is not a consistent measure of how much you value your current state. On the other hand, utilities measure value uniformly. We stipulate that each added unit of utility (sometimes called a **util**) is equally valuable to an agent. She is just as happy to go from −50 utils to −49 as she is to go from 1 util to 2, and so on.

Having introduced this uniform value scale, we can explain your preferences in the bus case using expectations. Admittedly, the coin flip gamble has a higher expected *monetary* payoff ($1.01) than the guaranteed dollar. But monetary value doesn't always translate neatly to utility, and utility reflects the values on which you truly make your decisions. Let's say that having no money is worth

0 utils to you in this case, receiving one dollar and being able to get on the bus is worth 100 utils, and receiving $2.02 is worth 102 utils. (The larger amount of money is still more valuable to you; just not *much* more valuable.) When we calculate the expected *utility* of the gamble, it only comes to 51 utils, which is much less than the 100 expected utils associated with the guaranteed dollar. So you prefer the dollar guarantee.

The setup of this example is somewhat artificial, because it makes the value of money change radically at a particular cutoff point. But economists think money generally has a **decreasing marginal utility** for agents. While an agent always receives some positive utility from each additional dollar (or peso, or yuan, or . . .), the more dollars she already has the less extra utility it will be. The first billion you earn makes your family comfortable; the second billion doesn't have as much significance for your life. Postulating an underlying locus of value distinguishable from monetary worth helps explain why we don't always chase the next dollar as hard as we chased the first.

With that said, quantifying value on a numerical scale introduces many of the same problems we found with quantifying confidence. First, it's not clear that a real agent's psychology will always be as nuanced as a numerical utility structure seems to imply. And second, the moment you assign numerical utilities to every arrangement of the world you make them all comparable; the possibility of incommensurable values is lost. (Compare Section 1.2.2.)

7.2 Expected utility theory

7.2.1 Preference rankings and money pumps

A **decision problem** presents an agent with a partition of **acts**, from which she must choose exactly one. Decision theory aims to lay down rational principles governing choices in decision problems. It does so by supposing that a rational agent's choice of acts tracks her preferences among those acts. If the available acts are A and B, and she prefers A to B (we write $A \succ B$), then the agent decides to perform action A. A similar point applies when $B \succ A$. Yet it might be that the agent is indifferent between A and B (we write $A \sim B$), in which case she is rationally permitted to choose either one.

Sometimes a decision among acts is easy. If the agent is certain how much utility will be generated by the performance of each act, the choice is simple— she prefers the act leading to the highest-utility result. Yet the utility resulting from an act often depends on features of the world beyond the agent's control (think, for instance, of the factors determining whether a particular career choice turns out well), and the agent may be uncertain how those features stand. In that case, the agent needs a technique for factoring uncertainty into her decision. She needs a technique for combining credences and utilities to generate preferences.

Decision theory responds to this problem by providing a **valuation function**, which combines credences and utilities to assign each act a numerical score. The agent's preferences are assumed to match these scores: $A \succ B$ just in case A receives a higher score than B, while $A \sim B$ when the scores are equal. Given a particular decision problem, a rational agent will select the available act with the highest score (or—if there are ties at the top—one of the acts with the highest score).

Here's an example of a valuation function, just to convey the idea: Suppose you assign each act a numerical score by considering all the possible worlds to which you assign nonzero credence, finding the one in which that act produces the lowest utility, and then assigning that minimal utility value as the act's score. This valuation function generates preferences satisfying the **maximin rule**, so called because it selects the act with the highest minimum utility payoff. Maximin attends to only the worst case scenario for each available act.

While maximin is just one valuation function (we'll see others later), any approach that ties preferences to numerical scores assigned over acts imposes a certain structure on an agent's preferences. For instance, it guarantees that her preferences will display:

Preference Transitivity: For any acts A, B, and C, if the agent prefers A to B and B to C, then the agent prefers A to C.

This follows from the simple fact that numerical inequalities are transitive: each act's score is a number, so if act A's score is greater than act B's, and B's is greater than C's, then A's must be greater than C's as well.

Preference Transitivity will be endorsed as a rational constraint by any decision theory that ties preferences to numerical valuation functions. One might object that an agent may prefer A to B and prefer B to C, but never have thought to compare A to C. In other words, one might think that such an agent's preference ranking could go silent on the comparison between A and C and still be rational. Yet by coordinating preference with a numerical valuation over the entire partition of acts, we have already settled this issue; we have required the agent's preferences to form a complete ranking. Since every act receives a score, every act is comparable, and our theory demands the agent assign a preference (or indifference) between any two acts. Decision theorists sometimes express this as:

Preference Completeness: For any acts A and B, exactly one of the following is true: the agent prefers A to B, the agent prefers B to A, or the agent is indifferent between the two.

Notice that Preference Completeness entails the following:

Preference Asymmetry: There do not exist acts A and B such that the agent both prefers A to B and prefers B to A.

To recap: Decision theory begins by requiring an agents' choices to reflect her preferences, then coordinates those preferences with a numerical valuation function combining credences and utilities. By making the latter move, decision theory requires preferences to satisfy Preference Transitivity and Asymmetry. Hopefully it's intuitive that rational preferences satisfy these two conditions. But we can do better than that: We can provide an *argument* that Preference Transitivity and Asymmetry are rational requirements.

Consider a situation in which some of us find ourselves frequently. On any given weeknight, I would prefer to do something else over washing the dishes. (Going to a movie? Great! Watching the game? Good idea!) But when the week ends and the dishes have piled up, I realize that I would've preferred foregoing one of those weeknight activites in order to avoid a disgusting kitchen. Each of my individual decisions was made in accordance with my preferences among the acts I was choosing between at the time, yet together those local preferences added up to a global outcome I disprefer.

A student once suggested to me that he prefers eating out to cooking for himself, prefers eating at a friend's to eating out, but prefers cooking for himself to eating at a friend's. Imagine one night my student is preparing himself dinner, then decides he'd prefer to order out. He calls up the takeout place, but before they pick up the phone he decides he'd rather drive to his friend's for dinner. He gets in his car and is halfway to his friend's, when he decides he'd rather cook for himself. At which point he turns around and goes home, having wasted a great deal of time and energy. Each of those choices reflects the student's preference between the two options he considers at the time, yet their net effect is to leave him right back where he started meal-wise and out a great deal of effort overall.

My student's preferences violate Transitivity; as a result he's susceptible to a **money pump**. In general, a money pump against intransitive preferences (preferring A to B, B to C, and C to A) can be constructed like this: Suppose you're about to perform act B, and I suggest I could make it possible to do A instead. Since you prefer A to B, there must be *some* amount of something (we'll just suppose it's money) you'd be willing to pay me for the option to perform A. So you pay the price, are about to perform A, but then I hold out the possibility of performing C instead. Since you prefer C to A, you pay me a small amount to make that switch. But then I offer you the opportunity to perform B rather than C—for a small price, of course. And now you're back to where you started with respect to A, B, and C, but out a few dollars for your trouble. To add insult to injury, I could repeat this set of trades again, and again, milking more and more money out of you until I decide to stop. Hence the "money pump" terminology.[3]

Violating Preference Transitivity leaves one susceptible to a money-pumping set of trades. (If you violate Preference Asymmetry, the money pump is even simpler.) In a money pump, the agent proceeds through a series of exchanges, each of which looks favorable given his preferences between the two acts involved. But when those exchanges are combined, the total package produces a net loss (which the agent would prefer to avoid). The money pump therefore seems to reveal an inconsistency between the agent's local and global preferences, as in my dishwashing example. (We will further explore this kind of inconsistency in our Chapter 9 discussion of Dutch Books.) The irrationality of being susceptible to a money pump has been taken as a strong argument against violating Preference Asymmetry or Transitivity.[4]

7.2.2 Savage's expected utility

Savage (1954) frames decision problems using a partition of acts available to the agent and a partition of **states** the world might be in. A particular act performed with the world in a particular state produces a particular **outcome**. Agents assign numerical utility values to outcomes; given partial information they also assign credences over states.[5]

Here's a simple example: Suppose you're trying to decide whether to carry an umbrella today, but you're uncertain whether it's going to rain. This table displays the utilities you assign various outcomes:

	rain	dry
take umbrella	0	−1
leave it	−10	0

You have two available acts, represented in the rows of the table. There are two possible states of the world, represented in the columns. Performing a particular act when the world is in a particular state produces a particular outcome. If you leave your umbrella behind and it rains, the outcome is you walking around wet. The cells in the table report your utilities for the outcomes produced by various act/state combinations. Your utility for walking around wet is −10 utils, while carrying an umbrella on a dry day is inconvenient but not nearly as unpleasant (−1 util).

How should you evaluate available acts and set your preferences among them? For a finite partition $\{S_1, S_2, \ldots, S_n\}$ of possible states of the world, Savage offers the following valuation function:

$$EU_{SAV}(A) = u(A \ \& \ S_1) \cdot cr(S_1) + u(A \ \& \ S_2) \cdot cr(S_2) \\ + \ldots + u(A \ \& \ S_n) \cdot cr(S_n) \tag{7.5}$$

Here A is the particular act being evaluated. Savage evaluates acts by calculating their expected utilities; $EU_{SAV}(A)$ represents the expected utility of act A calculated in the manner Savage prefers. (We'll see other ways of calculating expected utility later on.) $cr(S_i)$ is the agent's unconditional credence that the world is in state S_i; $u(A \ \& \ S_i)$ is the utility she assigns to the outcome that will eventuate should she perform act A in state S_i.[6] So EU_{SAV} calculates the weighted average of the utilities the agent might receive if she performs A, weighted by her credence that she will receive each one. Savage holds that given a decision among a partition of acts, a rational agent will set her preferences in line with her expected utilities. She will choose to perform an act with at least as great an expected utility as that of any act on offer.

Now suppose that in the umbrella case you have a 0.30 credence in rain. We can calculate expected utilities for each of the available acts as follows:

$$EU_{SAV}(take) = 0 \cdot 0.30 + -1 \cdot 0.70 = -0.7 \\ EU_{SAV}(leave) = -10 \cdot 0.30 + 0 \cdot 0.70 = -3 \tag{7.6}$$

Taking the umbrella has the higher expected utility, so Savage thinks that if you're rational you'll prefer to take the umbrella. You're more confident it'll be dry than rain, but this is outweighed by the much greater disutility of a disadvantageous decision in the latter case than the former.

EU_{SAV} is a valuation function that combines credences and utilities in a specific way to assign numerical scores to acts. As a numerical valuation function, it generates a preference ranking satisfying Preference Asymmetry, Transitivity, and Completeness. But calculating expected utilities this way also introduces new features not shared by all valuation functions. For example, Savage's expected utility theory yields preferences that satisfy the

Dominance Principle: If act *A* produces a higher-utility outcome than act *B* in each possible state of the world, then *A* is preferred to *B*.

The Dominance Principle[7] seems intuitively like a good rational principle. Yet, surprisingly, there are decision problems in which it yields very bad results. Since Savage's expected utility theory entails the Dominance Principle, it can be relied upon only when we don't find ourselves in decision problems like that.

7.2.3 Jeffrey's theory

To see what can go wrong with dominance reasoning, consider this example from (Weirich 2012):

A student is considering whether to study for an exam. He reasons that if he will pass the exam, then studying is wasted effort. Also, if he will not pass the exam, then studying is wasted effort. He concludes that because whatever will happen, studying is wasted effort, it is better not to study.

The student entertains two possible acts—study or don't study—and two possible states of the world—he either passes the exam or he doesn't. His utility table looks something like this:

	pass	fail
study	18	−5
don't study	20	−3

Because studying costs effort, passing having not studied is better than passing having studied, and failing having not studied is also better than failing having studied. So whether he passes or fails, not studying yields a higher utility. By the Dominance Principle, the student should prefer not studying to studying.

This is clearly a horrible argument; it ignores the fact that whether the student studies *affects whether he passes the exam*.[8] The Dominance Principle—and Savage's expected utility theory in general—breaks down when the state of the world is influenced by which act the agent performs. Savage recognizes this limitation, and so requires that the acts and states used in framing decision problems be independent of each other. Jeffrey (1965), however, notes that in real life we often analyze decision problems in terms of dependent acts and states. Moreover, he worries that agents might face decision problems in which they are unable to identify independent acts and states.[9] So it would be helpful to have a decision theory that didn't require acts and states to be independent.

Jeffrey offers just such a theory. The key innovation is a new valuation function that calculates expected utilities differently from Savage's. Given an act A and a finite partition $\{S_1, S_2, \ldots, S_n\}$ of possible states of the world,[10] Jeffrey calculates

$$\mathrm{EU_{EDT}}(A) = \mathrm{u}(A \,\&\, S_1) \cdot \mathrm{cr}(S_1 \,|\, A) + \mathrm{u}(A \,\&\, S_2) \cdot \mathrm{cr}(S_2 \,|\, A)$$
$$+ \ldots + \mathrm{u}(A \,\&\, S_n) \cdot \mathrm{cr}(S_n \,|\, A) \tag{7.7}$$

I'll explain the "EDT" subscript later on; for now, it's crucial to see that Jeffrey alters Savage's approach (Equation (7.5)) by replacing the agent's *unconditional* credence that a given state S_i obtains with the agent's *conditional* credence that S_i obtains given A. This incorporates the possibility that performing the act the agent is evaluating will change the probabilities of various states of the world.

To see how this works, consider Jeffrey's example of a guest deciding whether to bring white or red wine to dinner. The guest is certain his host will serve either chicken or beef, but doesn't know which. The guest's utility table is as follows:

	chicken	beef
white	1	−1
red	0	1

For this guest, bringing the right wine is always pleasurable. Red wine with chicken is merely awkward, while white wine with beef is a disaster.

At a typical dinner party, the entree for the evening is settled well before the guests arrive. But let's suppose that tonight's host is especially accommodating, and will select a meat in response to the wine provided. (Perhaps the host has a stocked pantry, and waits to prepare dinner until the wine has arrived.) The guest is 75% confident that the host will select the meat that best pairs with the wine provided. Thus the state (meat served) depends on the agent's act (wine chosen). This means the agent cannot assign a uniform unconditional credence to each state prior to his decision. Instead, the guest assigns one credence to chicken conditional on his bringing white, and another credence to chicken conditional on his bringing red. These credences are reflected in the following table:

	chicken	beef
white	0.75	0.25
red	0.25	0.75

It's important to read the credence table differently from the utility table. In the utility table, the entry in the white/chicken cell is the agent's utility assigned to the outcome of chicken served *and* white wine. In the credence table, the white/chicken entry is the agent's credence in chicken served *given* white wine. The probability axioms and Ratio Formula together require all the credences conditional on white wine sum to 1, so the values in the first row sum to 1. The values in the second row sum to 1 for a similar reason. (In this example the values in each column sum to 1 as well, but that won't always be the case.)

We can now use Jeffrey's formula to calculate the agent's expected utility for each act. For instance:

$$\begin{aligned}
EU_{EDT}(\text{white}) &= u(\text{white \& chicken}) \cdot cr(\text{chicken} \,|\, \text{white}) \\
&\quad + u(\text{white \& beef}) \cdot cr(\text{beef} \,|\, \text{white}) \\
&= 1 \cdot 0.75 + -1 \cdot 0.25 \\
&= 0.5
\end{aligned} \tag{7.8}$$

(We multiply the values in the first row of the utility table by the corresponding values in the first row of the credence table, then sum the results.) A similar calculation yields $EU_{EDT}(\text{red}) = 0.75$. Bringing red wine has a higher expected utility for the agent than bringing white, so the agent should prefer bringing red.

Earlier I said somewhat vaguely that Savage requires acts and states to be "independent"; Jeffrey's theory gives that notion a precise meaning. EU_{EDT}

revolves around an agent's conditional credences, so for Jeffrey the relevant notion of independence is probabilistic independence relative to the agent's credence distribution. That is, an act A and state S_i are independent for Jeffrey just in case

$$cr(S_i | A) = cr(S_i) \qquad (7.9)$$

In the special case where the act A being evaluated is independent of each state S_i, the $cr(S_i | A)$ expressions in Jeffrey's formula may be replaced with $cr(S_i)$ expressions. This makes Jeffrey's expected utility calculation identical to Savage's. When acts and states are probabilistically independent, Jeffrey's theory yields the same preferences as Savage's. And since Savage's theory entails the Dominance Principle, Jeffrey's theory will also embrace Dominance in this special case.

But what happens to Dominance when acts and states are *dependent*? Here Jeffrey offers a nuclear deterrence example. Suppose a nation is choosing whether to arm itself with nuclear weapons, and knows its rival nation will follow its lead. The possible states of the world under consideration are war versus peace. The utility table might be:

	war	peace
arm	−100	0
disarm	−50	50

Wars are worse when both sides have nuclear arms; peace is also better without nukes on hand (because of nuclear accidents, etc.). A dominance argument is available since whichever state obtains, disarming provides the greater utility. So applying Savage's theory to this example would yield a preference for disarming.

Yet the advocate of nuclear deterrence takes the states in this example to depend on the acts. The deterrence advocate's credence table might be:

	war	peace
arm	0.1	0.9
disarm	0.8	0.2

The idea of deterrence is that if both countries have nuclear arms, war becomes much less likely. If arming increases the probability of peace, the acts and states in this example are probabilistically dependent. Jeffrey's theory calculates the following expected utilities from these tables:

$$EU_{EDT}(\text{arm}) = -100 \cdot 0.1 + 0 \cdot 0.9 = -10$$
$$EU_{EDT}(\text{disarm}) = -50 \cdot 0.8 + 50 \cdot 0.2 = -30$$

(7.10)

Relative to the deterrence advocate's credences, Jeffrey's theory yields a preference for arming. Act/state dependence has created a preference ranking at odds with the Dominance Principle.[11] When an agent takes the acts and states in a decision problem to be independent, Jeffrey's and Savage's decision theories are interchangeable, and dominance reasoning is reliable. But Jeffrey's theory also provides reliable verdicts when acts and states are dependent, a case in which Savage's theory and the Dominance Principle may fail.

7.2.4 Risk aversion and Allais' Paradox

Different people respond to risks differently. Many agents are **risk-averse**; they would rather have a sure $10 than take a 50-50 gamble on $30, even though the expected dollar value of the latter is greater than that of the former.

Economists have traditionally explained this preference by appealing to the declining marginal utility of money. If the first $10 yields much more utility than the next $20 for the agent, then the sure $10 may in fact have a higher expected utility than the 50-50 gamble. This makes the apparently risk-averse behavior perfectly rational. But it does so by portraying the agent as only *apparently* risk-averse. On this explanation, the agent would be happy to take a risk if only it offered her a higher expectation of what she really values: utility. But might some agents be *genuinely* risk-averse—might they be willing to give up a bit of expected utility if it meant they didn't have to gamble? If we could offer agents a direct choice between a guaranteed 10 utils and a 50-50 gamble on 30, might some prefer the former? (Recall that utils are defined so as not to decrease in marginal value.) And might that preference be rationally permissible?

Let's grant for the sake of argument that simple risk-aversion cases involving monetary gambles can be explained by attributing to the agent a utility distribution with decreasing marginal utility over dollars. Other documented responses to risk cannot be explained by *any* kind of utility distribution. Suppose a fair lottery is to be held with 100 numbered tickets. You get to choose between two gambles, with the following payoffs should particular tickets be drawn:

	Ticket 1	Tickets 2–11	Tickets 12–100
Gamble A	$1M	$1M	$1M
Gamble B	$0	$5M	$1M

(Here "$1M" is short for 1 million dollars.) Which gamble would you prefer? After recording your answer somewhere, consider the next two gambles (on the same lottery) and decide which of them you would prefer if they were your only options:

	Ticket 1	Tickets 2–11	Tickets 12–100
Gamble C	$1M	$1M	$0
Gamble D	$0	$5M	$0

When subjects are surveyed, they often prefer Gamble D to C; they're probably not going to win anything, but if they do they'd like a serious shot at $5 million. On the other hand, many of the same subjects prefer Gamble A to B, because A guarantees them a payout of $1 million.

Yet anyone who prefers A to B while at the same time preferring D to C violates Savage's[12]

Sure-Thing Principle: If two acts yield the same outcome on a particular state, any preference between them remains the same if that outcome is changed.

In our example, Gambles A and B yield the same outcome for tickets 12 through 100: 1 million dollars. If we change that common outcome to 0 dollars, we get Gambles C and D. The Sure-Thing Principle requires an agent who prefers A to B also to prefer C to D. Put another way: if the Sure-Thing Principle holds, we can determine a rational agent's preferences between any two acts by focusing exclusively on the states for which those acts produce *different* outcomes. In both the decision problems here, tickets 12 through 100 produce the same outcome no matter which act the agent selects. So we ought to be able to determine her preferences by focusing exclusively on the outcomes for tickets 1 through 11. Yet if we focus exclusively on those tickets, A stands to B in exactly the same relationship as C stands to D. So the agent's preferences across the two decisions should be aligned.

The Sure-Thing Principle is a theorem of Savage's decision theory. It is therefore also a theorem of Jeffrey's decision theory for cases in which acts and states are independent, as they are in the present gambling example. Thus preferring A to B while preferring D to C—as real-life subjects often do—is incompatible with those two decision theories. And here we can't chalk up the problem to working with dollars rather than utils. There is no possible utility distribution over dollars on which Gamble A has a higher expected utility than Gamble B while Gamble D has a higher expected utility than Gamble C. (See Exercise 7.10.)

Jeffrey and Savage, then, must shrug off these commonly paired preferences as irrational. Yet Maurice Allais, the Nobel-winning economist who introduced the gambles in his (1953), thought that this combination of preferences could be perfectly rational. Because it's impossible to maintain these seemingly reasonable preferences while hewing to standard decision theory, the example is now known as **Allais' Paradox**. Allais thought the example revealed a deep flaw in the decision theories we've been considering.[13]

We have been discussing decision theories as *normative* accounts of how *rational* agents behave. Economists, however, often assume that decision theory provides an accurate *descriptive* account of *real* agents' market decisions. Real-life subjects' responses to cases like the Allais Paradox prompted economists to develop new descriptive theories of agents' behavior, such as Kahneman and Tversky's Prospect Theory (Kahneman and Tversky 1979; Tversky and Kahneman 1992). More recently, Buchak (2013) has proposed a generalization of standard decision theory that accounts for risk aversion without positing declining marginal utilities, and is consistent with the Allais preferences many real-life subjects display.

7.3 Causal Decision Theory

Although we have been focusing on the expected values of propositions describing acts, Jeffrey's valuation function can be applied to any sort of proposition. For example, suppose my favorite player has been out of commission for weeks with an injury, and I am waiting to hear whether he will play in tonight's game. I start wondering whether I would prefer that he play tonight or not. Usually it would make me happy to see him on the field, but there's the possibility that he will play despite his injury's not being fully healed. That would definitely be a bad outcome. So now I combine my credences about states of the world (is he fully healed? is he not?) with my utilities for the various possible outcomes (plays fully healed, plays not fully healed, etc.) to determine how happy I would be to hear that he's playing or not playing.

Having calculated expected utilities for both "plays" and "doesn't play", I decide whether I'd prefer that he play or not.

Put another way, I can use Jeffrey's expected utility theory to determine whether I would consider it good news or bad were I to hear that my favorite player will be playing tonight. And I can do so whether or not I have *any* influence on the truth of that proposition. Jeffrey's theory is sometimes described as calculating the "news value" of a proposition.

Even for propositions describing our own acts, Jeffrey's expected utility calculation assesses news value. I might be given a choice between a sure $1 and a 50-50 chance of $2.02. I would use my credences and utility distribution to determine expected values for each act, then declare which option I preferred. But notice that this calculation would go exactly the same if instead of my selecting among the options, someone else was selecting on my behalf. What's ultimately being compared are the proposition *that I receive a sure dollar* and the proposition *that I receive whatever payoff results from a particular gamble*. Whether I have the ability to make one of those propositions true rather than the other is irrelevant to Jeffrey's preference calculations.

7.3.1 Newcomb's Problem

Jeffrey's focus on news value irrespective of agency leads him into trouble with **Newcomb's Problem**. This problem was introduced to philosophy by Robert Nozick, who attributed its construction to the physicist William Newcomb. Here's how Nozick introduced the problem:

Suppose a being in whose power to predict your choices you have enormous confidence. (One might tell a science-fiction story about a being from another planet, with an advanced technology and science, who you know to be friendly, etc.) You know that this being has often correctly predicted your choices in the past (and has never, so far as you know, made an incorrect prediction about your choices), and furthermore you know that this being has often correctly predicted the choices of other people, many of whom are similar to you, in the particular situation to be described below. One might tell a longer story, but all this leads you to believe that almost certainly this being's prediction about your choice in the situation to be discussed will be correct.

There are two boxes. [The first box] contains $1,000. [The second box] contains either $1,000,000, or nothing.... You have a choice between two actions: (1) taking what is in both boxes (2) taking only what is in the second box.

Furthermore, and you know this, the being knows that you know this, and so on:

(I) If the being predicts you will take what is in both boxes, he does not put the $1,000,000 in the second box.

(II) If the being predicts you will take only what is in the second box, he does put the $1,000,000 in the second box.

The situation is as follows. First the being makes its prediction. Then it puts the $1,000,000 in the second box, or does not, depending upon what it has predicted. Then you make your choice. What do you do?   (1969, pp. 114–15)

Historically, Newcomb's Problem prompted the development of a new kind of decision theory, now known as Causal Decision Theory (sometimes just "CDT"). At the time of Nozick's discussion, extant decision theories (such as Jeffrey's) seemed to recommend taking just one box in Newcomb's Problem (so-called "one-boxing"). But many philosophers thought two-boxing was the rational act.[14] By the time you make your decision, the being has already made its prediction and taken its action. So the money is already either in the second box, or it's not—nothing you decide can affect whether the money is there. However much money is in the second box, you're going to get more money ($1,000 more) if you take both boxes. So you should two-box.

I've quoted Nozick's original presentation of the problem because in the great literature that has since grown up around Newcomb, there is often debate about what exactly counts as "a Newcomb Problem". Does it matter whether the agent is *certain* that the prediction will be correct? Does it matter *how* the predictor makes its predictions, and whether backward causation (some sort of information fed backwards from the future) is involved? Perhaps more importantly, who *cares* about such a strange and fanciful problem?

But our purpose is not generalized Newcombology—we want to understand why Newcomb's Problem spurred the development of Causal Decision Theory. That can be understood by working with just one version of the problem. Or better yet, it can be understood by working with a kind of problem that comes up in everyday life, and is much less fanciful:

I'm standing at the bar, trying to decide whether to order a third appletini. I reason through my decision as follows: Drinking a third appletini is the kind of act highly typical of people with addictive personalities. People with addictive personalities also tend to become smokers. I'd kind of like to have another drink, but I *really* don't want to become a smoker (smoking causes lung cancer, is increasingly frowned-upon in my social circle, etc.). So I don't order that next appletini.

Let's work through the reasoning just described using decision theory. First, stipulate that I have the following utility table:

	smoker	non
third appletini	−99	1
stop at two	−100	0

Ordering the third appletini is a dominant act. But dominance should dictate preference only when acts and states are independent, and my concern here is that they're not. My credence distribution has the following features (with A, S, and P representing the propositions that I order the appletini, that I become a smoker, and that I have an addictive personality, respectively):

$$\mathrm{cr}(S \,|\, P) > \mathrm{cr}(S \,|\, {\sim}P) \tag{7.11}$$
$$\mathrm{cr}(P \,|\, A) > \mathrm{cr}(P \,|\, {\sim}A) \tag{7.12}$$

I'm more confident I'll become a smoker if I have an addictive personality than if I don't. And having that third appletini is a positive indication that I have an addictive personality. Combining these two equations (and making a couple more assumptions I won't bother spelling out), we get:

$$\mathrm{cr}(S \,|\, A) > \mathrm{cr}(S \,|\, {\sim}A) \tag{7.13}$$

From my point of view, ordering the third appletini is positively correlated with becoming a smoker. Looking back at the utility table, I do not consider the states listed along the top to be probabilistically independent of the acts along the side. Luckily, Jeffrey's decision theory works even when acts and states are dependent. So I apply Jeffrey's valuation function to calculate expected utilities for the two acts:

$$\mathrm{EU}_{\mathrm{EDT}}(A) = -99 \cdot \mathrm{cr}(S \,|\, A) + 1 \cdot \mathrm{cr}({\sim}S \,|\, A)$$
$$\mathrm{EU}_{\mathrm{EDT}}({\sim}A) = -100 \cdot \mathrm{cr}(S \,|\, {\sim}A) + 0 \cdot \mathrm{cr}({\sim}S \,|\, {\sim}A) \tag{7.14}$$

Looking at these equations, you might think that A receives the higher expected utility. But I assign a considerably higher value to $cr(S \mid A)$ than $cr(S \mid \sim A)$, so the -99 in the top equation is multiplied by a significantly larger quantity than the -100 in the bottom equation. Assuming the correlation between S and A is strong enough, $\sim A$ receives the higher expected utility and I prefer to perform $\sim A$.

But this reasoning is all wrong! Whether I have an addictive personality is (let's say) determined by genetic factors, not anything I could possibly affect at this point in my life. The die is cast (so to speak); I either have an addictive personality or I don't; it's already determined (in some sense) whether an addictive personality is going to lead me to become a smoker. Nothing about this appletini—whether I order it or not—is going to change any of that. So I might as well enjoy the drink.[15]

Assuming the reasoning in the previous paragraph—rather than the reasoning originally presented in the example—is correct, it's an interesting question why Jeffrey's decision theory yields the wrong result. The answer is that on Jeffrey's theory, ordering the appletini gets graded down because it would be bad news about my future. If I order the drink, that's evidence that I have an addictive personality (as indicated in Equation (7.12)). Having an addictive personality is unfortunate because of its potential consequences for becoming a smoker. I expect the world in which I order another drink to be a worse world than the world in which I don't, and this is reflected in the EU_{EDT} calculation. Jeffrey's theory assesses the act of ordering a third appletini not in terms of outcomes it will *cause* to come about, but instead in terms of outcomes it provides *evidence* for. For this reason Jeffrey's theory is described as an Evidential Decision Theory (or "EDT").

The trouble with Evidential Decision Theory is that an agent's performing an act may be *evidence* of an outcome that it's too late for her to *cause* (or *prevent*). Even though the act indicates the outcome, it seems irrational to factor the value of that outcome into a decision about whether to peform the act. As Skyrms (1980a, p. 129) puts it, my not having the third drink in order to avoid becoming a smoker would be "a futile attempt to manipulate the cause by suppressing its symptoms." In making decisions we should focus on what we can control—the causal consequences of our acts. Weirich writes:

> Deliberations should attend to an act's causal influence on a state rather than an act's evidence for a state. A good decision aims to produce a good outcome rather than evidence of a good outcome. It aims for the good and not just signs of the good. Often efficacy and auspiciousness go hand in hand. When they come apart, an agent should perform an efficacious act rather than an auspicious act. (2012)

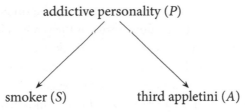

addictive personality (*P*)

smoker (*S*)   third appletini (*A*)

Figure 7.1 Third drink causal fork

7.3.2 A causal approach

The causal structure of our third drink example is depicted in Figure 7.1. As we saw in Chapter 3, correlation often indicates causation—but not *always*. Propositions on the tines of a causal fork will be correlated even though neither causes the other. This accounts for *A*'s being relevant to *S* on my credence distribution (Equation (7.13)) even though my ordering the third appletini has no causal influence on whether I'll become a smoker.

The causally spurious correlation in my credences affects Jeffrey's expected utility calculation because that calculation works with credences in states conditional on acts ($\mathrm{cr}(S_i \mid A)$). Jeffrey replaced Savage's $\mathrm{cr}(S_i)$ with this conditional expression to track dependencies between states and acts. The Causal Decision Theorist responds that while credal correlation is a *kind* of dependence, it's not the kind of dependence that decisions should track. Preferences should be based on *causal* dependencies. So the Causal Decision Theorist's valuation function is:

$$\mathrm{EU}_{\mathrm{CDT}}(A) = \mathrm{u}(A \ \& \ S_1) \cdot \mathrm{cr}(A \ \Box\!\!\rightarrow S_1) + \mathrm{u}(A \ \& \ S_2) \cdot \mathrm{cr}(A \ \Box\!\!\rightarrow S_2)$$
$$+ \ldots + \mathrm{u}(A \ \& \ S_n) \cdot \mathrm{cr}(A \ \Box\!\!\rightarrow S_n) \tag{7.15}$$

Here $A \ \Box\!\!\rightarrow S$ represents the subjunctive conditional "If the agent were to perform act *A*, state *S* would occur."[16] Causal Decision Theory uses such conditionals to track causal relations in the world.[17] Of course, an agent may be uncertain what consequences a given act *A* would cause. So $\mathrm{EU}_{\mathrm{CDT}}$ looks across the partition $\{S_1, \ldots, S_n\}$, and invokes the agent's credences that *A* would cause various states S_i to occur.

For many decision problems, Causal Decision Theory yields the same results as Evidential Decision Theory. In Jeffrey's wine example, it's plausible that

$$\mathrm{cr}(\text{chicken} \mid \text{white}) = \mathrm{cr}(\text{white} \ \Box\!\!\rightarrow \text{chicken}) = 0.75 \tag{7.16}$$

The guest's credence that chicken is served on the condition that she brings white wine is equal to her credence that if she were to bring white, chicken would be served. So one may be substituted for the other in expected utility calculations, and CDT's evaluations turn out the same as Jeffrey's.

But when conditional credences fail to track causal relations (as in cases involving causal forks), the two theories may yield different results. This is in part due to their differing notions of independence. EDT treats act A and state S as independent when they are *probabilistically* independent relative to the agent's credence distribution—that is, when $cr(S \mid A) = cr(S)$. CDT focuses on whether the agent takes A and S to be *causally* independent, which occurs just when

$$cr(A \mathbin{\square\!\!\rightarrow} S) = cr(S) \tag{7.17}$$

When an agent thinks A has no causal influence on S, her credence that S will occur if she performs A is just her credence that S will occur. In the third drink example my ordering another appletini may be evidence that I'll become a smoker, but I know it has no causal bearing on whether I take up smoking. So from a Causal Decision Theory point of view, the acts and states in that problem are independent. When acts and states are independent, dominance reasoning is appropriate, so CDT would have me prefer the dominant act and order the third appletini.

Now we can return to the Newcomb Problem, focusing on a version of it that distinguishes Causal from Evidential Decision Theory. Suppose that the "being" in Nozick's story makes its prediction by analyzing your brain state prior to your making the decision and applying a complex neuro-psychological theory. The being's track record makes you 99% confident that its predictions will be correct. And to simplify matters, let's suppose you assign exactly 1 util to each dollar, no matter how many dollars you already have. Then your utility and credence matrices for the problem are:

Utilities	P_1	P_2
T_1	1,000,000	0
T_2	1,001,000	1,000

Credences	P_1	P_2
T_1	0.99	0.01
T_2	0.01	0.99

where T_1 and T_2 represent the acts of taking one box or two boxes (respectively), and P_1 and P_2 represent the states of what the being predicted.

Jeffrey calculates expected values for the acts as follows:

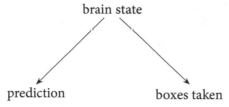

brain state

prediction   boxes taken

Figure 7.2 Newcomb Problem causal fork

$$EU_{EDT}(T_1) = u(T_1 \& P_1) \cdot cr(P_1 \mid T_1) + u(T_1 \& P_2) \cdot cr(P_2 \mid T_1) = 990,000$$
$$EU_{EDT}(T_2) = u(T_2 \& P_1) \cdot cr(P_1 \mid T_2) + u(T_2 \& P_2) \cdot cr(P_2 \mid T_2) = 11,000$$
$$(7.18)$$

So Evidential Decision Theory recommends one-boxing. Yet we can see from Figure 7.2 that this version of the Newcomb Problem contains a causal fork; the being's prediction is based on your brain state, which also has a causal influence on the number of boxes you take. This should make us suspicious of EDT's recommendations. The agent's act and the being's prediction are probabilistically correlated in the agent's credences, as the credence table reveals. But that's not because the number of boxes taken has any causal influence on the prediction.

Causal Decision Theory calculates expected utilities in the example like this:

$$EU_{CDT}(T_1) = u(T_1 \& P_1) \cdot cr(T_1 \mathrel{\Box\!\!\rightarrow} P_1) + u(T_1 \& P_2) \cdot cr(T_1 \mathrel{\Box\!\!\rightarrow} P_2)$$
$$= 1,000,000 \cdot cr(T_1 \mathrel{\Box\!\!\rightarrow} P_1) + 0 \cdot cr(T_1 \mathrel{\Box\!\!\rightarrow} P_2)$$

$$EU_{CDT}(T_2) = u(T_2 \& P_1) \cdot cr(T_2 \mathrel{\Box\!\!\rightarrow} P_1) + u(T_2 \& P_2) \cdot cr(T_2 \mathrel{\Box\!\!\rightarrow} P_2)$$
$$= 1,001,000 \cdot cr(T_2 \mathrel{\Box\!\!\rightarrow} P_1) + 1,000 \cdot cr(T_2 \mathrel{\Box\!\!\rightarrow} P_2)$$
$$(7.19)$$

It doesn't matter what particular values the credences in these expressions take, because the act has no causal influence on the prediction. That is,

$$cr(T_1 \mathrel{\Box\!\!\rightarrow} P_1) = cr(P_1) = cr(T_2 \mathrel{\Box\!\!\rightarrow} P_1) \qquad (7.20)$$

and

$$cr(T_1 \mathrel{\Box\!\!\rightarrow} P_2) = cr(P_2) = cr(T_2 \mathrel{\Box\!\!\rightarrow} P_2) \qquad (7.21)$$

With these causal independencies in mind, you can tell by inspection of Equation (7.19) that $EU_{CDT}(T_2)$ will be greater than $EU_{CDT}(T_1)$, and Causal Decision Theory endorses two-boxing.[18]

7.3.3 Responses and extensions

So is that it for Evidential Decision Theory? Philosophical debates rarely end cleanly; Evidential Decision Theorists have made a number of responses to the Newcomb Problem.

First, one might respond that one-boxing is the rationally mandated act. Representing the two-boxers, David Lewis once wrote:

> The one-boxers sometimes taunt us: if you're so smart, why ain'cha rich? They have their millions and we have our thousands, and they think this goes to show the error of our ways. They think we are not rich because we have irrationally chosen not to have our millions.   (1981b, p. 377)

Lewis's worry is this: Suppose a one-boxer and a two-boxer each go through the Newcomb scenario many times. As a highly accurate predictor, the being in the story will almost always predict that the one-boxer will one-box, and so place the $1,000,000 in the second box for him. Meanwhile, the two-boxer will almost always find the second box empty. The one-boxer will rack up millions of dollars, while the two-boxer will gain only thousands. Each agent has the goal of making as much money as possible, so one-boxing (and, by extension, EDT) seems to provide a better rational strategy for reaching one's goals than two-boxing (and CDT).

The Causal Decision Theorist's response (going at least as far back as Gibbard and Harper 1978/1981) is that some unfortunate situations reward agents monetarily for behaving irrationally, and the Newcomb Problem is one of them. The jury is still out on whether this response is convincing. In November 2009, the PhilPapers Survey polled over three thousand philosophers, and found that 31.4% of them accepted or leaned toward two-boxing in the Newcomb Problem, while 21.3% accepted or leaned toward one-boxing. (The remaining respondents were undecided or offered a different answer.) So not everyone considers EDT's embrace of one-boxing a fatal defect. Meanwhile, there are other cases in which EDT seems to give the intuitively rational result while CDT does not (Egan 2007).

Jeffrey, on the other hand, was convinced that two-boxing is rationally required in the Newcomb Problem. So he tried to reconcile Evidential Decision Theory with that verdict in a variety of ways. In the second edition of *The Logic of Decision* (1983), Jeffrey added a **ratifiability** condition to his EDT. Ratifiability holds that an act is rationally permissible only if the agent assigns it the highest expected utility conditional on the supposition that she chooses to perform it. Ratifiability avoids regret—if choosing to perform an act would make you wish you'd done something else, then you shouldn't choose it. In the Newcomb Problem, supposing that you'll choose to one-box makes you confident that the being predicted one-boxing, and so makes you confident that the $1,000,000 is in the second box. So supposing that you'll choose to one-box makes two-boxing seem the better choice. One-boxing is unratifiable, and so can be rationally rejected.

We won't cover the technical details of ratifiability here, in part because Jeffrey ultimately abandoned that response. Jeffrey eventually (1993, 2004) came to believe that the Newcomb Problem isn't really a decision problem, and therefore isn't the kind of thing against which a decision theory (like EDT) should be tested. Suppose that in the Newcomb Problem the agent assigns the credences we described earlier because she takes the causal structure of her situation to be something like Figure 7.2. In that case, she will see her physical brain state as having such a strong influence on how many boxes she takes that whether she one-boxes or two-boxes will no longer seem like a free choice. Jeffrey held that in order to make a genuine decision, an agent must see her choice as the cause of the act (and ultimately the outcome) produced. Read in this light, the Newcomb case seemed to involve too much causal influence on the agent's act from factors beyond her choice. In the final sentences of his last work, Jeffrey wrote, "I now conclude that in Newcomb problems, 'One box or two?' is not a question about how to choose, but about what you are already set to do, willy-nilly. Newcomb problems are not decision problems" (2004, p. 113).

7.4 Exercises

Unless otherwise noted, you should assume when completing these exercises that the credence distributions under discussion satisfy the probability axioms and Ratio Formula. You may also assume that whenever a conditional credence expression occurs, the needed proposition has nonzero unconditional credence so that conditional probabilities are well defined.

Problem 7.1. 🎵 When you play craps in a casino there are a number of different bets you can make at any time. Some of these are "proposition bets" on the outcome of the next roll of two fair dice. Below is a list of some proposition bets, and the odds at which casinos offer them:

Name of bet	Wins when	Odds paid
Big red	Dice total 7	4 : 1
Any craps	Dice total 2, 3, or 12	7 : 1
Snake eyes	Dice total 2	30 : 1

Suppose you place a $1 bet on each proposition at the odds listed above. Rank the three bets from highest expected net return to lowest.

Problem 7.2. 🎵 Suppose you're guarding Stephen Curry in an NBA game and he is about to attempt a three-point shot. You have to decide whether to foul him in the act of shooting.
  (a) If you don't foul him, he will attempt the shot. During the 2014–15 NBA season, Steph Curry made 44.3% of his three-point shot attempts. What is the expected number of points you will yield on Curry's shot attempt if you decide not to foul him?
  (b) Suppose that if you decide to foul Curry, you can ensure he doesn't get a three-point shot attempt off. However, your foul will send him to the free-throw line, where he will get three attempts, each worth one point if he makes it. During the 2014–15 NBA season, Curry made 91.4% of his free-throw attempts. Assuming that the result of each free-throw attempt is probabilistically independent of the results of all the others, what is the expected number of points you will yield on Curry's free throws if you foul him?
  (c) Given your calculations from parts (a) and (b), should you foul Steph Curry when he attempts a three-pointer?

Problem 7.3. The St. Petersburg game is played as follows: A fair coin is flipped repeatedly until it comes up heads. If the coin comes up heads on the first toss, the player wins $2. Heads on the second toss pays $4, heads on the third toss pays $8, etc.[19]
  (a) 🎵 If you assign fair prices equal to expected monetary payouts (and credences equal to objective chances), how much should you be willing to pay to play the St. Petersburg game?

(b) 🖊 If you were confronted with this game in real life, how much would you be willing to pay to play it? Explain your answer.

Problem 7.4. 🌶 Asked to justify his decision to bring along his chicken-replace-inator, Dr. Doofenshmirtz replies: "I'd rather have it and not need it than need it and not have it."

(a) Supposing the relevant states of the world are N (need-replace-inator) and $\sim N$, and the relevant acts are B (bring-replace-inator) and $\sim B$, what two outcomes is Doofenshmirtz referencing, and how is he claiming their utilities compare for him?

(b) Explain why on Savage's utility theory, this fact about Doofenshmirtz's utilities does not necessarily make his decision rationally permissible.

Problem 7.5. 🌶 Consider once again the utility table for the umbrella decision problem on page 255. Given this utility distribution, how confident would you need to be in rain for Savage's decision theory to recommend that you take your umbrella?

Problem 7.6. 🌶 Imagine there's some proposition P in which I'm highly interested (and whose truth I view as probabilistically independent of my behavior). Learning of my interest, a nefarious character offers to sell me the following betting ticket for $0.70:

> This ticket entitles the bearer
> to $1 if P is true,
> and nothing otherwise.

(a) For my fair betting price for this ticket to be exactly $0.70, what would my credence in P have to be?

(b) The nefarious character also has a second ticket available, which he offers to sell me for $0.70 as well:

> This ticket entitles the bearer
> to $1 if $\sim P$ is true,
> and nothing otherwise.

For my fair betting price in this second ticket to be exactly $0.70, what would my credence in $\sim P$ have to be?

(c) Suppose I throw caution to the wind and purchase both tickets, each at a price of $0.70. Without knowing my actual credences in P and $\sim P$, can you nevertheless calculate my expected *total* monetary value for the two tickets combined—taking into account both what I spent to get them and what they might pay out?

Problem 7.7. 🎵🎵
(a) Suppose an agent is indifferent between two gambles with the following utility outcomes:

	P	~P
Gamble 1	x	y
Gamble 2	y	x

where P is a proposition about the state of the world, and x and y are utility values with $x \neq y$. Assuming this agent maximizes EU_{SAV}, what can you determine about the agent's cr(P)?

(b) Suppose the same agent is also indifferent between these two gambles:

	P	~P
Gamble 3	d	w
Gamble 4	m	m

where $cr(P) = cr(\sim P)$, $d = 100$, and $w = -100$. What can you determine about m?

(c) Finally, suppose the agent is indifferent between these two gambles:

	Q	~Q
Gamble 5	r	s
Gamble 6	t	t

where $r = 100$, $s = 20$, and $t = 80$. What can you determine about cr(Q)?

Problem 7.8. 🎵🎵 You are confronted with a decision problem involving two possible states of the world (S and ~S) and three available acts (A, B, and C).
(a) Suppose that of the three S-outcomes, B & S does not have the highest utility for you. Also, of the three ~S-outcomes, B & ~S does not have the highest utility. Applying Savage's decision theory, does it follow that you should not choose act B? Defend your answer.
(b) Suppose that of the S-outcomes, B & S has the *lowest* utility for you. Also, of the three ~S-outcomes, B & ~S has the *lowest* utility. Still applying Savage's decision theory, does it follow that you should not choose act B? Defend your answer.

(c) Suppose now that you apply Jeffrey's decision theory to the situation in part (b). Do the same conclusions necessarily follow about whether you should choose act B? Explain.[20]

Problem 7.9. 🎵🎵 Suppose an agent faces a decision problem with two acts A and B and finitely many states.

(a) Prove that if the agent sets her preferences using EU_{SAV}, those preferences will satisfy the Dominance Principle.

(b) If the agent switches from EU_{SAV} to EU_{EDT}, exactly where will your proof from part (a) break down?

Problem 7.10. 🎵🎵 Referring to the payoff tables for Allais' Paradox in Section 7.2.4, show that no assignment of values to $u(\$0)$, $u(\$1M)$, and $u(\$5M)$ that makes $EU_{EDT}(A) > EU_{EDT}(B)$ will also make $EU_{EDT}(D) > EU_{EDT}(C)$. (You may assume that the agent assigns equal credence to each numbered ticket's being selected, and this holds regardless of which gamble is made.)

Problem 7.11. 🎵🎵 Having gotten a little aggressive on a routine single to center field, you're now halfway between first base and second base. You must decide whether to proceed to second base or run back to first.

The throw from the center fielder is in midair, and given the angle you can't tell whether it's headed to first or second base. But you do know that this center fielder has a great track-record at predicting where runners will go—your credence in his throwing to second conditional on your going there is 90%, while your credence in his throwing to first conditional on your going to first is 80%.

If you and the throw go to the same base, you will certainly be out, but if you and the throw go to different bases you'll certainly be safe. Being out has the same utility for you no matter where you're out. Being safe at first is better than being out, and being safe at second is better than being safe at first by the same amount that being safe at first is better than being out.

(a) Of the two acts available (running to first or running to second), which should you prefer according to Evidential Decision Theory (that is, accoring to Jeffrey's decision theory)?

(b) Does the problem provide enough information to determine which act is preferred by Causal Decision Theory? If so, explain which act is preferred. If not, explain what further information would be required and how it could be used to determine a preference.

Problem 7.12. ✍ In the Newcomb Problem, do you think it's rational to take just one box or take both boxes? Explain your thinking.

7.5 Further reading

INTRODUCTIONS AND OVERVIEWS

Michael D. Resnik (1987). *Choices: An Introduction to Decision Theory*. Minneapolis: University of Minnesota Press

Martin Peterson (2009). *An Introduction to Decision Theory*. Cambridge Introductions to Philosophy. Cambridge: Cambridge University Press

Each of these provides a book-length general introduction to decision theory, including chapters on game theory and social choice theory.

CLASSIC TEXTS

Leonard J. Savage (1954). *The Foundations of Statistics*. New York: Wiley

Savage's classic book laid the foundations for modern decision theory and much of contemporary Bayesian statistics.

Richard C. Jeffrey (1983). *The Logic of Decision*. 2nd edition. Chicago: University of Chicago Press

In the first edition, Jeffrey's Chapter 1 introduced a decision theory capable of handling dependent acts and states. In the second edition, Jeffrey added an extra section to this chapter explaining his "ratifiability" response to the Newcomb Problem.

EXTENDED DISCUSSION

Lara Buchak (2013). *Risk and Rationality*. Oxford: Oxford University Press

Presents a generalization of the decision theories discussed in this chapter that is consistent with a variety of real-life agents' responses to risk. For instance, Buchak's theory accommodates genuine risk-aversion, and allows agents to simultaneously prefer Gamble *A* to Gamble *B* and Gamble *D* to Gamble *C* in Allais' Paradox.

James M. Joyce (1999). *The Foundations of Causal Decision Theory.* Cambridge: Cambridge University Press

A systematic explanation and presentation of causal decision theory, unifying that approach under a general framework with evidential decision theory and proving a representation theorem that covers both.

Notes

1. The law of large numbers comes in many different forms, each of which has slightly different conditions and a slightly different conclusion. Most versions require the repeated trials to be independent and identically distributed (IID), meaning that each trial has the same probability of yielding a given result and the result on a given trial is independent of all previous results. (In other words, you think our batter is consistent across games and unaffected by previous performance.) Most versions also assume Countable Additivity for their proof. Finally, since we are dealing with results involving the infinite, we should remember that in this context credence 1 doesn't necessarily mean certainty. An agent who satisfies the probability axioms, the Ratio Formula, and Countable Additivity will assign credence 1 to the average's approaching the expectation in the limit, but that doesn't mean she *rules out* all possibilities in which those values don't converge. (For Countable Additivity and cases of credence-1 that don't mean certainty, see Section 5.4. For more details and proofs concerning laws of large numbers, see Feller 1968, Ch. X.)

2. See Bernoulli (1738/1954) for both his discussion and a reference to Cramer.

3. The first money pump was presented by Davidson, McKinsey, and Suppes (1955, p. 146), who attributed the inspiration for their example to Norman Dalkey. I don't know who introduced the "money pump" terminology.

   By the way, if you've ever read Dr. Seuss's story "The Sneetches", the Fix-it-Up Chappie (Sylvester McMonkey McBean) gets a pretty good money pump going before he packs up and leaves.

4. Though Quinn (1990) presents a case ("the puzzle of the self-torturer") in which it may be rational for an agent to have intransitive preferences.

5. While Savage thought of acts as functions from states to outcomes, it will be simpler for us to treat acts, states, and outcomes as propositions—the proposition that the agent will perform the act, the proposition that the world is in a particular state, and the proposition that a particular outcome occurs.

6. For simplicity's sake we set aside cases in which some S_i make particular acts impossible. Thus $A \& S_i$ will never be a contradiction.

7. The Dominance Principle I've presented is sometimes known as the Strong Dominance Principle. The Weak Dominance Principle says that if *A* produces *at least as good* an

outcome as B in each possible state of the world, plus a better outcome in at least one possible state of the world, then A is preferred to B. The names of the principles can be a bit confusing—it's not that Strong Dominance is a stronger *principle*; it's that it involves a stronger kind of dominance. In fact, the Weak Dominance Principle is logically stronger than the Strong Dominance Principle, in the sense that the Weak Dominance Principle entails the Strong Dominance Principle. (Thanks to David Makinson for suggesting this clarification.)

Despite being a logically stronger principle, Weak Dominance is also a consequence of Savage's expected utility theory, and has the same kinds of problems as Strong Dominance.

8. In a similar display of poor reasoning, Shakespeare's Henry V (Act 4, Scene 3) responds to Westmoreland's wish for more troops on their side of the battle—"O that we now had here but one ten thousand of those men in England, that do no work today"—with the following:

> If we are marked to die, we are enough to do our country loss;
> and if to live, the fewer men, the greater share of honor.
> God's will, I pray thee wish not one man more.

9. For a brief discussion and references, see Jeffrey (1983, § 1.8).
10. Instead of referring to "acts", "states", "outcomes", and "utilities", Jeffrey speaks of "acts", "conditions", "consequences", and "desirabilities" (respectively). As in my presentation of Savage's theory, I have made some changes to Jeffrey's approach for the sake of simplicity, and consistency with the rest of the discussion.
11. The decision-theoretic structure here bears striking similarities to Simpson's Paradox. We saw in Section 3.2.3 that while DeMar DeRozan had a better overall field-goal percentage than James Harden during the 2016–17 NBA season, from each distance (two-pointer versus three-pointer) Harden was more accurate. This was because a much higher proportion of DeRozan's shot attempts were two-pointers, which are much easier to make. So if you selected a DeRozan attempt at random, it was much more likely than a Harden attempt to have been a two-pointer, and so much more likely to have been made. Similarly, the deterrence utility table shows that disarming yields better outcomes than arming on each possible state of the world. Yet arming is much more likely than disarming to land you in the peace state (the right-hand column of the table), and so get you a desirable outcome.
12. While Savage coined the phrase "Sure-Thing Principle", it's actually a bit difficult to tell from his text exactly what he meant by it. I've presented a contemporary cleaning-up of Savage's discussion, inspired by the Sure-Thing formulation in Eells (1982, p. 10). It's also worth noting that the Sure-Thing Principle is intimately related to decision-theoretic principles known as Separability and Independence, but we won't delve into those here.
13. Heukelom (2015) provides an accessible history of the Allais Paradox, and of Allais' disputes with Savage over it. Another well-known counterexample to Savage's decision theory based on risk aversion is the Ellsberg Paradox, which we'll discuss in Section 14.1.3.

14. In case you're looking for a clever way out of Newcomb's Problem, Nozick specifies in a footnote that if the being predicts you will decide what to do via some random process (like flipping a coin), he does not put the $1,000,000 in the second box.

15. Eells (1982, p. 91) gives a parallel example from theology: "Calvinism is sometimes thought to involve the thesis that election for salvation and a virtuous life are effects of a common cause: a certain kind of soul. Thus, while leading a virtuous life does not cause one to be elected, still the probability of salvation is higher conditional on a virtuous life than conditional on an unvirtuous life. Should one lead a virtuous life?"

16. It's important for Causal Decision Theory that $A \mathbin{\Box\!\!\rightarrow} S$ conditionals be "causal" counterfactuals rather than "backtracking" counterfactuals; we hold facts about the past fixed when assessing A's influence on S. (See Lewis 1981a for the distinction and some explanation.)

17. There are actually many ways of executing a causal decision theory; the approach presented here is that of Gibbard and Harper (1978/1981), drawing from Stalnaker (1972/1981). Lewis (1981a) thought Causal Decision Theory should instead return to Savage's unconditional credences and independence assumptions, but with the specification that acts and states be *causally* independent. For a comparison of these approaches along with various others, plus a general formulation of Causal Decision Theory that attempts to cover them all, see Joyce (1999).

18. If you feel like Newcomb's Problem is too fanciful and our appletini example too frivolous to merit serious concern, consider that Gallo et al. (2018) found substantial evidence that smoking more cigarettes or smoking for a longer time is correlated with a decreased risk of developing Parkinson's disease. If Reichenbach's Principle of the Common Cause (Section 3.2.4) is true, then either smoking has a causal effect on whether one develops Parkinson's, Parkinson's somehow affects whether one smokes, or some other cause makes one both more likely to smoke and less likely to develop Parkinson's. Pursuing this third, causal-fork option, the researchers speculated that a dopamine shortage in the brain might contribute both to Parkinson's and to a "low-risk-taking personality trait" that makes people less likely to smoke or more likely to quit. If that's right, then should you take up smoking to avoid Parkinson's? Other studies have found that high levels of education correlate positively with developing Parkinson's (Frigerio et al. 2005). Should you cut short your education to avoid the disease?

19. This game was invented by Nicolas Bernoulli in the eighteenth century.

20. This problem was inspired by a problem of Brian Weatherson's.

PART IV
ARGUMENTS FOR BAYESIANISM

To my mind, the best argument for Bayesian epistemology is the uses to which it can be put. In the previous part of this book we saw how the Bayesian approach interacts with confirmation and decision theory, two central topics in the study of theoretical and practical rationality (respectively). The five core normative Bayesian rules grounded formal representations of how an agent should assess what her evidence supports and how she should make decisions in the face of uncertainty. These are just two of the many applications of Bayesian epistemology, which have established its significance in the minds of contemporary philosophers.

Nevertheless, Bayesian history also offers more direct arguments for the normative Bayesian rules. The idea is to *prove* from premises plausible on independent grounds that, say, a rational agent's unconditional credences satisfy Kolmogorov's probability axioms. The three most prominent kinds of arguments for Bayesianism are those based on representation theorems, Dutch Books, and accuracy measurements. This part of the book will devote one chapter to each type of argument.

Some of these argument-types can be used to establish more than just the probability axioms as requirements of rationality; the Ratio Formula, Conditionalization, Countable Additivity, and other norms we have discussed may be argued for. Each argument-type has particular norms it can and can't be used to support; I'll mention these applications as we go along. But they *all* can be used to argue for probabilism.

As I mentioned in Chapter 2, probabilism is the thesis that a rational agent's unconditional credence distribution at a given time satisfies Kolmogorov's three axioms. (I sometimes call a distribution that satisfies the axioms a "probabilistic" distribution.) Among the probability axioms, by far the most difficult to establish as a rational requirement is Finite Additivity. We'll see why as we dig into the arguments' particulars, but it's worth a quick reminder at this point what Finite Additivity does.

Chapter 2 introduced three characters: Mr. Prob, Mr. Weak, and Mr. Bold. We imagine there is some single proposition P for which the three of them assign the following credences:

Mr. Prob: $cr(F) = 0$ $cr(P) = 1/6$ $cr(\sim P) = 5/6$ $cr(T) = 1$
Mr. Weak: $cr(F) = 0$ $cr(P) = 1/36$ $cr(\sim P) = 25/36$ $cr(T) = 1$
Mr. Bold: $cr(F) = 0$ $cr(P) = 1/\sqrt{6}$ $cr(\sim P) = \sqrt{5}/\sqrt{6}$ $cr(T) = 1$

All three of these characters satisfy the Non-Negativity and Normality axioms. They also satisfy such intuitive credence norms as Entailment—the rule that a proposition must receive at least as much credence as any proposition that entails it. We could easily introduce conditional credences that have them satisfy the Ratio Formula as well. Yet of the three, only Mr. Prob satisfies Finite Additivity. This demonstrates that Finite Additivity is logically independent of these other norms; they can be satisfied even if Finite Additivity is not.

Mr. Weak's credences are obtained by squaring each of Mr. Prob's. This makes Mr. Weak's levels of confidence in logically contingent propositions $(P, \sim P)$ lower than Mr. Prob's. Mr. Weak is comparatively conservative, unwilling to be very confident in contingent claims. So while Mr. Weak is certain of $P \vee \sim P$, his individual credences in P and $\sim P$ sum to less than 1. Mr. Bold's distribution, on the other hand, is obtained by square-rooting Mr. Prob's credences. Mr. Bold is highly confident of contingent propositions, to the point that his credences in P and $\sim P$ sum to more than 1.

When we argue for Finite Additivity as a rational norm, we are arguing that Mr. Weak and Mr. Bold display a rational flaw not present in Mr. Prob. It's worth wondering in exactly what respect Mr. Weak and Mr. Bold make a rational mistake. This is especially pressing because empirical findings suggest that real humans consistently behave like Mr. Bold: they assign credences to mutually exclusive disjuncts that sum to more than their credence in the disjunction. Tversky and Koehler (1994) summarize a great deal of evidence on this front. In one particularly striking finding, subjects were asked to write down the last digit of their phone number and then estimate the percentage of American married couples with exactly that many children. The subjects with numbers ending in 0, 1, 2, and 3 each assigned their digit a value greater than 25%. If we suppose the reported values express estimates common to all of these subjects, then each of them assigns estimates summing to more than 100% before we even get to families with more than three kids!

Each of the three argument-types we consider will explain what's wrong with violating Finite Additivity in a slightly different way. And for each argument, I will ultimately have the same complaint. Finite Additivity is a linearity constraint; it requires a disjunction's credence to be a linear combination of the credences in its mutually exclusive disjuncts. In order to support Finite Additivity, each of the arguments assumes some other linearity constraint, whose normative credentials are no more clear than those of Finite Additivity. I call this the Linearity In, Linearity Out problem: We want to establish that treating credences additively is required by more fundamental rational principles. But ultimately we rely on principles that are additive (or intimately bound up with additive distributions) themselves. Instead of demonstrating that some deep aspect of rationality demands linearity, we have snuck in the linearity through our premises.

If traditional *arguments* for probabilism are question-begging in this manner, then probabilism's *applications* become all the more significant as reasons to endorse probabilistic norms. Near the end of Chapter 10 I'll ask whether Finite Additivity is really necessary for those applications. We'll briefly examine whether Bayesian epistemology's successes in confirmation and decision theory could be secured without such a strong commitment to probabilism.

Further Reading

Alan Hájek (2009a). Arguments for—or against—Probabilism? In: *Degrees of Belief*. Ed. by Franz Huber and Christoph Schmidt-Petri. Vol. 342. Synthese Library. Springer, pp. 229–51

Excellent introduction to, and assessment of, all the arguments for probabilism discussed in this part of the book.

8

Representation Theorems

Decision theory aligns a rational agent's preferences among available acts with her credence and utility distributions. It does so in two steps: first, a valuation function combines the agent's credences and utilities to assign each act a numerical score; second, the rational agent prefers acts with higher scores.

Savage's decision theory evaluates an act A by calculating its expected utility as follows:

$$EU(A) = u(A \& S_1) \cdot cr(S_1) + u(A \& S_2) \cdot cr(S_2) + \ldots + u(A \& S_n) \cdot cr(S_n) \quad (8.1)$$

where u represents the agent's utilities, cr represents her credences, and states S_1 through S_n form a finite partition. A rational agent has

$A \succ B$ just in case

$$u(A \& S_1) \cdot cr(S_1) + \ldots + u(A \& S_n) \cdot cr(S_n) \quad (8.2)$$
$$> u(B \& S_1) \cdot cr(S_1) + \ldots + u(B \& S_n) \cdot cr(S_n)$$

where "$A \succ B$" indicates that the agent prefers act A to act B.

Equation (8.2) relates three types of attitudes that impact an agent's practical life: her preferences among acts, her credences in states, and her utilities over outcomes.[1] It's a bit like an equation with three variables; if we know two of them we can solve for the third. For instance, if I specify a rational agent's full utility and credence distributions, you can determine her preference between any two acts using Equation (8.2). Going in a different direction, de Finetti (1931/1989) showed that if you know an agent's utilities and her preferences among certain kinds of betting acts, you can determine her credences. Meanwhile von Neumann and Morgenstern (1947) showed that given an agent's preferences over risky acts with specified credal profiles (called "lotteries"), one can determine her utilities. (See Figure 8.1.)[2]

Yet at some point during the 1920s, Frank Ramsey discovered how to do something remarkable: given only one of the variables in Equation (8.2), he figured out how to determine the other two. (The relevant paper, (Ramsey

Fundamentals of Bayesian Epistemology 2: Arguments, Challenges, Alternatives. Michael G. Titelbaum,
Oxford University Press. © Michael G. Titelbaum 2022. DOI: 10.1093/oso/9780192863140.003.0008

Author	Preferences	Utilities	Credences
(straightforward)	determine	given	given
de Finetti (1931)	given	given	determine
von Neumann and Morgenstern (1947)	given	determine	given
Ramsey (1931)	given	determine	determine

Figure 8.1 Results deriving some decision-theoretic attitudes from others

1931), was published after Ramsey's death in 1930 at age 26.) Given an agent's full preference ranking over acts, Ramsey showed how to determine both that agent's credences and her utilities. Ramsey's method laid the groundwork for representation theorems later proven by Savage and others. And these representation theorems ground an important argument for probabilism.

This chapter begins with an overview of Ramsey's method for determining credences and utilities from preferences. I will then present Savage's representation theorem and discuss how it is taken to support probabilism. Finally, I will present contemporary criticisms of the representation theorem argument for probabilism. Especially eager readers may skip over the Ramsey section; strictly speaking one needn't know how Ramsey pulled the trick to understand representation theorems and their relation to probabilism. Yet I will not be presenting any proof of the representation theorem, so if you want to know how it's possible to get both credences and utilities from preferences it may be worth studying Ramsey's approach. Ramsey's process also illustrates why certain structural assumptions are necessary for the theorems that came later.

One note before we begin: Readers familiar with decision theory (perhaps from Chapter 7) will know that many contemporary decision theorists have found fault with Savage's expected utility formula as a valuation function. But since we will mainly be discussing Savage's representation theorem, I will use Savage-style expected utilities (as defined in Equation (8.1)) throughout this chapter. One can find similar representation theorems for Jeffrey-style Evidential Decision Theory in (Jeffrey 1965), and for Causal Decision Theory in (Joyce 1999).

8.1 Ramsey's four-step process

Here's how Ramsey's process works. We imagine we are given an agent's complete preference ranking over acts, some of which are acts of accepting various "gambles" (which provide one outcome if a proposition is true, another outcome if that proposition is false). We assume that the agent assigns finite numerical utilities, credences satisfying the probability axioms, and preferences in line with her (Savage-style) expected utilities. Yet we are given no further information about which credence and utility values she assigns to particular propositions. That's what we want to determine.

Ramsey's process works by sorting through the agent's preference rankings until we find preferences that fit certain patterns. Those patterns allow us to determine particular features of the agent's credences and utilities, which we then leverage to determine further features, until we can set a utility and credence value for each proposition in the agent's language $\mathcal{L}$.

Step One: Find ethically neutral propositions

Ramsey defines a proposition P as **ethically neutral** for an agent if the agent is indifferent between any two gambles whose outcomes differ only in replacing P with $\sim P$. The intuitive idea is that an agent just doesn't care how an ethically neutral proposition comes out, so she values any outcome in which P occurs just as much as she values an otherwise-identical outcome in which $\sim P$ occurs. (Despite the terminology, Ramsey is clear that a proposition's "ethical neturality" has little to do with ethics at all.) For instance, a particular agent might care not one whit about hockey teams and how they fare; this lack of caring will show up in her preferences among various acts (including gambles). Suppose this agent is confronted with two acts: one will make the Blackhawks win the Stanley Cup and also get her some ice cream, while another will make the Blackhawks lose but still get her the same ice cream. If propositions about hockey results are ethically neutral for the agent, she will be indifferent between performing those two acts.

In Step One of Ramsey's process, we scour the agent's preferences to find a number of propositions that are ethically neutral for her. We can tell an ethically neutral proposition P because every time P appears in the outcomes of a gamble, she will be indifferent between that gamble and another gamble in which every P in an outcome has been replaced by a $\sim P$.

Step Two: Find ethically neutral P, $\sim P$ with equal credence

We now examine the agent's preferences until we find three propositions X, Y, and P such that P is ethically neutral for the agent and the agent is indifferent between these two gambles:

	P	$\sim P$
Gamble 1	$X \,\&\, P$	$Y \,\&\, \sim P$
Gamble 2	$Y \,\&\, P$	$X \,\&\, \sim P$

In this decision table the possible states of the world are listed across the top row, while the acts available to the agent are listed down the first column. Since we don't know the agent's utility values, we can't put them in the cells. So I've listed there the outcome that will result from each act-state pair. For instance, Gamble 1 yields outcome X & P if P is true, Y & $\sim P$ if P is false. If we've established in Step One that hockey results are ethically neutral for our agent, then Gamble 1 might make it the case that the agent receives chocolate ice cream (X) if the Blackhawks win and vanilla (Y) if they lose, while Gamble 2 gives her vanilla if they win and chocolate if they lose.

   If the agent is indifferent between the acts of making Gamble 1 and Gamble 2, and if the agent's preferences reflect her expected utilities, then we have

$$u(X \& P) \cdot cr(P) + u(Y \& \sim P) \cdot cr(\sim P) = EU(\text{Gamble 1})$$
$$= EU(\text{Gamble 2}) = u(Y \& P) \cdot cr(P) + u(X \& \sim P) \cdot cr(\sim P) \tag{8.3}$$

But we've already ascertained that P is ethically neutral for the agent—she doesn't care whether P is true or false. So

$$u(X \& P) = u(X \& \sim P) = u(X) \tag{8.4}$$

Since the agent gets no utility advantage from P's being either true or false, her utility for X & P is just her utility for X, which is also her utility for X & $\sim P$.[3] A similar equation holds for Y. Substituting these results into Equation (8.3), we obtain

$$u(X) \cdot cr(P) + u(Y) \cdot cr(\sim P) = u(Y) \cdot cr(P) + u(X) \cdot cr(\sim P) \tag{8.5}$$

which we can rewrite as

$$u(X)[cr(P) - cr(\sim P)] = u(Y)[cr(P) - cr(\sim P)] \tag{8.6}$$

One way to make this equation true is to have $u(X) = u(Y)$. How can we determine whether those utilities are equal strictly from the agent's preferences? We might offer her a gamble that produces X no matter what—sometimes called a **constant act**—and another gamble that produces Y no matter what. If the agent is indifferent between the constant act that produces X and the constant act that produces Y, she must assign X and Y the same utilities.

   But now suppose we offer the agent a choice between those constant acts and she turns out to have a preference between X and Y. With $u(X) \neq u(Y)$, the only way to make Equation (8.6) true is to have $cr(P) = cr(\sim P)$, so that $cr(P) -$ $cr(\sim P) = 0$. If the agent is indifferent between Gambles 1 and 2, considers P ethically neutral, and assigns distinct utilities to X and Y, she must be equally confident in P and $\sim P$.

Intuitively, here's how this step works: If you prefer one outcome to another then you'll lean toward gambles that make you more confident you'll receive the preferred result. The only way you'll be indifferent between a gamble that gives you the preferred outcome on P and a gamble that gives you that preferred outcome on $\sim P$ is if your confidence in P is equal to your confidence in $\sim P$. To return to our earlier example: Suppose hockey propositions are ethically neutral for our agent, she prefers chocolate ice cream to vanilla, and she is offered two gambles. The first gamble provides chocolate on a Blackhawks win and vanilla on a loss; the second provides vanilla on a Blackhawks win and chocolate on a loss. If she thinks the Blackhawks are likely to win she'll prefer the first gamble (because she wants that chocolate); if she thinks the Blackhawks are likely to lose she'll prefer the second. Being indifferent between the gambles makes sense only if she thinks a Blackhawks loss is just as likely as a win.

Step Three: Determine utilities
We've now found at least one ethically neutral proposition P that the agent takes to be as likely as not. Next we survey the agent's preferences until we find three propositions D, M, and W satisfying the following two conditions: First, $u(D) > u(M) > u(W)$. (We can determine this by examining the agent's preferences among constant acts involving D, M, and W.) Second, the agent is indifferent between these two gambles:

	P	$\sim P$
Gamble 3	$D \& P$	$W \& \sim P$
Gamble 4	$M \& P$	$M \& \sim P$

Because P is ethically neutral for the agent, $u(D) = u(D \& P)$, $u(W) = u(W \& \sim P)$, and $u(M) = u(M \& P) = u(M \& \sim P)$. So the agent's indifference between these gambles tell us that

$$u(D) \cdot cr(P) + u(W) \cdot cr(\sim P) = u(M) \cdot cr(P) + u(M) \cdot cr(\sim P) \qquad (8.7)$$

We've also selected a P such that $cr(P) = cr(\sim P)$. So we can just divide through by this value, leaving

$$u(D) + u(W) = u(M) + u(M) \tag{8.8}$$

$$u(D) - u(M) = u(M) - u(W) \tag{8.9}$$

In other words, the gap between the agent's utilities in D and M must equal the gap between her utilities in M and W.

Intuitively: The agent prefers D to M, so if P is true then the agent would rather have Gamble 3 than Gamble 4. On the other hand, the agent prefers M to W, so if $\sim P$ then the agent would rather have Gamble 4. If the agent considered P much more likely than $\sim P$, then a small preference for D over M could balance a much stronger preference for M over W. But we've chosen a P that the agent finds just as likely as $\sim P$. So if the agent is indifferent between Gambles 3 and 4, the advantage conferred on Gamble 3 by its potential to provide D instead of M must precisely balance out the advantage conferred on Gamble 4 by its potential to provide M rather than W. The agent must value D over M by the exact same amount that she values M over W.

This kind of gamble allows us to establish equal utility gaps among various propositions. In this case, the utility gap between D and M must equal that between M and W. Suppose we stipulate that $u(D) = 100$ and $u(W) = -100$. (As we'll see in the next section, any finite values would've worked equally well here as long as $u(D) > u(W)$.) Equation (8.9) then tells us that $u(M) = 0$.

By repeatedly applying this technique, we can find a series of benchmark propositions for the agent's utility scale. For example, we might find a proposition C such that the utility gap between C and D is equal to that between D and M. In that case we know that $u(C) = 200$. On the other hand, we might find a proposition I whose utility is just as far from M as it is from D; I has utility 50. Then we find proposition G with utility 75. As we find more and more of these propositions with special utility values, we can use them to establish the utilities of other propositions (even propositions that don't enter into convenient Gambles like 3 and 4 between which the agent is indifferent). If the agent prefers the constant act that produces E to the constant act that produces G, her utility for E must be greater than 75. But if she prefers D's constant act to E's, $u(E)$ must be less than 100. By drawing finer and finer such distinctions, we can specify the agent's utility for an arbitrary proposition to as narrow an interval as we like. Repeated applications of this step will determine the agent's full utility distribution over a propositional language to any desired level of precision.

Step Four: Determine credences

We've now determined the agent's utilities for every proposition in her language; the final step is to determine her credences. To determine the agent's credence in an arbitrarily selected proposition Q, we find propositions R, S, and T such that the agent is indifferent between a constant act providing T and the following gamble:

	Q	$\sim Q$
Gamble 5	$R \& Q$	$S \& \sim Q$

We then have

$$u(T) = u(R \& Q) \cdot cr(Q) + u(S \& \sim Q) \cdot cr(\sim Q) \qquad (8.10)$$

We assumed at the outset that the agent's credence distribution satisfies the probability axioms. So we can replace $cr(\sim Q)$ with $1 - cr(Q)$, yielding

$$u(T) = u(R \& Q) \cdot cr(Q) + u(S \& \sim Q) \cdot [1 - cr(Q)] \qquad (8.11)$$

We then apply a bit of algebra to obtain

$$cr(Q) = \frac{u(T) - u(S \& \sim Q)}{u(R \& Q) - u(S \& \sim Q)} \qquad (8.12)$$

Since we already know the agent's utilities for every proposition in her language, we can fill out all the values on the right-hand side and calculate her credence in Q. And since this method works for arbitrarily selected Q, we can apply it repeatedly to determine the agent's entire credence distribution over her language.[4]

8.2 Savage's representation theorem

The previous section didn't flesh out *all* the details of Ramsey's process for determining credences and utilities from preferences. But Savage (1954) proved a representation theorem which guarantees that the necessary details can be provided. I'll start by presenting the theorem, then explain some of its individual parts.

Representation Theorem: If an agent's preferences satisfy certain constraints, then there exists a unique probabilistic credence distribution and unique utility distribution (up to positive affine transformation) that yield those preferences when the agent maximizes expected utility.

We saw the basic idea in Ramsey's four-step process: The Representation Theorem says that starting from an agent's preferences, we'll always be able to construct a unique probabilistic credence distribution and unique utility distribution (up to positive affine transformation, which I'll explain shortly) that generate those preferences through expected utility maximization.

In order for this to work, the preferences must satisfy certain constraints, often called the **preference axioms**. These constraints are called "axioms" because we take the agent's satisfaction of them as given in applying the representation theorem; calling them "axioms" does *not* mean they cannot be independently argued for.

For example, Savage assumes the preferences under discussion will satisfy these two constraints (introduced in Chapter 7):

Preference Asymmetry: There do not exist acts *A* and *B* such that the agent both prefers *A* to *B* and prefers *B* to *A*.

Preference Transitivity: For any acts *A*, *B*, and *C*, if the agent prefers *A* to *B* and *B* to *C*, then the agent prefers *A* to *C*.

Section 7.2.1's money pump argument tries to show that these axioms will be satisfied by the preferences of any *rational* agent.[5]

Preference Asymmetry and Transitivity are substantive constraints on an agent's preferences—the kinds of things we might rationally fault her for failing to meet. Yet many of Savage's axioms merely require the agent's preference structure to display a certain level of richness; Suppes (1974) calls these "structure axioms". We saw one good example in Chapter 7:

Preference Completeness: For any acts *A* and *B*, exactly one of the following is true: the agent prefers *A* to *B*, the agent prefers *B* to *A*, or the agent is indifferent between the two.

Even more demanding assumptions[6] popped up in Ramsey's four-step process: At various stages, we had to assume that if we combed through enough of the agent's preferences, we'd eventually find propositions falling into a very specific preference pattern. In Step Four, for example, we assumed that for any arbitrary proposition *Q* there would be propositions *R*, *S*, and *T* such that the agent was indifferent between *T*'s constant act and a gamble that generated *R* on *Q* and *S* otherwise. More generally, we assumed a large supply of propositions the agent treated as ethically neutral, and among those some propositions the agent took to be as likely as not.

It's doubtful that *any* agent has *ever* had preferences rich enough to satisfy all of these assumptions. And we wouldn't want to rationally fault agents for failing to do so. Yet decision theorists tend to view the structure axioms as harmless assumptions added in to make the math come out nicely. Since Savage's work, a number of alternative representation theorems have been proven, many of which relax his original structural assumptions.[7]

If an agent's preferences satisfy the preference axioms, Savage's Representation Theorem guarantees the existence of a unique probabilistic credence distribution for the agent and a unique utility distribution "up to positive affine transformation". Why can't we determine a unique utility distribution for the agent full stop? Recall that in Step Three of Ramsey's process—the step in which we determined the agent's utility distribution—we stipulated that proposition D had a utility of 100 and proposition W a utility of -100. I chose those values because they were nice round numbers; they had no special significance, and we easily could have chosen other values (as long as D came out more valuable than W). Stipulating other utilities for these propositions would have affected our utility assignments down the line. For example, if we had chosen $u(D) = 100$ and $u(W) = 0$ instead, the proposition M that we proved equidistant between D and W would have landed at a utility of 50 (rather than 0).

Yet I hope it's clear that differing utility scales resulting from different utility stipulations for D and W would have many things in common. This is because they measure the same underlying quantity: the extent to which an agent values a particular arrangement of the world. Different numerical scales that measure the same underlying quantity may be related in a variety of ways; we will be particularly interested in measurement scales related by scalar and affine transformations.

Two measurement scales are related by a **scalar transformation** when values on one scale are constant multiples of values on the other. A good example are the kilogram and pound measurement scales for mass. An object's mass in pounds is its mass in kilograms times 2.2. Thus the kilogram and pound scales are related by a scalar transformation. In this case the multiplying constant (2.2) is positive, so we call it a *positive* scalar transformation. Scalar transformations maintain zero points and ratios, and positive scalar transformations maintain ordinal ranks. Taking these one at a time: Anything that weighs 0 kilograms also weighs 0 pounds; the pound and kilogram scales have the same zero point. Moreover, if I'm twice as heavy as you in kilograms then I'll be twice as heavy as you in pounds. Scalar transformations preserve ratios among values. Finally, since it's a positive scalar transformation, ordering a group of people from lowest to highest weight in kilograms will also order them from lowest to highest weight in pounds.

Affine transformations are a bit more complex: the conversion not only multiplies by a constant but also adds a constant. Celsius and Fahrenheit temperatures are related by an affine transformation; to get Fahrenheit from Celsius you multiply by 1.8 then add 32. This is a *positive* affine transformation (determined again by the sign of the multiplying constant). Positive affine transformations maintain ordinal ranks, but not necessarily zero points or ratios among values. Again, one at a time: Tahiti is hotter than Alaska whatever temperature scale you use; a positive affine transformation keeps things in the same order from lowest to highest. While 0°C is the (usual) freezing point of water, 0°F is a much colder temperature. So a value of 0 does not indicate the same worldly situation on both temperature scales. Positive affine transformations may also distort ratios: 20°C is twice 10°C, but their equivalents 68°F and 50°F (respectively) do not fall in a ratio of 2 to 1. Affine transformations do, however, preserve facts about gap equality. Suppose I tell you, "Tomorrow will be hotter than today by the same number of degrees that today was hotter than yesterday." This will be true on the Fahrenheit scale just in case it's true on the Celsius scale. (Scalar transformations preserve gap equality as well, since scalar transformations are the special case of affine transformations in which the constant added is 0.)

Savage's representation theorem guarantees that if an agent's preferences satisfy the preference axioms, we will be able to find a probabilistic credence distribution and a utility distribution that match those preferences via expected utility maximization. In fact, we will be able to find *many* such utility distributions, but all the utility distributions that match this particular agent's preferences will be related by positive affine transformation. Decision theorists tend to think that if two utility distributions are related by a positive affine transformation, there is no real underlying difference between an agent's having one and the agent's having another. Each distribution will rank states of affairs in the same order with respect to value, and when put into an expected utility calculation with the same credence distribution, each will produce the same preferences among acts.[8] The real difference between such utility distributions lies in the particular utility values we—the utility measurers— stipulate to set up our measurement scale. Whether *we* choose to represent an agent's love for chocolate ice cream with a utility score of one hundred or one million, the *agent* will still prefer chocolate ice cream to vanilla, and vanilla ice cream to none. And our choice of measurement scale will not change the facts about whether she prefers chocolate to vanilla by exactly the same amount that she prefers vanilla to none. (Affine transformations preserve the equivalence of gaps; establishing equivalent utility gaps was the main business of Ramsey's third step.)

This equanimity among utility scales related by positive affine transformation does lose us absolute zero points and ratios among utility assignments. Different utility scales will yield different results about whether our agent likes chocolate ice cream *twice as much as* vanilla. If we're going to treat each of those measurement scales as equally accurate, we'll have to deny that there's any fact of the matter about the ratio between the agent's utility for chocolate and utility for vanilla. But it's unclear what it would even *mean* for an agent to value chocolate twice as much as vanilla (especially since such facts could have no bearing on her preferences among acts). So decision theorists tend not to mourn their inability to make utility ratio claims.

8.3 Representation theorems and probabilism

It's time to pause and ask what decision theory—and representation theorems—are really *for*. Kenny Easwaran writes:

> Naive applications of decision theory often assume that it works by taking a specification of probabilities and utilities and using them to calculate the expected utilities of various acts, with a rational agent being required to take whichever act has the highest (or sufficiently high) expected utility. However, justifications of the formal framework of expected utility theory generally work in the opposite way—they start with an agent's preferences among acts, and use them to calculate an implied probability and utility function . . . .
>
> The orthodox view of decision theory endorsed by Savage (1954) and Jeffrey (1965) takes preferences over acts with uncertain outcomes to be the fundamental concept of decision theory, and shows that if these preferences satisfy a particular set of axioms, then they can be represented by a probability function and a utility function . . . . This conflicts with a naive reading of the concept of expected utility, which was perhaps the dominant understanding of theories that were popular in the 17th to 19th centuries. One often assumes that utilities and probabilities are prior to preference, and that decision theory says that you *should* prefer an act with a higher expected utility over any act with a lower expected utility. And this is how the theory of expected utility is often applied in practical contexts.   (2014a, pp. 1–2, emphasis in original)

Decision theory is often presented (and was largely presented in Chapter 7) as a first-personal guide to decision-making and problem-solving. (Blaise Pascal initiated his famous calculations of expected returns because he and Pierre Fermat wanted to find the proper way of settling up the payouts for a casino game (Fermat and Pascal 1654/1929)). Once an agent has assigned her credences that various states of the world obtain, and her utilities to all the relevant outcomes, she can combine them via a valuation function to determine which available act she ought rationally to prefer.

Representation theorems belong to a fairly different approach to decision theory—what we might call a third-personal approach. Suppose an economist has been studying a particular subject, noting the decisions she has made when confronted by various choices in the past. This reveals *some* of the subject's preferences to the economist, but certainly not all. Suppose the economist also assumes that the subject is rational in the sense that her total set of preferences (both revealed and as-yet-unrevealed) together satisfies the preference axioms (Transitivity, Asymmetry, Completeness, etc.). A representation theorem then guarantees that the agent can be understood *as if* her past and future preferences are the result of maximizing expected utility relative to some utility distribution and probabilistic credence distribution. So the economist can take the subject's past preferences and deduce features of credence and utility distributions that would generate those preferences were she maximizing expected utility. The economist then uses what's known about those (imagined) credences and utilities to predict preferences not yet observed. The subject's future preferences must match these predictions, on pain of violating the preference axioms. (In part (b) of Exercise 8.3 you'll use Ramsey's four-step process to make a prediction in this way.) To the extent that decision theorists and economists can assume real agents satisfy the preference axioms, this makes decision theory a powerful predictive tool.

The third-personal approach can also be applied in a more abstract fashion. Since any agent who is rational in the sense of satisfying the preference axioms is representable as maximizing expected utility, we can prove results about the preferences of rational agents by proving that maximizing expected utility requires certain kinds of preference relationships. For instance, we could argue that any agent displaying the preferences in the Allais Paradox (Section 7.2.4) must be irrational by showing that no possible utility distribution would generate those preferences for an expected-utility maximizer.

All of these third-personal results—both abstract and particular—suppose at some point that the agent sets her preferences by maximizing expected utility relative to a utility distribution and probabilistic credence distribution.

But this supposition is a kind of bridge, taking the theorist from premises about the agent's preferences to conclusions that concern her preferences as well. To get the relevant results, we need not demonstrate that the agent *actually* sets her preferences using utilities and probabilistic credences. Expected-utility maximization acts as a mathematical model, making the acquisition of preference results more tractable.[9] Resnik (1987, p. 99) writes, "the [representation] theorem merely takes information already present in facts about the agent's preferences and reformulates it in more convenient numerical terms."

Could we do more? Some Bayesian epistemologists have used representation theorems to argue that rational agents must have probabilistic degrees of belief. At a first pass, the argument runs something like this:

Representation Theorem Argument for Probabilism

(Premise)   Every rational agent has preferences satisfying the preference axioms.

(Theorem)   Any agent whose preferences satisfy the preference axioms can be represented as maximizing expected utility relative to a probabilistic credence distribution.

(Conclusion) Every rational agent has a probabilistic credence distribution.

The rest of this chapter investigates whether any argument in the vicinity of this one is sound. I want to emphasize, though, that even if representation theorems cannot be employed in a convincing argument for probabilism, that does not undermine their usefulness for the third-person modeling projects just described.

8.3.1 Objections to the argument

As usual, a first approach to refuting this Representation Theorem Argument would be to deny its premise. Whatever representation theorem one uses in the argument (Savage's or one of its descendents), that theorem will assume that every rational agent satisfies some particular set of preference axioms. One might then deny that rationality requires satisfying those axioms. For example, some philosophers have argued that Preference Transitivity is not a rational requirement. (See Chapter 7, note 4.)

On the other hand one might accept the premise, but read it in a way that doesn't generate the desired conclusion. Chapter 1 distinguished practical

298 REPRESENTATION THEOREMS

rationality, which concerns connections between attitudes and action, from theoretical rationality, which assesses representational attitudes considered as such. The preference axioms, being constraints on preferences between *acts*, are requirements of practical rationality. So if it's successful, the Representation Theorem Argument demonstrates that any agent who satisfies the requirements of practical rationality has probabilistic credences. As Ramsey put it after laying out his four-step process,

> Any definite set of degrees of belief which broke [the probability rules] would be inconsistent in the sense that it violated the laws of preference between options, such as that preferability is a transitive asymmetrical relation.
>
> (1931, p. 84)

Yet we have offered probabilism as a thesis about *theoretical* rationality. The Representation Theorem Argument seems to show that an agent with non-probabilistic credences will make irrational decisions about how to behave in her life. But we wanted to show that non-probabilistic credences are flawed as representations in themselves, independently of how they lead to action. Adding the word "practically" before the word "rational" in the argument's Premise forces us to add "practically" before "rational" in its Conclusion as well, but that isn't the Conclusion we were hoping to obtain.[10]

Setting aside concerns about the Premise, one might worry about the validity of the Representation Theorem Argument as I've reconstructed it. Logically, the Premise and Theorem together entail that every rational agent can be *represented* as maximizing expected utility relative to a probabilistic credence distribution. How does it follow that every rational agent *has* probabilistic credences?

To establish that a rational agent has probabilistic credences, we need to establish two claims: (1) that the agent has numerical credences to begin with; and (2) that those credences satisfy the probability axioms. It's unclear that the Representation Theorem Argument can establish even the first of these claims. Alan Hájek explains the trouble as follows:

> The concern is that for all we know, the mere *possibility* of representing you one way or another might have less force than we want; your acting *as if* the representation is true of you does not make it true of you. To make this concern vivid, suppose that I represent your preferences with *Voodooism*. My voodoo theory says that there are warring voodoo spirits inside you. When you prefer A to B, then there are more A-favouring spirits inside

you than *B*-favouring spirits. I interpret all of the usual rationality axioms in voodoo terms. Transitivity: if you have more *A*-favouring spirits than *B*-favouring spirits, and more *B*-favouring spirits than *C*-favouring spirits, then you have more *A*-favouring spirits than *C*-favouring spirits .... And so on. I then 'prove' Voodooism: if your preferences obey the usual rationality axioms, then there exists a Voodoo representation of you. That is you act *as if* there are warring voodoo spirits inside you in conformity with Voodooism. Conclusion: rationality requires you to have warring Voodoo spirits in you. Not a happy result.   (2009a, p. 238, emphases in original)

It's possible to defend the representation theorem approach—and close this gap in the argument—by adopting a metaphysically thin conception of the attitudes in question. Voodoo (or Voudou) is a complex set of cultural traditions involving a variety of ontological and metaphysical commitments. Demonstrating that an agent behaves *as if* there were voodoo spirits inside her seems insufficient to establish such metaphysical claims. On the other hand, one might *define* the notion of a credence such that all there is to possessing a particular credence distribution is acting according to preferences with a particular structure. At one point Ramsey writes, "I suggest that we introduce as a law of psychology that [an agent's] behaviour is governed by what is called the mathematical expectation .... We ... define degree of belief in a way which presupposes the use of the mathematical expectation" (1931, p. 76). Bruno de Finetti (1937/1964) employs a similar definitional approach. On such a metaphysically thin behaviorist or functionalist account,[11] an agent's acting as if she has probabilistic credences may be tantamount to her having such credences. (This point of view also makes it less worrisome that the argument's Premise invokes constraints of practical rationality.)

The prominence of such operationalist views during parts of the twentieth century explains why so little distance was perceived between conclusions that follow uncontroversially from Savage's Representation Theorem and the more controversial claims of the Representation Theorem Argument. Yet such straightforward, metaphysically thin operationalisms have fallen out of favor, for a variety of reasons. For example, even if we identify mental states using their functional roles, it's too restrictive to consider only roles related to preferences among acts. Doxastic attitudes have a variety of functions within our reasoning, and may even be directly introspectible. Christensen (2004) also notes that even if they aren't directly introspectible, degrees of belief affect other mental states that can be introspected, such as our emotions. (Consider how confidence that you will perform well affects your feelings upon taking the stage.) With a thicker conception of credences in place, it's difficult to imagine that merely observing an agent's preferences among acts would suffice to attribute such doxastic attitudes to her.[12]

Perhaps, then, we can scale back what the Representation Theorem Argument is meant to show. Suppose we have convinced ourselves on independent grounds that agents have numerical degrees of belief, perhaps via considerations about comparisons of confidence like those adduced in Chapter 1. With the existence of credences already established, could the Representation Theorem Argument show that rationality requires those credences to satisfy the probability axioms? Unfortunately the logic of the argument prevents it from achieving even that.

Go back and carefully reread the Representation Theorem on page 291. The phrase "there exists a unique probabilistic credence distribution" contains a key ambiguity. One might be tempted to read it as saying that given an agent's full preference ranking of acts, there will be exactly one credence distribution that matches those preferences via expected utility maximization, and *moreover* that credence distribution will be probabilistic. But that's not how the theorem works. The proof begins by assuming that we're looking for a probabilistic credence distribution, and then showing that out of all the probabilistic distributions, there is exactly one that will match the agent's preferences. (If you look closely at Step Four of Ramsey's process—the step that determines credence values—you'll notice that halfway through we had to *assume* those values satisfy the probability axioms.) In the course of an argument for probabilism, this is extremely question-begging—how do we know that there isn't some other, non-probabilistic distribution that would lead to the same preferences by expected utility maximization? What if it turns out that any agent who can be represented as if she is maximizing expected utility with respect to a *probabilistic* distribution can also be represented as maximizing expected utility with respect to a *non-probabilistic* distribution? This would vitiate the argument's ability to privilege probabilistic distributions.[13]

8.3.2   Reformulating the argument

We can address these questions by proving a new version of the Representation Theorem:[14]

Revised Representation Theorem: If an agent's preferences satisfy certain constraints, then there exists a unique credence distribution (up to positive scalar transformation) and unique utility distribution (up to positive affine transformation) that yield those preferences when the agent maximizes expected utility. Moreover, one of those scalar credence transforms satisfies the probability axioms.

The revised theorem shows that if an agent maximizes expected utility, her full set of preferences narrows down her credences to a very narrow class of possible distributions. These distributions are all positive scalar transformations of each other; they all satisfy Non-Negativity and Finite Additivity; and they all satisfy something like Normality. The main difference between them is in the particular numerical value they assign to tautologies. For instance, one of the distributions will assign a credence of 100 to every logical truth (we can think of this distribution as working on a percentage scale), while another distribution will assign tautologies degree of belief 1. For any particular proposition P, the former distribution will assign a value exactly 100 times that assigned by the latter. If we *stipulate* that our credence scale tops out at 1, there will be only one credence distribution matching the agent's preferences, and that distribution will satisfy the Kolomogorov axioms. Given the stipuation of maximal credence 1, the Revised Representation Theorem demonstrates that any agent who generates preferences satisfying the preference axioms by maximizing expected utilities has a probabilistic credence distribution.

The revised theorem provides a revised argument:

Revised Representation Theorem Argument for Probabilism

(Premise 1)   Every rational agent has preferences satisfying the preference axioms.

(Premise 2)   Every rational agent has preferences that align with her credences and utilities by expected utility maximization.

(Theorem)   If an agent's preferences satisfy the preference axioms and align with her credences and utilities by expected utility maximization, then that agent has probabilistic credences (up to a positive scalar transformation).

(Conclusion) Every rational agent has a probabilistic credence distribution (or a positive scalar transformation thereof).

This argument has the advantage of being valid. Its conclusion isn't quite probabilism, but if we treat the maximum numerical credence value as a stipulated matter, it's close enough to probabilism to do the trick. So this version of the argument makes real progress: if an agent assigns rational preferences that align with her credences and utilities by maximizing expected utility, then we can show not merely that she's represent*able* as satisfying certain constraints; we can show that she *must* assign essentially probabilistic credences.

Yet by clearing out some of our earlier problems, this version of the argument highlights the truly central assumption of representation theorems, enshrined in Premise 2 of the argument and the antecedent of the conditional that ended the previous paragraph. Why should we assume that rational preferences maximize expected utility? Savage's expected utility equation is just one of many valuation functions that could be used to combine credences and utilities into preferences. In Chapter 7 we considered other valuation functions endorsed by Jeffrey and by causal decision theorists. But those valuation functions all maximized expected utilities in some sense—they worked by calculating linear averages. An agent might instead determine her preferences using a "squared credence" rule instead:

$A \succ B$ just in case

$$u(A \And S_1) \cdot cr(S_1)^2 + \ldots + u(A \And S_n) \cdot cr(S_n)^2 \qquad (8.13)$$
$$> u(B \And S_1) \cdot cr(S_1)^2 + \ldots + u(B \And S_n) \cdot cr(S_n)^2$$

This valuation function behaves differently than the expected utility rules. For example, if an agent has a 2/3 credence that a particular bet will pay her \$4 and a 1/3 credence that it will pay her nothing, applying the squared credence rule will lead her to prefer a guaranteed \$2 to this gamble. Expected utility maximization, on the other hand, recommends taking the bet. Squared-credence valuations are more risk-averse than expected utilities.

Here's another interesting feature of the squared credence rule: Return to our friends Mr. Prob and Mr. Bold from the introduction to this part of the book. Mr. Prob's credences are probabilistic, while Mr. Bold's credence in any proposition is the square-root of Mr. Prob's. Mr. Bold satisfies Non-Negativity and Normality, but not Finite Additivity. His credence in any contingent proposition is higher than Mr. Prob's. Now suppose that while Mr. Prob determines his preferences by maximizing Savage-style expected utilities, Mr. Bold's preferences are generated using Equation (8.13). In that case, Mr. Prob and Mr. Bold have the *exact same* preferences between any two acts.[15] It's easy to see why: Mr. Bold's credence in a given state S_i is the square-root of Mr. Prob's, but Mr. Bold squares his credence values in the process of calculating his valuation function. Mr. Bold's aggressive attitude assignments and risk-averse act selections cancel out precisely, leaving him with preferences identical to Mr. Prob's. This means that if Mr. Prob's preferences satisfy the preference axioms, Mr. Bold's do as well.[16]

If all you know about an arbitrary agent is that her preferences satisfy the preference axioms, it will be impossible to tell whether she has probabilistic credences and maximizes expected utility, or has non-probabilistic credences and a different valuation function. If I assure you that this agent is fully rational, does that break the tie? Why does rationality require maximizing expected utility—what's rationally *wrong* with the way Mr. Bold proceeds?

For the (revised) Representation Theorem Argument to succeed, we need a convincing argument that maximizing expected utility is rationally required.[17] The argument cannot be that any agent who fails to maximize expected utility will adopt intuitively unappealing preferences among acts. (This was the strategy of our money pump argument for some of the preference axioms.) The alternative to maximizing expected utility we're considering here is capable of generating all the same act preferences—intuitively appealing or otherwise— that expected utilities can.

I said in the introduction to this part of the book that Finite Additivity is the most difficult to establish of Kolmogorov's axioms. The revised Representation Theorem Argument shows that if we can assume rational agents set their preferences by maximizing expected utility, then Finite Additivity is entailed by the preference axioms. But now we have a Linearity In, Linearity Out problem. We want to demonstrate that rational agents satisfy Finite Additivity, which calculates the credence of a disjunction as a linear combination of the credences of its mutually exclusive disjuncts. To make that demonstration, we assume that rational agents maximize expected utility, which calculates a valuation as a linear combination of state credences. We can criticize Mr. Bold for being non-linear in his credences only if it's antecedently permissible to criticize him for making non-linear valuations.

To be clear: I have no problem with a decision theory that lists both probabilism and expected-utility maximization among its rationally required norms. We saw earlier how these norms complement each other and allow rational choice theorists to derive interesting and substantive results. My complaint is with expected-utility maximization as a premise in what's meant to be an independent argument for probabilism. Representation theorem arguments for probabilism rely on an assumption that looks just as nonobvious and in need of independent support as probabilism did.[18]

8.4 Exercises

Problem 8.1. 🖋 Show that any real-number measurement scale with finite upper and lower bounds can be converted by a positive affine transformation into a scale with the bounds 0 and 1.

Problem 8.2. 🖋

  (a) List three different real-life examples of two distinct measuring scales that measure the same quantity and are related by a positive scalar transformation. (Measurements of mass cannot be one of your examples.)

  (b) List three different real-life examples of two distinct measuring scales that measure the same quantity and are related by a positive affine transformation that is not a scalar transformation. (Measurements of temperature cannot be one of your examples.)

Problem 8.3. 🖋🖋 Shane is a graduate student who doesn't care about the outcomes of sporting events (though he has opinions about which outcomes will occur). Assume the propositions *the Heat win the NBA Finals* and *the Blackhawks win the Stanley Cup* are ethically neutral for Shane. Among Shane's preferences between various acts and gambles are the following:

Go to movie

—preferred to—

Read book

—indifferent with—

Go to movie if Heat win the NBA Finals,
work on dissertation if Heat don't win

—indifferent with—

Go to movie if Heat don't win, dissertate if Heat win

—preferred to—

Go to gym

—indifferent with—

Read book if Heat win, dissertate if Heat don't win

—indifferent with—

Go to movie if Blackhawks win the Stanley Cup,
dissertate if Blackhawks don't win

—preferred to—

Dissertate

For the sake of definiteness, suppose Shane assigns a utility of 100 to going to a movie and a utility of 0 to working on his dissertation. Suppose also that Shane's preferences satisfy the preference axioms, his credences satisfy the probability axioms, and he determines his preferences by maximizing expected utilities as prescribed by Savage.

(a) Use Ramsey's four-step process to determine as much about Shane's utility and credence values as you can. Be sure to explain your method.

(b) Imagine Shane is offered a gamble on which he reads a book if the Blackhawks win the Stanley Cup, but dissertates if they don't win. Would Shane prefer to accept this gamble or go to the gym?

Problem 8.4. 🌙

(a) Suppose an agent assigns $cr(P) = 1/3$ and sets her preferences according to Savage-style expected utilities. Explain how she might nevertheless prefer a guaranteed \$10 to a gamble that pays \$40 on P and nothing otherwise, if dollars have declining marginal utility for her.

(b) Now suppose the agent doesn't have declining marginal utility for money—in fact, she assigns exactly 1 util per dollar gained or lost, no matter how many she already has. Show that such an agent could still prefer a guaranteed \$10 to a gamble that pays \$40 on P and nothing otherwise, if she assigns preferences using the "squared credences" valuation function of Equation (8.13).

Problem 8.5. ✐ Suppose I've been cross-examining an agent for some time about her preferences, and all the preferences I've elicited satisfy the preference axioms. Mr. Prob comes along and calculates a utility distribution and probabilistic credence distribution that would generate the elicited preferences if the agent is an expected-utility maximizer. Mr. Bold then claims that Mr. Prob is wrong about the agent's credence values—according to Mr. Bold, the agent's nonextreme credences are actually the square-root of what Mr. Prob has suggested, but the agent is a squared-credence maximizer.

Do you think there's any way to tell if Mr. Prob or Mr. Bold is correct about the agent's credence values? Could there even *be* a fact of the matter to the effect that one of them is right and the other is wrong?

8.5 Further reading

INTRODUCTIONS AND OVERVIEWS

Richard C. Jeffrey (1965). *The Logic of Decision*. 1st edition. McGraw-Hill Series in Probability and Statistics. New York: McGraw-Hill

Chapter 3 carefully explains techniques for drawing out credences and utilities from preferences, including a step-by-step walkthrough of Ramsey's approach with examples.

CLASSIC TEXTS

Frank P. Ramsey (1931). Truth and Probability. In: *The Foundations of Mathematics and Other Logic Essays*. Ed. by R. B. Braithwaite. New York: Harcourt, Brace and Company, pp. 156–98

Ramsey inspired all future representation theorems with his four-step process for determining an agent's credences and utilities from her preferences.

Leonard J. Savage (1954). *The Foundations of Statistics*. New York: Wiley

Though the proof is spread out over the course of the book, this work contains the first general representation theorem.

EXTENDED DISCUSSION

Patrick Maher (1993). *Betting on Theories*. Cambridge Studies in Probability, Induction, and Decision Theory. Cambridge: Cambridge University Press
Mark Kaplan (1996). *Decision Theory as Philosophy*. Cambridge: Cambridge University Press
David Christensen (2004). *Putting Logic in its Place*. Oxford: Oxford University Press

Each of these authors explains and defends some version of a representation theorem argument for rational constraints on degrees of belief: Maher in his Chapter 8, Kaplan in his Chapter 1, and Christensen in his Chapter 5.

Lyle Zynda (2000). Representation Theorems and Realism about Degrees of Belief. *Philosophy of Science* 67, pp. 45–69

Demonstrates that rational preferences representable as maximizing expected utilities based on probabilistic credences can also be represented as maximizing some other quantity based on non-probabilistic credences, then explores the consequences for realism about probabilistic credence.

Christopher J.G. Meacham and Jonathan Weisberg (2011). Representation Theorems and the Foundations of Decision Theory. *Australasian Journal of Philosophy* 89, pp. 641–63

A critical examination of representation theorem arguments, assessing their potential to establish both descriptive and normative claims about degrees of belief.

Notes

1. As in Chapter 7, we will read "*A*" as the proposition that the agent performs a particular act and "*S_i*" as the proposition that a particular state obtains. Thus preferences, credences, and utilities will all be propositional attitudes.
2. Alan Hájek was the first person to explain to me the significance of Ramsey's result by comparing it with these other results as in Figure 8.1.
3. To be slightly more careful about Equation (8.4): Standard expected utility theories (such as Savage's) entail a principle for utilities similar to the Conglomerability principle for credences we saw in Section 5.4—for any X and P, u(X & P) and u(X & $\sim P$) set the bounds for u(X). If u(X & P) = u(X & $\sim P$), u(X) must equal this value as well.
4. Ramsey points out that this approach will not work for propositions Q with extreme unconditional credences. But those can be ferreted out easily: for instance, if the agent is certain of Q then she will be indifferent between receiving D for certain and a gamble that yields D on Q and M on $\sim Q$.
   Also, we have to be sure in Step 4 that we've selected Q, R, S, T such that u(T) $\neq$ u(S & $\sim Q$) $\neq$ u(R & Q).
5. Maher (1993, Chs 2 and 3) provides an excellent general defense of the preference axioms as rational requirements.
6. As I pointed out in Chapter 7, Preference Completeness actually has some substantive consequences when considered on its own. For example, it entails Preference Asymmetry. But decision theorists often think of these principles in the following order: If we first take on substantive constraints such as Preference Asymmetry and Transitivity, then *adding* Preference Completeness is just a matter of requiring preference rankings to be complete.

7. See Fishburn (1981) for a useful survey. One popular move is to replace the requirement that an agent's preference ranking *actually* satisfy a particular richness constraint with a requirement that the ranking be *extendable* into a fuller ranking satisfying that constraint. For more on this move of constraining incomplete rankings by asking whether they can be extended to complete rankings with particular features—and the justification for it—see Section 14.1.3.

8. The Hypothetical Priors Theorem presented in Section 4.3.3 may be thought of as a representation theorem of sorts. It says that any agent whose credence distributions over time satisfy the core Bayesian norms can be represented as if those distributions are all generated by conditionalizing an underlying hypothetical prior. The full details of that hypothetical prior are usually underspecified by the agent's history of credence assignments, but it doesn't matter because all the hypothetical priors suited to the job will make the same predictions about her future responses to possible evidence streams. The family of hypothetical priors matching an agent's sequence of credence distributions is rather like the family of utility distributions matching her preferences.

9. In every case where a result about rational preferences is obtained by assuming those preferences align with numerical utilities and probabilistic credences, the same result could have been derived directly from the preference axioms and the subject's given preferences, without invoking utilities and credences at all. Yet the "direct" derivation may be extended and difficult, while the utility-maximization representation saves a great deal of work.

10. Of course, if one holds the position (mentioned in Section 1.1.2) that all requirements of theoretical rationality ultimately boil down to requirements of practical rationality, this objection may not be a concern.

11. Historically, it's interesting to consider exactly what philosophy of mind Ramsey was working with. On the one hand, he was surrounded by a very positivist, behaviorist milieu. (Among other things, Ramsey was Wittgenstein's supervisor at Cambridge and produced the first English translation of *Tractatus Logico-Philosophicus*.) On the other hand, Ramsey's writings contain suggestions of an early functionalism. Brian Skyrms writes that, "Ramsey thinks of personal probabilities as theoretical parts of an imperfect but useful psychological model, rather than as concepts given a strict but operational definition" (1980b, p. 115).

12. That's not to say that the links between credence and rational preference become useless once operationalism about doxastic attitudes is abandoned. Savage (1954, pp. 27–8) has a nice discussion of the advantages of determining an agent's numerical credence values by observing her preferences over determining them by asking her to introspect.

13. This concern is forcefully put by Meacham and Weisberg (2011).

14. This revised theorem was proven for a relevant set of preference axioms by myself and Lara Buchak. For a sketch of the proof, see the appendix to Titelbaum (2019). The result is a fairly straightforward extension of proofs used in standard Dutch Book arguments for probabilism.

15. The idea of mimicking a probabilistic agent's preference structure by giving a non-probabilistic agent a non-standard valuation function comes from Zynda (2000) (with a precursor in Arrow 1951).

16. One might have thought that I dismissed our original Representation Theorem Argument for Probabilism too quickly, for the following reason: Even if we're not operationalists about degrees of belief, we might think that if probabilistic degrees of belief figure in the simplest, most useful *explanation* of observed rational agent preferences then that's a good reason to maintain that rational agents possess them. In this vein, Patrick Maher writes:

> I suggest that we understand attributions of probability and utility as essentially a device for interpreting a person's preferences. On this view, an attribution of probabilities and utilities is correct just in case it is part of an overall interpretation of the person's preferences that makes sufficiently good sense of them and better sense than any competing interpretation does.... If a person's preferences all maximize expected utility relative to some cr and u, then it provides a perfect interpretation of the person's preferences to say that cr and u are the person's probability and utility functions. Thus, having preferences that all maximize expected utility relative to cr and u is a sufficient (but not necessary) condition for cr and u to be one's probability and utility functions. (1993, p. 9)

Suppose we accept Maher's criterion for correct attitude attribution. The trouble is that even if the interpretation of a rational agent's preferences based on probabilistic credences and maximizing expected utilities is a perfect one, the alternate interpretation based on Bold-style credences and the squared credence valuation function looks just as perfect as well. Thus the probabilist interpretation fails to make better sense than competing interpretations, and a representation theorem argument for probabilism cannot go through.

17. Notice that in our earlier quote from Ramsey (1931, p. 76), Ramsey simply introduces "as a law of psychology" that agents maximize expected utility.

18. In Section 7.1 we suggested that the law of large numbers provides one reason to use expectations in estimating values. The idea is that one's expectation of a numerical quantity equals the average value one anticipates that quantity will approach in the limit. Why doesn't this provide an argument for making decisions on the basis of expected utilities?

One might worry here that using the long-run *average* smuggles in a linear bias. But there's an even deeper problem with the proposed argument: The law of large numbers says that *if* you satisfy the probability axioms, *then* you'll have credence 1 that the average in the limit equals your expectation. A result that assumes probabilism cannot be used to support maximizing expected utility if we hope to use the latter as part of our argument for probabilism.

9

Dutch Book Arguments

Chapter 8 presented the Representation Theorem Argument for probabilism. In its best form, this argument shows that any agent who satisfies certain preference axioms and maximizes expected utility assigns credences satisfying Kolmogorov's probability rules.[1] Contraposing, an agent who maximizes expected utility but fails to assign probabilistic credences will violate at least one of the preference axioms.

But why should rationality require satisfying the preference axioms? In Chapter 7 we argued that an agent who violates certain of the preference axioms (such as Preference Asymmetry or Preference Transitivity) will be susceptible to a money pump: a series of decisions, each of which is recommended by the agent's preferences, but which together leave her back where she started with less money on her hands. It looks irrational to leave yourself open to such an arrangement, and therefore irrational to violate the preference axioms.

While money pumps may be convincing, it's an awfully long and complicated road from them to probabilism. This chapter assesses a set of arguments that are similar to money pump arguments, but which constrain credences in a much more direct fashion. These arguments show that if an agent's credences violate particular constraints, we can construct a Dutch Book against her: a set of bets, each of which the agent views as fair, but which together guarantee that she will lose money come what may. Dutch Books can be constructed not only against agents whose credences violate Kolmogorov's probability axioms, but also against agents whose credences violate the Ratio Formula, updating by Conditionalization, or some of the other credal constraints we considered in Chapter 5.

Historically, Dutch Book Arguments were a very important motivation for probabilism. de Finetti (1937/1964) called Dutch-Bookable credence distributions "incoherent", and thought them rationally inconsistent.[2] When a Bayesian proposed a new constraint on credences, she would often vindicate that new requirement by showing that agents who violated it were susceptible to a Dutch Book.

In this chapter we will work through a variety of Bayesian norms, showing how to construct Dutch Books against agents who violate each one. We will

Fundamentals of Bayesian Epistemology 2: Arguments, Challenges, Alternatives. Michael G. Titelbaum, Oxford University Press. © Michael G. Titelbaum 2022. DOI: 10.1093/oso/9780192863140.003.0009

then ask whether the fact that a Dutch Book can be constructed against agents who violate a particular norm can be turned into an argument that that norm is rationally required. After offering the most plausible version of a Dutch Book Argument we can find, we will canvass a number of traditional objections to that argument.

9.1 Dutch Books

Dutch Book Arguments revolve around agents' betting behavior, so we'll begin by discussing how an agent's credences influence the bets she'll accept. For simplicity's sake we will assume throughout this chapter that an agent assigns each dollar the same amount of utility (no matter how many dollars she already has). That way we can express bets in dollar terms instead of worrying about the logistics of paying off a bet in utils.

Suppose I offer to sell you the following ticket:

> This ticket entitles the bearer
> to $1 if P is true,
> and nothing otherwise.

for some particular proposition P. If you're rational, what is your fair price for that ticket—how much would you be willing to pay to possess it? It depends how confident you are that P is true. If you think P is a long shot, then you think this ticket is unlikely to be worth anything, so you won't pay much for it. The more confident you are in P, however, the more you'll pay for the ticket. For example, if P is the proposition that a fair coin flip comes up heads, you might be willing to pay anything up to $0.50 for the ticket. If you pay exactly $0.50 for the ticket, then you've effectively made a bet on which you net $0.50 if P is true (coin comes up heads) but lose $0.50 if P is false (coin comes up tails). Seems like a fair bet.

A ticket that pays off on P is worth more to a rational agent the more confident she is of P. In fact, we typically assume that a rational agent's fair betting price for a $1 ticket on P is $cr(P)$—she will purchase a ticket that pays $1 on P for any amount *up to* $cr(P)$ dollars.

For example, suppose neither you nor I know anything about the day of the week on which Frank Sinatra was born. Nevertheless, I offer to sell you the following ticket:

> This ticket entitles the bearer
> to $1 if Sinatra was born on a weekend,
> and nothing otherwise.

If you spread your credences equally among the days of the week, then $2/7—or about $0.29—is your fair betting price for this ticket. To buy the ticket at that price is to place a particular kind of bet that the selected day fell on a weekend. If you lose the bet, you're out the $0.29 you paid for the ticket. If you win the bet, it cost you $0.29 to buy a ticket which is now worth $1, so you're up $0.71. Why do you demand such a premium—why do you insist on a higher potential payout for this bet than the amount of your potential loss? Because you think you're more likely to lose than win, so you'll only make the bet if the (unlikely) payout is greater than the (probable) loss.

Now look at the same transaction from my point of view—the point of view of someone who's selling the ticket, and will be on the hook for $1 if Sinatra was born on a weekend. You spread your credences equally among the days, and are willing to *buy* the ticket for *up to* $0.29. If I spread my credences in a similar fashion, I should be willing to *sell* you this ticket for *at least* $0.29. On the one hand, I'm handing out a ticket that may entitle you to $1 from me once we find out about Sinatra's birthday. On the other hand, I don't think it's very likely that I'll have to pay out, so I'm willing to accept as little as $0.29 in exchange for selling you the ticket. In general, an agent's fair betting price for a gambling ticket is both the *maximum* amount she would *pay* for that ticket and the *minimum* amount for which she would *sell* it.

All the tickets we've considered so far pay out $1 if a particular proposition is true. Yet tickets can be bought or sold for other potential payoffs. We generally assume that the rational fair betting price for a ticket that pays S if P is true and nothing otherwise is $S \cdot cr(P)$.[3] (You might think of this as the fair betting price for S tickets, each of which pays $1 on P.) This formula works both for run-of-the-mill betting cases and for cases in which the agent has extreme opinions. For instance, consider an agent's betting behavior when her credence in P is 0. Our formula sets her fair betting price at $0, whatever the stakes S. Since the agent doesn't think the ticket has *any* chance of paying off, she will not pay any amount of money to possess it. On the other hand, she will be willing to sell such a ticket for any amount you like, since she doesn't think she's incurring any liability in doing so.

Bayesians (and bookies) often quote bets using odds instead of fair betting prices. For instance, a bet that Sinatra was born on a weekend would typically go off at 5 to 2 odds. This means that the ratio of your potential net payout to your potential net loss is 5:2 (0.71 : 0.29). A rational agent will accept a bet on P at her odds against P (that is, cr($\sim P$) : cr(P)) or better. Yet despite the ubiquity of odds talk in professional gambling, we will use fair betting prices going forward.

9.1.1 Dutch Books for probabilism

Suppose we have an agent who violates the probability axioms by assigning both cr(P) = 0.7 and cr($\sim P$) = 0.7 for some particular proposition P. (Perhaps he's a character like Mr. Bold.) Given his credence in P, this agent's fair betting price for a ticket that pays \$1 if P is true will be \$0.70. Given his credence in $\sim P$, his fair betting price for a ticket that pays \$1 if $\sim P$ is true will also be \$0.70. So let's sell him both of these tickets, at \$0.70 each.

Our agent is now in trouble. He has paid a total of \$1.40 for the two tickets, and there's no way he can make all that money back. If P is true, his first ticket is worth \$1 but his second ticket is worth nothing. If P is false, his first ticket is worth nothing and his second ticket pays only \$1. Either way, he'll be out \$0.40.

We can summarize this agent's situation with the following table:

	P	$\sim P$
Ticket pays on P	0.30	−0.70
Ticket pays on $\sim P$	−0.70	0.30
TOTAL	**−0.40**	**−0.40**

The columns of this table partition the possible states of the world. In this case, our partition contains the propositions P and $\sim P$. The agent purchases two tickets; each ticket is recorded on one row. The entries in the cells report the agent's *net* payout for that ticket in that state; all values are in dollars, and negative numbers indicate a loss. So, for instance, the upper-right cell reports that if $\sim P$ is true then the agent loses \$0.70 on his P ticket (the ticket cost him \$0.70, and doesn't win him anything in that state). The upper-left cell records that the P ticket cost the agent \$0.70, but he makes \$1 on it if P is true, for a net profit of \$0.30. The final row reports the agent's total payout for all his tickets in a given state of the world. As we can see, an agent who purchases both tickets will lose \$0.40 no matter which state the world is in. Purchasing this set of tickets guarantees him a net loss.

A **Dutch Book** is a set of bets, each placed with an agent at her fair betting price (or better), that together guarantee her a sure loss come what may.[4] The idea of a Dutch Book is much like that of a money pump (Section 7.2.1): we make a series of exchanges with the agent, each of which individually looks fair (or favorable) from her point of view, but which together yield an undesirable outcome. In a Dutch Book, each bet is placed at a price the agent considers fair given her credence in the proposition in question, but when all the bets are added up she's guaranteed to lose money no matter which possible world is actual.

Ramsey (1931, p. 84) recognized a key point about Dutch Books, which was proven by de Finetti (1937/1964):

Dutch Book Theorem: If an agent's credence distribution violates at least one of the probability axioms (Non-Negativity, Normality, or Finite Additivity), then a Dutch Book can be constructed against her.

We will prove this theorem by going through each of the axioms one at a time, and showing how to make a Dutch Book against an agent who violates it.

Non-Negativity and Normality are relatively easy. An agent who violates Non-Negativity will set a negative betting price for a ticket that pays $1 on some proposition P. Since the agent assigns a negative betting price to that ticket, she is willing to sell it at a negative price. In other words, this agent is willing to pay *you* some amount of money to take a ticket which, if P is true, entitles you to a further $1 from her beyond what she paid you to take it. Clearly this is a losing proposition for the agent.

Next, suppose an agent violates Normality by assigning credence greater than 1 to a tautology. Say, for instance, an agent assigns $cr(P \vee \sim P) = 1.4$. This agent will pay $1.40 for a ticket that pays $1 if $P \vee \sim P$ is true. The agent will definitely win that $1, but will still have lost money overall. On the other hand, if an agent assigns a credence less than 1 to a tautology, she will sell for less than $1 a ticket that pays $1 if the tautology is true. The tautology will be true in every possible world, so in every world the agent will lose money on this bet.

Now let's turn to Finite Additivity. Suppose that for mutually exclusive P and Q, an agent violates Finite Additivity by assigning $cr(P) = 0.5$, $cr(Q) = 0.5$, and $cr(P \vee Q) = 0.8$. Because of these credences, the agent is willing to pay $0.50 for a ticket that pays $1 on P, and $0.50 for a ticket that pays $1 on Q. Then we have her sell us for $0.80 a ticket that pays $1 if $P \vee Q$.

At this point, the agent has collected $0.80 from us and paid a total of $1 for the two tickets she bought. So she's down $0.20. Can she hope to make this money back? Well, the tickets she's holding will be worth $1 if either P or Q is true. She can't win on both tickets, because P and Q were stipulated to be mutually exclusive. So at most, the agent's tickets are going to earn her $1. But if either P or Q is true, $P \vee Q$ will be true as well, so she will have to pay out $1 on the ticket she sold to us. The moment she earns $1, she'll have to pay it back out. There's no way for the agent to make her money back, so no matter what happens she'll be out a net $0.20.

The situation is summed up in this table:

	$P \& \sim Q$	$\sim P \& Q$	$\sim P \& \sim Q$
Ticket pays on P	0.50	−0.50	−0.50
Ticket pays on Q	−0.50	0.50	−0.50
Ticket pays on $P \vee Q$	−0.20	−0.20	0.80
TOTAL	**−0.20**	**−0.20**	**−0.20**

Since P and Q are mutually exclusive, there is no possible world in which $P \& Q$ is true, so our partition has only three members. On the first row, the P-ticket for which the agent paid $0.50 nets her a positive $0.50 in the state where P is true. Similarly for Q on the second row. The third row represents a ticket the agent *sold*, so she makes $0.80 on it unless $P \vee Q$ is true, in which case she suffers a net loss. The final row sums the rows above it to show that each possible state guarantees the agent a $0.20 loss from her bets. A similar Book can be constructed for any agent who assigns $cr(P \vee Q) < cr(P) + cr(Q)$. For a Book against agents who violate Finite Additivity by assigning $cr(P \vee Q) > cr(P) + cr(Q)$, see Exercise 9.1.

9.1.2 Further Dutch Books

We can also construct Dutch Books against agents who violate other credence requirements. For example, suppose an agent has the probabilistic unconditional credence distribution specified by the following probability table:

P	Q	cr
T	T	1/4
T	F	1/4
F	T	1/4
F	F	1/4

But now suppose that this agent violates the Ratio Formula by assigning $cr(P|Q) = 0.6$. To construct a Dutch Book against this agent, we first need to understand conditional bets. Suppose we sell this agent the following ticket:

> If Q is true, this ticket entitles the bearer
> to \$1 if P is true and nothing otherwise.
> If Q is false, this ticket may be returned to
> the seller for a full refund of its purchase price.

If Q turns out to be false, the agent's full purchase price for this ticket will be refunded to her. So if Q is false, it doesn't matter what she pays; the ticket will net her \$0 no matter what. That means the agent's purchase price for this ticket should be dictated by her opinion of P in worlds where Q is true. In other words, the agent's purchase price for this ticket should be driven by $cr(P|Q)$. We say that this ticket creates a **conditional bet** on P given Q. A conditional bet on P given Q wins or loses money for the agent only if Q is true; if the payoff on P (given Q) is \$1, the agent's fair betting price for such a bet is $cr(P|Q)$. In general, conditional bets are priced using conditional credences.

Our agent who assigns $cr(P|Q) = 0.6$ will purchase the ticket above for \$0.60. We then have her sell us two more tickets:

1. We pay the agent \$0.25 for a ticket that pays us \$1 if $P \& Q$.
2. We pay the agent \$0.30 for a ticket that pays us \$0.60 if $\sim Q$.

Notice that Ticket 2 is for stakes other than \$1; we've calculated the agent's fair betting price for this ticket (\$0.30) by multiplying her credence in $\sim Q$ (1/2) by the ticket's payoff (\$0.60).

The agent has received \$0.55 from us, but she's also paid out \$0.60 for the conditional ticket. So she's down \$0.05. If Q is false, she'll get a refund of \$0.60 for the conditional ticket, but she'll also owe us \$0.60 on Ticket 2. If Q is true and P is true, she gets \$1 from the conditional ticket but owes us \$1 on Ticket 1. And if Q is true and P is false, she neither pays nor collects on any of the tickets and so is still out \$0.05. No matter what, the agent loses \$0.05. The following table summarizes the situation:

	$P \& Q$	$\sim P \& Q$	$\sim Q$
Ticket 1	−0.75	0.25	0.25
Ticket 2	0.30	0.30	−0.30
Conditional ticket	0.40	−0.60	0
TOTAL	**−0.05**	**−0.05**	**−0.05**

A similar Dutch Book can be constructed against any agent who violates the Ratio Formula.

David Lewis figured out how to turn this Dutch Book against Ratio Formula violators into a strategy against anyone who fails to update by Conditionalization.[5] Suppose we have an agent who assigns the unconditional credence distribution described above at t_i, satisfies the Ratio Formula, but will assign $cr_j(P) = 0.6$ if she learns Q between t_i and t_j. Since she satisfies the Ratio Formula, this agent assigns $cr_i(P \mid Q) = 0.5$. That means the $cr_j(P)$ value she will assign upon learning Q does not equal her $cr_i(P \mid Q)$ value, leaving her in violation of Conditionalization.

We take advantage of this agent's Conditionalization violation by first purchasing the Tickets 1 and 2 described above from her at t_i. The prices on these tickets match the agent's unconditional t_i credences, so she will be willing at t_i to sell them at the prices listed. We then implement the following strategy: If the agent learns Q between t_i and t_j, we sell her a ticket at t_j that pays \$1 on P. We know that this agent will assign $cr_j(P) = 0.6$ if she learns Q, so in that circumstance she'll be willing to buy this ticket for \$0.60. If the agent doesn't learn Q between t_i and t_j, we make no transactions with her beyond Tickets 1 and 2.[6]

Putting all this together, the agent's payoffs once more are:

	$P \& Q$	$\sim P \& Q$	$\sim Q$
Ticket 1	−0.75	0.25	0.25
Ticket 2	0.30	0.30	−0.30
Ticket if Q learned	0.40	−0.60	0
TOTAL	**−0.05**	**−0.05**	**−0.05**

(Because the agent purchases the third ticket only if Q is true, it neither costs nor pays her anything if Q is false.)

This agent received \$0.55 from us for selling two tickets at t_i. If Q is false, no more tickets come into play, but she owes us \$0.60 on Ticket 2, so she's out a total of \$0.05. If Q is true, she purchases the third ticket, and so is out \$0.05. If P is also true, she wins \$1 on that third ticket but has to pay us \$1 on Ticket 1, so she's still down \$0.05. If P is false (while Q is true), none of the tickets pays, and her net loss remains at \$0.05. No matter what, the agent loses money over the course of t_i to t_j.

A quick terminological remark: A Dutch Book is a specific set of bets guaranteed to generate a loss. Strictly speaking, we haven't just built a Dutch Book against Conditionalization violators, because we haven't described a single set of bets that can be placed against the agent to guarantee a sure loss in every case. Instead, we've specified *two* sets of bets, one to be placed if the agent learns Q and the other to be placed if not. (The former set contains three bets, while the latter contains two.) We've given the bookie a *strategy* for placing different sets of bets in different circumstances, such that each potential set of bets is guaranteed to generate a loss in the circumstances in which it's placed. For this reason, Lewis's argument supporting Conditionalization is usually known as a **Dutch Strategy** argument rather than a Dutch Book argument.[7]

Dutch Books or Strategies have been constructed to punish violators of many of the additional Bayesian constraints we considered in Chapter 5: the Principal Principle (Howson 1992), the Reflection Principle (van Fraassen 1984), Regularity (Kemeny 1955; Shimony 1955), Countable Additivity (Adams 1962), and Jeffrey Conditionalization (Armendt 1980; Skyrms 1987b). I will not work through the details here. Instead, we will consider the normative consequences of these Dutch constructions.

9.2 The Dutch Book Argument

A Dutch Book is a set of bets, not an argument—and neither is the Dutch Book Theorem on its own. But once we know that a Dutch Book can be constructed for a particular kind of situation, we can use that fact to argue for a norm. For example, there's the

Dutch Book Argument for Probabilism

(Premise)   No Dutch Book can be constructed against a rational agent.
(Theorem)   If an agent's credence distribution violates at least one of the probability axioms, then a Dutch Book can be constructed against her.
(Conclusion) Every rational agent has a probabilistic credence distribution.

The key premise is that no rational agent is susceptible to being Dutch Booked; just as rational preferences help one avoid money pumps, so should rational credences save us from Dutch Books. Once we have this premise, similar Dutch Book Arguments can be constructed for the Ratio Formula, Conditionalization, and all the other norms mentioned in the previous section. But is the premise plausible? What if it turns out that a Dutch Book can be constructed against *any* agent, no matter what rules her credences do or don't satisfy? In other words, imagine that Dutch Books were just a rampant, unavoidable fact of life. That would render the Dutch Book Argument's premise false.

To reassure ourselves that we don't live in this dystopian Book-plagued world, we need a series of what Hájek (2009a) calls **Converse Dutch Book Theorems**. The usual Dutch Book Theorem tells us that if an agent violates the probability axioms, she is susceptible to a Dutch Book. A Converse Dutch Book Theorem would tell us that if an agent *satisfies* the probability axioms, she is *not* susceptible to a Dutch Book. If we had a Converse Dutch Book Theorem, then we wouldn't need to worry that whatever credences we assigned, we could be Dutch Booked. The Converse Dutch Book Theorem would guarantee us safety from Dutch bookies as long as we maintained probabilistic credences; together with the standard Dutch Book Theorem, the Converse Theorem would constitute a powerful consideration in favor of assigning probabilistic over non-probabilistic credence distributions.

Unfortunately we can't get a Converse Dutch Book Theorem of quite the kind I just described. Satisfying the probability axioms with her unconditional credences does not suffice to innoculate an agent against Dutch Books. An agent whose unconditional credence distribution satisfies the probability axioms might still violate the Ratio Formula with her conditional credences, and as we've seen this would leave her open to Book. So it can't be a theorem that no agent with probabilistic credences can ever be Dutch Booked (because that isn't true!). Instead, our Converse Dutch Book Theorem has to say that as long as an agent's credences satisfy the probability axioms, she can't be Dutch Booked *with the kind of Book we deployed against agents who violate the axioms*. For instance, if an agent satisfies Non-Negativity, there won't be any propositions to which she assigns a negative credence, so we won't be able to construct a Book against her that requires selling a ticket at a negative fair betting price (as we did against the Non-Negativity violator). Lehman (1955) and Kemeny (1955) each independently proved that if an agent's credences satisfy the probability axioms, she isn't susceptible to Books of the sort we considered in Section 9.1.1.[8] This clears the way for the Dutch Book Argument's premise to be at least plausible.

Converse Dutch Book Theorems can also help us beat back another challenge to Dutch Book Arguments. Hájek (2009a) defines a **Czech Book** as a set of bets, each placed with an agent at her fair betting price (or better), that together guarantee her a sure *gain* come what may. It's easy to see that whenever one can construct a Dutch Book against an agent, one can also construct a Czech Book. We simply take each ticket contained in the Dutch

Book, leave its fair betting price intact, but have the agent *sell* it rather than *buy* it (or vice versa). In the betting tables associated with each Book, this will flip all the negative payouts to positive and positive payouts to negative. So the total payouts on the bottom row of each column in the table will be positive, and the agent will profit come what may.

According to the Dutch Book Theorem, a Dutch Book can be constructed against any agent who violates a probability axiom. We now know that whenever a Dutch Book can be constructed, a Czech Book can be as well. This gives us the

Czech Book Theorem: If an agent's credence distribution violates at least one of the probability axioms, then a Czech Book can be constructed in her favor.

Violating the probability axioms leaves an agent susceptible to Dutch Books, which seems to be a disadvantage. But violating the probability axioms also opens up the possibility that an agent will realize Czech Books, which seems to be an advantage. Perhaps a rational agent would leave herself susceptible to Dutch Books in order to be ready for Czech Books, in which case the premise of our argument fails once more. In general, Hájek worries that any argument for probabilism based on Dutch Books will be canceled by Czech Books, leaving the Dutch Book Theorem normatively inert.

At this point, converse theorems become significant. A Converse Dutch Book Theorem says that satisfying the probability axioms protects an agent from the disadvantage of susceptibility to particular kinds of Dutch Book. But there is no Converse Czech Book Theorem—it's just not true that any agent who satisfies the probability axioms must forgo Czech Books. That's because a rational agent will purchase betting tickets at anything *up to* her fair betting price (and sell at anything *at least* that number). For instance, an agent who satisfies the axioms will assign credence 1 to any tautology and so set $1 as her fair betting price for a ticket that pays $1 if some tautology is true. But if we offer her that ticket for, say, $0.50 instead, she will be perfectly happy to take it off our hands. Since the ticket pays off in every possible world, the agent will make a profit come what may. So here we have a Czech Book available to agents who satisfy the probability axioms.

Non-probabilistic agents are susceptible to certain kinds of Dutch Books, while probabilistic agents are not. Non-probabilistic agents can take advantage of Czech Books, but probabilistic agents can too. The advantage goes to probabilism.

9.2.1 Dutch Books depragmatized

Recent authors have reformulated the Dutch Book Argument in response to two objections. First, like the Representation Theorem Argument of Chapter 8, the Dutch Book Argument seems to move from a practical premise to a theoretical conclusion. The argument establishes that an agent with non-probabilistic credences may *behave* in ways that are practically disadvantageous—buying and selling gambling tickets that together guarantee a loss. But this shows only that it's *practically* irrational to assign credences violating the probability axioms. (As with the Representation Theorem Argument, it feels like the word "practically" should be inserted before the word "rational" in both the Dutch Book Argument's premise and conclusion.) We wanted to establish the axioms as requirements of *theoretical* rationality (see Chapter 1), and the argument seems unable to do that.

The distinction between requirements of practical and theoretical rationality might disappear if one understood doxastic attitudes purely in terms of their effects on action. de Finetti, for example, explored a position that *defines* an agent's credences in terms of her betting behavior:

> Let us suppose that an individual is obliged to evaluate the rate p at which he would be ready to exchange the possession of an arbitrary sum S (positive or negative) dependent on the occurrence of a given event E, for the possession of the sum pS; we will say by definition that this number p is the measure of the degree of probability attributed by the individual considered to the event E, or, more simply, that p is the probability of E (according to the individual considered).   (1937/1964, pp. 101–2)

As we discussed in Chapter 8, this kind of behaviorism about mental states is fairly unpopular these days. And even if we take this definitional approach, a second objection to the Dutch Book Argument remains: As a practical matter, susceptibility to Book doesn't seem that significant. Few of us are surrounded by bookies ready to press gambles on us should we violate the probability axioms. If the Dutch Book Argument is supposed to talk us into probabilistic credences on the grounds that failing to be probabilistic will lead to bad practical consequences, those practical consequences had better be a realistic threat.

This second objection is a bit unfair as it stands. As Julia Staffel writes:

> Dutch book arguments start from the idea that one central function of our credences is guiding our actions. Using bets as stand-ins for actions more

generally, they are supposed to demonstrate that probabilistic credences are suitable for guiding action, whereas incoherent credences are not.... Dutch book arguments show us that coherent thinkers avoid guaranteed betting losses, hence, given that we understand bets as stand-ins for actions more generally, coherent credences are suitable for guiding our actions.

(2019, pp. 57–8)

The idea is that Dutch Books aren't the only kind of practical trouble into which non-probabilistic credences might lead an agent; they're just a particularly straightforward and vivid example. While none of us may ever face a Dutch bookie, and some of us may avoiding betting entirely, there are nevertheless a variety of practical situations in which non-probabilistic credences will drive us to make sub-optimal decisions.

Still, if the concern behind Dutch Books is a practical one, shouldn't we investigate how often such situations really arise, and how much our non-probabilistic credences are liable to cost us? Maybe it's not worth the trouble to correct our degrees of belief...

To avoid these sorts of objections (and others we'll see later), recent authors have recast the Dutch Book Argument as establishing a requirement of theoretical rationality. They suggest that despite the Dutch Book's pragmatic appearance, the bookie and his Books are merely a device for dramatizing an underlying doxastic inconsistency. These authors take their inspiration from the original passage in which Ramsey mentioned Dutch Books:

These are the laws of probability, which we have proved to be necessarily true of any consistent set of degrees of belief.... If anyone's mental condition violated these laws, his choice would depend on the precise form in which the options were offered him, which would be absurd. He could have a book made against him by a cunning better and would then stand to lose in any event.   (1931, p. 84)

Interpreting this passage, Skyrms writes that "for Ramsey, the cunning bettor is a dramatic device and the possibility of a dutch book a striking symptom of a deeper incoherence" (1987a, p. 227). Since the bookie is only a device for revealing this deeper incoherence, it doesn't matter whether we are actually surrounded by bookies or not. (The fact that a child's current medical condition would lead her to break 100°F on a thermometer indicates an underlying problem, whether or not any thermometers are around.) As Brad Armendt puts it:

We should resist the temptation to think that a Dutch book argument demonstrates that the violations (violations of probability, for the synchronic argument) are bound to lead to dire outcomes for the unfortunate agent. The problem is not that violators are bound to suffer, it is that their action-guiding beliefs exhibit an inconsistency. That inconsistency can be vividly depicted by imagining the betting scenario, and what would befall the violators were they in it. The idea is that the irrationality lies in the inconsistency, when it is present; the inconsistency is portrayed in a dramatic fashion when it is linked to the willing acceptance of certain loss. The value of the drama lies not in the likelihood of its being enacted, but in the fact that it is made possible by the agent's own beliefs, rather than a harsh, brutal world.   (1992, p. 218)

To argue that Dutch Book vulnerability reveals a deeper rational inconsistency, we start by relating credences to betting behavior in a more nuanced manner than de Finetti's. Howson and Urbach (2006), for instance, say that an agent who assigns credence $cr(P)$ to proposition P won't necessarily *purchase* a \$1 ticket on P at price \$$cr(P)$, but will *regard* such a purchase as fair. (This ties the doxastic attitude of credence to another attitude—regarding as fair—rather than tying credences directly to behavior.) The Dutch Book Theorem then tells us that an agent with nonprobabilistic credences will regard each of a set of bets as fair that together guarantee a sure loss. Since such a set of bets is clearly *un*fair, a nonprobabilistic agent's degrees of belief are theoretically inconsistent because they regard as fair something that is guaranteed not to be.

Christensen (2004, Ch. 5) attenuates the connection between credences and betting rates even further. As a purely descriptive matter, an agent with particular degrees of belief may or may not regard any particular betting arrangement as fair (perhaps she makes a calculation error; perhaps she doesn't have any views about betting arrangements; etc.). Nevertheless, Christensen argues for a *normative* link between credences and fair betting prices. If an agent assigns a particular degree of belief to P, that degree of belief *sanctions as fair* purchasing a ticket for \$$cr(P)$ that pays \$1 on P; it *justifies* the agent's evaluating such a purchase as fair; and it makes it *rational* for the agent to purchase such a ticket at (up to) that price.

Christensen then argues to probabilism from three premises:[9]

Depragmatized Dutch Book Argument for Probabilism

(Premise)   An agent's degrees of belief sanction as fair monetary bets at odds matching her degrees of belief. (Christensen calls this premise "Sanctioning".)

(Premise)	A set of bets that is logically guaranteed to leave an agent monetarily worse off is rationally defective. ("Bet Defectiveness")
(Premise)	If an agent's beliefs sanction as fair each of a set of bets, and that set of bets is rationally defective, then the agent's beliefs are rationally defective. ("Belief Defectiveness")
(Theorem)	If an agent's degrees of belief violate the probability axioms, there exists a set of bets at odds matching her degrees of belief that is logically guaranteed to leave her monetarily worse off.
(Conclusion)	If an agent's degrees of belief violate the probability axioms, that agent's degrees of belief are rationally defective.

The theorem in this argument is, once more, the Dutch Book Theorem, and the argument's conclusion is a version of probabilism. Christensen assesses this kind of Dutch Book Argument ("DBA") as follows:

> This distinctively non-pragmatic version of the DBA allows us to see why its force does not depend on the real possibility of being duped by clever bookies. It does not aim at showing that probabilistically incoherent degrees of belief are unwise to harbor for practical reasons. Nor does it locate the problem with probabilistically incoherent beliefs in some sort of preference inconsistency. Thus it does not need to identify, or define, degrees of belief by the ideally associated bet evaluations. Instead, this DBA aims to show that probabilistically incoherent beliefs are rationally defective by showing that, in certain particularly revealing circumstances, they would provide *justification* for bets that are rationally defective in a particularly obvious way. The fact that the diagnosis can be made *a priori* indicates that the defect is not one of fitting the beliefs with the way the world happens to be: it is a defect internal to the agent's belief system.   (2004, p. 121, emphasis in original)

9.3 Objections to Dutch Book Arguments

If we can construct both a Dutch Book Theorem and a Converse Dutch Book Theorem for a particular norm (probabilism, the Ratio Formula, updating by Conditionalization, etc.), then we have a Dutch Book Argument that rationality requires honoring that norm. I now want to review various objections to Dutch Book Arguments that have arisen over the years; these objections apply equally well to depragmatized versions of such arguments.

It's worth beginning with a concern that is often overlooked. Dutch Book Arguments assume that a rational agent's fair betting price for a bet that pays $1 on P is $cr(P)$. An author like de Finetti who *identifies* an agent's credence in P with the amount she's willing to pay for a ticket that yields $1 on P is free to make this move. But contemporary authors unwilling to grant that identification need some argument that these betting prices are rationally required.

A simple argument comes from expected value calculations. As we saw in Chapter 7, Equation (7.3), an agent's expected monetary payout for a ticket that pays $1 on P is

$$\$1 \cdot cr(P) + \$0 \cdot cr(\sim P) = \$cr(P) \tag{9.1}$$

So an agent whose preferences are driven by expected value calculations will assign that ticket a fair betting price of $cr(P)$. (The calculation can be generalized to bets at other stakes.)

But this argument for the fair betting prices we've been assuming takes as a premise that rational agents maximize expected utility. (Or expected monetary return—recall that we assumed for the duration of this chapter that agents assign constant marginal utility to money.) If we had *that* premise available, we could argue much more directly for probabilism via the Revised Representation Theorem of Chapter 8.[10]

So what else can we try? At the beginning of Section 9.1, I tried to motivate the typical formula for fair betting prices without appealing to expectations. I invoked intuitions about how an agent's fair betting price for a ticket should rise and fall as her credences and the stakes change. Unfortunately, those intuitive motivations can't take us quite far enough. An agent could assign fair betting prices that rise and fall in the manner described without setting those fair betting prices equal to her credences.

Recall Mr. Bold, who assigns to each proposition the square-root of the credence assigned by Mr. Prob. Mr. Prob's credences satisfy the probability axioms, while Mr. Bold's violate Finite Additivity. Now suppose that Mr. Bold sets his fair betting prices for various gambling tickets equal not to his credences, but instead to the *square* of his credences. Mr. Bold's fair betting prices (for tickets on contingent propositions) will still rise and fall in exactly the ways that intuition requires. In fact, he will be willing to buy or sell any gambling ticket at exactly the same prices as Mr. Prob. And since Mr. Prob isn't susceptible to various kinds of Dutch Book, Mr. Bold won't be either. In general, an agent who assigns nonprobabilistic credences may be able to avoid

Book by assigning his betting prices in nonstandard fashion. Without a strong assumption about how rational agents use their credences to set betting prices, the Dutch Book Argument cannot show that nonprobabilistic credences are irrational.[11]

9.3.1 The Package Principle

The objection just raised applies to any Dutch Book Argument, because it questions how fair betting prices are set for the bets within a Book. Another, more traditional objection applies only to Books involving more than one gambling ticket; for instance, it applies to the Dutch Book against Finite Additivity violators but not to the Books against Non-Negativity and Normality offenders. (As I keep saying, Finite Additivity is the most difficult of the three axioms to establish as a rational rule.)

This traditional objection begins with **interference effects** that may be generated by placing a series of bets in succession. Interference effects occur when the initial bets in a series interfere with an agent's willingness to accept the remaining bets. While she might have accepted the remaining bets as fair had they been offered to her in isolation, the bets she's made already turn her against them. For example, the agent might have a personal policy never to tie up more than a certain total amount of money in gambles at one time. Or the third bet she's offered might be on the proposition "I will never make more than two bets in my life." More to the point, suppose we have an agent whose credences violate the probability axioms; we carefully construct a set of bets guaranteeing a sure loss, each of which will be placed at odds matching her degree of belief in the relevant proposition. We offer these bets to her one at a time. There's no guarantee that placing the first few wagers won't interfere with the agent's willingness to accept the remainder. Besides the interference effects just mentioned, the agent might see her sure loss coming down the pike, and simply refuse to place any more bets past some point! Interference effects undermine the claim that any agent with nonprobabilistic credences can be trapped into placing a sure-loss set of bets.

Interference effects are often introduced (as I've just done) by talking about placing bets with an agent one at a time. A Dutch Book defender might respond by suggesting that the bookie place his bets with the agent all at once—as a package deal. Yet the agent might still reject this package on the grounds that she doesn't like to tie up so much money in gambles, or that she can see a sure loss on the way. The sequential offering of the bets over time is ultimately irrelevant to the dialectic. A more promising response to interference effects points out how heavily they rely on the transactional pragmatics of betting. Depragmatized Dutch Book Arguments indict a nonprobabilistic agent on the grounds that her credences *sanction* a sure-loss set of bets; whether interference effects would impede her actually *placing* those bets is neither here nor there.

Yet there's a problem in the vicinity even for depragmatized arguments. Howson and Urbach's and Christensen's arguments contend that Dutch Bookability reveals an underlying doxastic inconsistency. What's the *nature* of that inconsistency? Earlier we saw Ramsey suggesting of the probability rules that, "If anyone's mental condition violated these laws, his choice would depend on the precise form in which the options were offered him, which would be absurd." From this suggestion, Skyrms takes the principle that for a rational agent, "A betting arrangement gets the same expected utility no matter how described" (1987a, p. 230). Similarly, Joyce writes that an agent's nonprobabilism "leads her to commit both the prudential sin of squandering happiness and the epistemic sin of valuing prospects differently depending on how they happen to be described"(1998, p. 96).

The rational inconsistency revealed by Dutch Bookability seems to be that the agent evaluates *one and the same entity* differently depending on how it is presented. Skyrms calls the entity being evaluated a "betting arrangement". To illustrate how one and the same betting arrangement might be presented in two different ways, let's return to our Dutch Book against a Finite Additivity violator (Section 9.1.1). That Book begins by selling the agent a ticket that pays $1 if P is true and another ticket that pays $1 on Q (call these the "P-ticket" and the "Q-ticket", respectively). The agent assigns $cr(P) = cr(Q) = 0.5$, so she will buy these tickets for $0.50 each. At that point the agent has purchased a package consisting of two tickets:

This ticket entitles the bearer to $1 if P is true, and nothing otherwise.	This ticket entitles the bearer to $1 if Q is true, and nothing otherwise.

Call this the "PQ-package". We assume that since the agent is willing to pay $0.50 for each of the two tickets in the package, she will pay $1 for the package as a whole.

In the next step of the Finite Additivity Dutch Book, we buy the following ticket from the agent (which we'll call the "$P \vee Q$-ticket"):

> This ticket entitles the bearer
> to $1 if $P \vee Q$ is true,
> and nothing otherwise.

Our agent assigns $cr(P \vee Q) = 0.8$, so she sells us this ticket for $0.80.

Now compare the PQ-package with the $P \vee Q$-ticket, and keep in mind that in this example P and Q are mutually exclusive. If either P or Q turns out to be true, the PQ-package and the $P \vee Q$-ticket will each pay exactly $1. Simlarly, each one pays $0 if neither P nor Q is true. So the PQ-package pays out the same amount as the $P \vee Q$-ticket in every possible world. This is the sense in which they represent the same "betting arrangement". When we offer the agent that betting arrangement as a package of two bets on atomic propositions, she values the arrangement at $1. When we offer that arrangement as a single bet on a disjunction, she values it at $0.80. She values the same thing—the same betting arrangement—differently under these two presentations. If she's willing to place bets based on those evaluations, we can use them to take money from her. (We sell the arrangement to her in the form she holds dear, then buy it back in the form she'll part with for cheap.) But even if the agent won't actually place the bets, the discrepancy in her evaluations reveals a rational flaw in her underlying credences.

The general idea is that any Dutch Book containing multiple bets reveals a violation of

Extensional Equivalence: If two betting arrangements have the same payoff as one another in each possible world, a rational agent will value them equally.

I certainly won't question Extensional Equivalence. But the argument above sneaks in another assumption as well. How did we decide that our agent valued the PQ-package at $1? We assumed that since she was willing to pay $0.50 for the P-ticket on its own and $0.50 for the Q-ticket as well, she'd pay $1 for these two tickets bundled together as a package. We assumed the

Package Principle: A rational agent's value for a package of bets equals the sum of her values for the individual bets it contains.

Our argument needed the Package Principle to get going (as does every Dutch Book consisting of more than one bet). We wanted to indict the set of credences our agent assigns to P, Q, and $P \vee Q$. But bets based on those individual propositions would not invoke Extensional Equivalence, because no two such bets have identical payoffs in each possible world. So we *combined* the P- and Q-tickets into the PQ-package, a betting arrangement extensionally equivalent to the P ∨ Q-ticket. We then needed a value for the PQ-package, a new object not immediately tied to any of the agent's credences. So we applied the Package Principle.[12]

Is it legitimate to assume the Package Principle in arguing for Finite Additivity? I worry that we face a Linearity In, Linearity Out problem once more. In order to get a Dutch Book Argument for Finite Additivity, we need to assume that a rational agent's value for a package of bets on mutually exclusive propositions is a linear combination of her values for bets on the individual propositions. Schick (1986, p. 113) calls this "the unspoken assumption . . . of value additivity" in Dutch Book Arguments; it seems to do exactly for bet valuations what Finite Additivity does for credences.[13] Without an independent argument for this Package Principle, the Dutch Book Argument for probabilism cannot succeed.[14]

9.3.2 Dutch Strategy objections

The first objection discussed in this section—concerning fair betting prices—applies to any Dutch Book. The Package Principle objection applies to Books containing multiple bets. But even beyond those objections, special problems arise for Dutch arguments involving credences assigned at different times. I will focus here on Lewis's Dutch Strategy Argument for Conditionalization; similar points apply to Strategies supporting Jeffrey Conditionalization and other potential diachronic norms.

To get the concern going, we need to focus on an aspect of Dutch Books and Strategies that I haven't mentioned yet. I keep saying that a Dutch Book guarantees the agent will lose money in every possible world. What set of possible worlds am I talking about? It can't be the set of logically possible worlds; after all, there are logically possible worlds in which no bets are ever placed. When we say that a Dutch Book guarantees the agent a sure loss in every world, we usually mean something like the agent's doxastically possible worlds—the worlds she entertains as a live option.

It makes sense to construct Dutch Books around worlds the agent considers possible. Dutch Book susceptibility is supposed to be a *rational* flaw, and rationality concerns how things look from the agent's own point of view. Imagine a bookie sells you for $0.50 a bet that pays $1 if a particular fair coin flip comes up heads. The bookie then claims he has Dutch Booked you, because he's already seen the flip outcome and it came up tails! Your willingness to purchase that bet didn't reveal any rational flaw in your credences. Admittedly there's *some* sense in which the bookie sold you a bet that's a loser in every live possibility. But that's a sense of "live possibility" to which you didn't have access when you placed the bet; relative to *your* information the bet wasn't a sure loss. To constrain our attention to Dutch Books or Strategies capable of revealing *rational* flaws, we usually require them to generate a sure loss across the agent's space of doxastically possible worlds. A convenient way to do this is to stipulate that the bookie in a Dutch Book or Strategy must be capable of constructing the Book or implementing the Strategy without employing any contingent information the agent lacks.

With that in mind, let's return to Lewis's Dutch Strategy against Conditionalization violators. Here's the particular set of bets we used in Section 9.1.2:

	$P \& Q$	$\sim P \& Q$	$\sim Q$
Ticket 1	−0.75	0.25	0.25
Ticket 2	0.30	0.30	−0.30
Ticket if Q learned	0.40	−0.60	0
TOTAL	−0.05	−0.05	−0.05

This Strategy was constructed against an agent who assigns equal unconditional credence to each of the four P/Q state-descriptions at t_i, assigns $cr_i(P \mid Q) = 0.5$, yet assigns $cr_j(P) = 0.6$ if she learns that Q between t_i and t_j.

At a first pass, this Strategy seems to meet our requirement that the bookie need not know more than the agent. Tickets 1 and 2 are purchased from the agent at t_i using betting prices set by her t_i credences. The third ticket is sold to the agent at t_j only if she learns Q between t_i and t_j. But by t_j the agent (and the bookie) know whether she has learned Q, so the bookie needn't know more than the agent to decide whether to sell that ticket.

Yet matters turn out to be more subtle than that. To see why, I'd suggest that the reader construct a Dutch Strategy against an agent who assigns the same t_i credences as in our example but assigns $cr_j(P) = 0.4$ if she learns that Q between t_i and t_j. Obviously the change in $cr_j(P)$ value changes the bet made if Q is learned; that bet must be keyed to the agent's fair betting prices at the later time. More interestingly, though, you'll find that while the bets placed at t_i have much the same structure as Ticket 1 and Ticket 2, the bookie needs to *sell* them at t_i—rather than *buying* them—in order to guarantee the agent a sure loss.

Now imagine a bookie is confronted at t_i by an agent who assigns equal credence to all four P/Q state-descriptions and satisfies the probability axioms and Ratio Formula. The bookie wants to initiate a Dutch Strategy that will cost the agent money should she fail to update by Conditionalization at t_j. But the bookie doesn't know which Strategy to pursue: the Strategy against agents who assign $cr_j(P) > 0.5$, or the Strategy against $cr_j(P) < 0.5$. These Strategies require the bookie to take different sides on his t_i bets, so in order to pursue a course that will definitely cost the agent, the bookie must know at t_i what the agent's credences will be at t_j. Given our stipulation that the bookie knows only what the agent does, this means that a Dutch Strategy can be constructed against an agent who violates Conditionalization only if that agent knows in advance how she'll be violating it.

How might an agent know in advance what credences she'll assign in the future? One possibility is if the agent has a standing policy, or plan, for updating in response to evidence. If the agent has such a plan, and it recommends different credences than Conditionalization, then the agent (and bookie) will be able to tell at t_i that she'll violate Conditionalization at t_j. Moreover, if the agent knows at t_i that, say, her updating plan would lead her to assign $cr_j(P) = 0.6$ upon learning that Q, the bookie can take advantage of this information to set up a Dutch Strategy by placing the appropriate bets at t_i.

By itself, violating Conditionalization doesn't make an agent susceptible to a Dutch Strategy. A Strategy can be implemented against an agent only if she violates Conditionalization *after planning to do so*. To dramatize the point, consider an agent who plans at t_i to conditionalize, but when t_j comes around actually violates Conditionalization. No Dutch Strategy can be implemented against such an agent; since at t_i the bookie won't know the details of the violation, he won't be able to place the requisite t_i bets.

Once we see this point, we might wonder whether the rational fault lies in the plan, or in the implementation. van Fraassen writes:

> Let us emphasize especially that these features are demonstrable *beforehand*, without appeal to any but logical considerations, and the strategy's implementation requires no information inaccessible to the agent himself. The general conclusion must be that an agent vulnerable to such a Dutch strategy has an initial state of opinion and practice of changing his opinion, which together constitute a demonstrably bad guide to life.
>
> (1984, p. 240, emphasis in original)

According to van Fraassen, it's the initial state of opinion plus the *initial practice of changing opinion* that together constitute a bad guide. If that's right, then the

Dutch Strategy targets a synchronic inconsistency among stances adopted at t_i, not any truly diachronic inconsistency.

One might respond that it's irrational for an agent to make one plan then carry out another, so a rational agent plans to conditionalize just in case she does so. (If that's right, then any argument that a rational agent plans to conditionalize is also an argument that she does conditionalize.) Notice first that this biconditional deems it irrational to update without first planning to do so—must rationality bar spontaneity? But setting that point aside, there's a legitimate question about how to *argue* for the diachronic rational requirement that an agent update as she had planned. For the previous few paragraphs, I've been trying to establish that this requirement cannot be established using a Dutch Strategy. But if you still aren't convinced, the following analogy may help.

Suppose my sister and I each have credences satisfying the probability axioms and Ratio Formula, but I assign $cr(P) = 0.7$, while she assigns $cr(\sim P) = 0.7$. A clever bookie could place bets with each of us that together guaranteed him a sure profit. Should this bother me? Or my sister? Not unless we antecedently think there's something wrong with our having differing opinions.

Now consider an agent who, at t_j, takes her credences to stand in the same relation to her t_i assignments that I take my credences to stand in to my sister's. This t_j agent doesn't see anything rationally pressing about lining up her credences with the credences or plans she made at t_i. A bookie might place a series of bets with the agent's t_i and t_j selves that together guarantee a sure loss. But the t_j agent will find that no more impressive than the sure-loss contract constructible against me and my sister. If the t_j agent doesn't antecedently feel any rational pressure to coordinate her current attitudes with those of her t_i self, pointing out that their combined activities result in a guaranteed loss will not create that pressure.

A Dutch Strategy may establish that it's rational for an agent to *plan* to Conditionalize. But it cannot establish that an agent is rationally required at a later time to do what she planned earlier. There may of course be *other* arguments for such a diachronic rational requirement, but they must be independently established before a Dutch Strategy can have any diachronic bite. As Christensen puts it:

> Without some independent reason for thinking that an agent's present beliefs must cohere with her future beliefs, her potential vulnerability to the Dutch strategy provides no support at all for [conditionalization].   (1991, p. 246)

A Dutch Strategy Argument may fill out the details of rational updating norms should any exist. But it is ill-suited to establish the existence of such norms in the first place.[15]

9.4 Exercises

Problem 9.1. 🎵 In Section 9.1.1 we constructed a Dutch Book against an agent like Mr. Bold whose credences are subadditive. Now construct a Dutch Book against an agent whose credences are *super*additive: for mutually exclusive P and Q, he assigns $cr(P) = 0.3, cr(Q) = 0.3$, but $cr(P \lor Q) = 0.8$. Describe the bets composing your Book, say why the agent will find each one acceptable, and show that the bets guarantee him a loss in every possible world.

Problem 9.2. 🎵🎵 Roxanne's credence distribution at a particular time includes the following values:

$$cr(A \& B) = 0.5 \qquad cr(A) = 0.1 \qquad cr(B) = 0.5 \qquad cr(A \lor B) = 0.8$$

(Do *not* assume that A and B are mutually exclusive!)
  (a) Show that Roxanne's distribution violates the probability axioms.
  (b) Construct a Dutch Book against Roxanne's credences. Lay out the bets involved, then show that those bets actually constitute a Dutch Book against Roxanne.
     Note: The Book must be constructed using only the credences described above; since Roxanne is non-probabilistic you may not assume anything about the other credences she assigns. However, the Book need not take advantage of all four credences.
  (c) Construct a Czech Book in Roxanne's favor. Lay out the bets involved and show that they guarantee her a profit in every possible world.
  (d) Does the success of your Dutch Book against Roxanne require her to satisfy the Package Principle? Explain.

Problem 9.3. 🎵🎵 You are currently certain that you are not the best singer in the world. You also currently satisfy the probability axioms and the Ratio Formula. Yet you assign credence 0.5 that you will go to a karaoke bar tonight, and are certain that if you *do* go to the bar, cheap beer and persuasive friends will make you certain that you *are* the best singer in the world. Suppose a bookie offers you the following two betting tickets right now:

> This ticket entitles you
> to $20 if you go to the bar,
> and nothing otherwise.

> If you go to the bar, this ticket entitles you
> to $40 if you are not the world's best singer,
> and nothing if you are.
> If you don't go to the bar,
> this ticket may be returned to the seller
> for a full refund of its purchase price.

(a) Suppose that right now, a bookie offers to sell you the first ticket above for $10 and the second ticket for $30. Explain why, given your current credences, you will be willing to buy the two tickets at those prices. (Remember that the second ticket involves a conditional bet, so its fair betting price is determined by your current *conditional* credences.)

(b) Describe a Dutch Strategy the bookie can plan against you right now. In particular, describe a third bet that he can plan to place with you later tonight only if you're at the bar and certain of your singing prowess, such that he's guaranteed to make a net profit from you come what may. Be sure to explain why you'll be willing to accept that third bet later on, and how it creates a Dutch Strategy against you.[16]

Problem 9.4. 🖋 Do you think there is any kind of Dutch Book or Strategy that reveals a rational flaw in an agent's attitudes? If so, say why, and say which *kinds* of Dutch Books/Strategies you take to be revealing. If not, explain why not.

9.5 Further reading

INTRODUCTIONS AND OVERVIEWS

Susan Vineberg (2011). Dutch Book Arguments. In: *The Stanford Encyclopedia of Philosophy*. Ed. by Edward N. Zalta. Summer 2011

Covers all the topics discussed in this chapter in much greater depth, with extensive citations.

CLASSIC TEXTS

Frank P. Ramsey (1931). Truth and Probability. In: *The Foundations of Mathematics and Other Logic Essays*. Ed. by R. B. Braithwaite. New York: Harcourt, Brace and Company, pp. 156–98

Bruno de Finetti (1937/1964). Foresight: Its Logical Laws, its Subjective Sources. In: *Studies in Subjective Probability*. Ed. by Henry E. Kyburg Jr and H.E. Smokler. New York: Wiley, pp. 94–158. Originally published as "La prévision; ses lois logiques, ses sources subjectives" in *Annales de l'Institut Henri Poincaré*, Volume 7, 1–68

On p. 84, Ramsey notes that any agent whose degrees of belief violated the laws of probability "could have a book made against him by a cunning better and would then stand to lose in any event." de Finetti goes on to prove it.

Frederic Schick (1986). Dutch Bookies and Money Pumps. *The Journal of Philosophy* 83, pp. 112–19

Compares Dutch Book and money pump arguments, then offers a Package Principle objection to each.

Paul Teller (1973). Conditionalization and Observation. *Synthese* 26, pp. 218–58

First presentation of Lewis's Dutch Strategy Argument for Conditionalization; also contains a number of other interesting arguments for Conditionalization.

EXTENDED DISCUSSION

David Christensen (2001). Preference-based Arguments for Probabilism. *Philosophy of Science* 68, pp. 356–76

Presents depragmatized versions of both the Representation Theorem and Dutch Book Arguments for probabilism, then responds to objections.

Notes

1. Or something very close to Kolmogorov's rules—the agent may assign a maximal credence other than 1.
2. Over the years, Bayesians have used the term **coherent** in a variety of ways. de Finetti defined a "coherent" credence distribution as one that avoided Dutch Book, then proved

that coherence requires probabilistic credences. Other Bayesian authors take satisfying the probability axioms as the definition of "coherence". Some use "coherent" simply as a synonym for "rationally consistent", whatever such consistency turns out to require.

3. This formula is easy to derive if we assume that the agent selects her fair betting prices so as to maximize expected dollar return, as we did in Section 7.1. Yet I've been scrupulously avoiding making that assumption here, for reasons I'll explain in Section 9.3.

4. A "book" is a common term for a bet placed with a "bookmaker" (or "bookie"), but why *Dutch*? Hacking (2001, p. 169) attributes the term to Ramsey, and suggests it may have been English betting slang of Ramsey's day. Concerned to avoid using what is possibly a slur, Hacking prefers to speak of "sure-loss contracts". Yet I haven't been able to find the "Dutch" terminology anywhere in Ramsey's text, so its origins remain a mystery to me. Since the phrase "Dutch Book" is near-ubiquitous in the Bayesian literature, and we have no evidence that (or how) it gives offence, I will continue to use it here. But I'd be happy for Bayesians to move in another direction if more information came to light.

5. Lewis didn't publish this argument against non-Conditionalizers. Instead it was reported by Teller (1973), who attributed the innovation to Lewis.

6. Notice that the ticket sold to the agent at t_j does not constitute a conditional bet. It's a normal ticket paying out \$1 on P no matter what, and we set the agent's fair betting price for this ticket using her unconditional credences at t_j. It's just that we decide whether to sell her this (normal, unconditional) ticket on the basis of what she learns between t_i and t_j.

   Notice also that for purposes of this example we're assuming that the agent learns Q between t_i and t_j just in case Q is true.

7. Because I like to keep things tidy, I have constructed all of my Dutch Books and Dutch Strategies above so that the agent loses the same amount in each possible world. That isn't a requirement for Dutch Books or Strategies, and you certainly don't have to follow it when constructing your own—as long as the net payout in each world is negative, the goal has been achieved.

8. The phrase "Books of the sort we considered in Section 9.1.1" underspecifies the precise sets of Books in question; Lehman and Kemeny are much clearer about what kinds of betting packages their results concern. See also an early, limited Converse result in (de Finetti 1937/1964, p. 104).

9. I have slightly altered Christensen's premises to remove his references to a "simple agent". Christensen uses the simple agent to avoid worries about the declining marginal utility of money and about the way winning one bet may alter the value of a second's payout. Inserting the simple agent references back into the argument would not protect it from the objections I raise in the next section.

10. For more on the question of whether rationality requires agents to set their fair betting prices using expected utilities, see Hedden (2013), Wroński and Godziszewski (2017), and Pettigrew (2021).

11. All this should sound very reminiscent of my criticisms of Representation Theorem Arguments in Chapter 8. Mathematically, the proof of the Revised Representation Theorem from that chapter is very similar to standard proofs of the Dutch Book Theorem.

12. A similar move is hidden in Christensen's "Belief Defectiveness" principle. The principle says that if an agent's degrees of belief sanction as fair each bet in a set of bets, and that set of bets is rationally defective, then the agent's beliefs are rationally defective. The intuitive idea is that beliefs which sanction something defective are themselves defective. Yet without the Package Principle, an agent's beliefs might sanction as fair each of the bets in a particular set without sanctioning the *entire set* as fair. And it's the entire set of bets that is the rationally defective object—it's the *set* that guarantees the agent a sure loss.

13. Schick contends that money pump arguments for the Preference Axioms (Section 7.2.1) also assume something like the Package Principle.

14. If we assume that rational agents maximize expected utility, we can generate straightforward arguments for both Extensional Equivalence and the Package Principle. But again, if we are allowed to make *that* assumption, then probabilism follows quickly. (In fact, Extensional Equivalence is a key lemma in proving the Revised Representation Theorem.)

15. For what it's worth, one could make a similar point about Dutch Book Arguments for *synchronic* norms relating distinct degrees of belief. To dramatize the point, imagine that an agent's propositional attitudes were in fact little homunculi, each assigned its own proposition to tend to and adopt a degree of belief toward. If we demonstrated to one such homunculus that combining his assignments with those of other homunculi would generate a sure loss, he might very well not care.

   The point of this fanciful scenario is that while Dutch Books may fill out the details of rational relations among an agent's degrees of belief at a given time, they are ill-suited to establish that rationality requires such synchronic relations in the first place. Absent an antecedent rational pressure to coordinate attitudes adopted at the same time, the fact that such attitudes could be combined into a sure loss would be of little normative significance. No one ever comments on this point about Dutch Books, because we all assume in the background that degrees of belief assigned by the same agent at the same time are required to stand in some relations of rational coherence.

16. I owe this entire problem to Sarah Moss.

10

Accuracy Arguments

The previous two chapters considered arguments for probabilism based on Representation Theorems and Dutch Books. We criticized both types of argument for beginning with premises about practical rationality—premises about how a rational agent views certain acts (especially acts of placing bets). We want to establish the probability axioms as requirements of *theoretical* rationality on an agent's credences, and it's difficult to move from practical premises to a theoretical conclusion.

This chapter builds arguments for probabilism from explicitly epistemic premises. The basic idea is that, as representational attitudes, credences can be assessed for **accuracy**. We often assess other doxastic attitudes, such as binary beliefs, for accuracy: a belief in the proposition P is accurate if P is true; disbelief in P is accurate if P is false. A traditional argument moves from such accuracy assessments to a rational requirement that agents' belief sets be logically consistent (Chapter 1's Belief Consistency norm). The argument begins by noting that if a set of propositions is logically inconsistent, then by definition there is no (logically) possible world in which all of those propositions are true. So if an agent's beliefs are logically inconsistent, she's in a position to know that at least some of them are inaccurate. Moreover, she can know this a priori—without invoking any contingent truths. (Since an inconsistent set contains falsehoods in *every* possible world, no matter which world is actual her inconsistent belief set misrepresents how things are.) Such unavoidable, a priori inaccuracy reveals a rational flaw in any logically inconsistent set of beliefs.[1]

There are plenty of potential flaws in this argument—starting with its assumption that beliefs have a teleological "aim" of being accurate. But the argument is a good template for the arguments for probabilism to be discussed in this chapter. Whatever concerns you have about the Belief Consistency argument above, keep them in mind as you consider accuracy arguments for probabilism.

Assessing credences for accuracy isn't as straightforward as assessing binary beliefs: a credence of, say, 0.6 in proposition P doesn't say that P is true, but neither does it say that P is false. So we can't assess the accuracy of this credence

Fundamentals of Bayesian Epistemology 2: Arguments, Challenges, Alternatives. Michael G. Titelbaum, Oxford University Press. © Michael G. Titelbaum 2022. DOI: 10.1093/oso/9780192863140.003.0010

by asking whether it assigns a truth-value to P matching P's truth-value in the world. Nor can we say that $\text{cr}(P) = 0.6$ is accurate just in case P is true "to degree 0.6"; we've assumed that propositions are wholly true or wholly false, full stop. So just as we moved from classificatory to quantitative doxastic attitudes in Chapter 1, we need to move from a classificatory to a quantitative concept of accuracy. This chapter will begin by considering various numerical measures of just *how* accurate a credence (or set of credences) is. We'll start with historical "calibration" approaches that measure credal accuracy by comparing credences with frequencies. Then we'll reject calibration in favor of contemporary "gradational accuracy" approaches.

The most commonly used gradational accuracy measure is known as the Brier score. Using the Brier score, we will construct an argument for probabilism similar to the Belief Consistency argument above: violating the probability axioms impedes a credence set's accuracy in every possible world. It will then turn out that an argument like this can be constructed using not just the Brier score, but any gradational accuracy measure in a class known as the "strictly proper scoring rules".

Which leads to the question of why strictly proper scoring rules are superior to other accuracy-measurement options—especially options that rule *out* probabilism. The spectre will arise once more that our argument for probabilism has a question-begging, Linearity-In, Linearity-Out structure. This will force us to ask something you may have started wondering over the last couple of chapters: How important is it, *really*, that rational credences satisfy Finite Additivity, as opposed to other norms with similar consequences for thought and behavior?

Besides arguing for probabilism, Bayesian epistemologists have offered accuracy-based arguments for other norms such as the Principal Principle (Pettigrew 2013a), the Principle of Indifference (Pettigrew 2014), Reflection (Easwaran 2013), and Conglomerability (Easwaran 2013). We'll close this chapter with an argument for Conditionalization based on minimizing expected future inaccuracy. Unfortunately this argument has the same drawback as Dutch Strategy arguments for Conditionalization: it is insufficient on its own to establish truly *diachronic* norms.

10.1 Accuracy as calibration

In Section 5.2.1 we briefly considered a putative rational principle for matching one's credence that a particular outcome will occur to the frequency with which that kind of outcome occurs. In that context, the match was supposed to be between one's credence that outcome B will occur and the frequency with which *one's evidence* suggests B-type outcomes occur. But we might instead assess an agent's credences relative to *actual frequencies* in the world: If events of type A actually produce outcomes of type B with frequency x, an agent's credence that a particular A-event will produce a B-outcome is more accurate the closer it is to x.

Now imagine that an agent managed to be perfectly accurate with respect to the actual frequencies. In that case, she would assign credence 2/3 to types of outcomes that occurred 2/3 of the time, credence 1/2 to outcome-types that occurred 1/2 of the time, etc. Or—flipping this around—propositions to which she assigned credence 2/3 would turn out to be true 2/3 of the time, propositions to which she assigned credence 1/2 would turn out to be true 1/2 of the time, etc. This conception of accuracy—getting the frequencies right, as it were—motivates assessing credences with respect to their

Calibration: A credence distribution over a finite set of propositions is perfectly calibrated when, for any real x, out of all the propositions to which the distribution assigns x, the fraction that turn out to be true is x.

For example, suppose your weather forecaster comes on television every night and reports her degree of confidence that it will snow the next day. You might notice that every time she says she's 20% confident of snow, it snows the next day. In that case she's not a very accurate forecaster. But if it snows on just about 20% of those days, we'd say she's doing her job well. If exactly 20% of the days on which she's 20% confident of snow turn out to have snow (and exactly 30% of the days on which she's 30% confident . . . etc.), we say the forecaster is perfectly calibrated.[2] Calibration seems a plausible way to gauge accuracy.[3]

I've defined only what it means to be *perfectly* calibrated; there are also numerical measures for assessing degrees of calibration short of perfection (see Murphy 1973).[4] But all the good and bad consequences of reading accuracy as calibration can be understood by thinking solely about perfect calibration. First, the good: van Fraassen (1983) and Abner Shimony (1988) both argued for probabilism by showing that in order for a credence distribution to be embeddable in larger and larger distributions with calibration scores approaching perfection, that original credence distribution must satisfy the probability axioms. This seems a powerful argument for probabilism—if we're on board with calibration as a measure of accuracy.

Here's why we might not be. Consider two agents, Sam and Diane, who assign the following credence distributions over propositions X_1 through X_4:

	X_1	X_2	X_3	X_4
Sam	1/2	1/2	1/2	1/2
Diane	1	1	1/10	0

Now suppose that propositions X_1 and X_2 are true, while X_3 and X_4 are false. Look at the table and ask yourself whose credences intuitively seem more accurate.[5]

I take it the answer is Diane. Yet Sam's credences are perfectly calibrated—he assigns credence 1/2 to all four propositions, exactly half of which are true—while Diane's credences are not. This is an intuitive flaw with measuring accuracy by calibration.

A similar point can be made by considering the following (real life!) example: On the morning of February 1, 2015, I looked outside and found it was snowing heavily. At least four inches had accumulated during the night, the snow was still coming down, and it showed no signs of stopping. The online weather report on my smartphone, however, showed an at-the-moment 90% probability of snow. Why hadn't the forecaster simply looked out her window and updated the report to 100%?

I was suddenly struck by a possible explanation. Let's imagine (what's probably not true) that the forecaster posts to the online weather report her current credence that it will snow on the current day. Suppose also that weather forecasting sites are graded for accuracy, and promoted on search engines based on how well they score. Finally, suppose this accuracy scoring is done by measuring calibration. What if, up to February 1, it had snowed every time the forecaster reported a 100% credence, but it had snowed on only eight of the nine occasions on which she had expressed a 90% credence? The snow on February 1 would then present her with an opportunity. She could report her true, 100% confidence in snow for February 1 on the website. Or she could post a 90% probability of snow. Given that it was clearly snowing on February 1, the latter option would bring her up to a perfect calibration score, and shoot her website to the top of the search rankings. Calibration gives the forecaster an incentive to misreport her own credences—and the content of her own evidence.

Calibration is one example of a **scoring rule**; a procedure for rating distributions with respect to accuracy. James M. Joyce reports that "the term 'scoring rule' comes from economics, where values of [such rules] are seen as imposing penalties for making inaccurate probabilistic predictions" (2009, p. 266). Done right, the imposition of such penalties can be a good way of finding out what experts really think—what's known as **credence elicitation**. If you reward (or punish) an expert according to the accuracy of her reports, you incentivize her to gather the best evidence she can, consider it carefully, and then report to you her genuine conclusions. Seen through this lense of credence elicitation, calibration fails as a scoring rule. As we've just seen, rewarding a forecaster according to her level of calibration can incentivize her to misreport her true opinions, and what she takes to be the import of her evidence.

Yet perhaps it's unfair to criticize calibration on the grounds that it perversely incentivizes credence *reports*; norms for assertion can be messy, and anyway probabilism is a norm on agents' thoughts, not their words. So let's consider calibration as a direct accuracy measure of our forecaster's credences. Prior to February 1 it has snowed whenever the forecaster was certain of snow, but on the days when she assigned 0.9 credence to snow, it has snowed eight of nine times. Looking out her window and seeing snow, the forecaster assigns credence 1 to snow.[6] Yet if her goal is to be as accurate as possible with her credences, and if accuracy is truly measured by calibration, then the forecaster will *wish* that her credence in snow was 0.9. After all, that would make her pefectly calibrated!

Assessing the forecaster's credences by calibration makes those credences *unstable*. By the forecaster's own lights—given the credences she has formed in light of her evidence—she thinks she'd be better off with different credences. Such instability is an undesirable feature in a credence distribution, and is generally thought to be a hallmark of irrationality. David Lewis offers the following analogy:

> It is as if *Consumer Bulletin* were to advise you that *Consumer Reports* was a best buy whereas *Consumer Bulletin* itself was not acceptable; you could not possibly trust *Consumer Bulletin* completely thereafter.   (1971, p. 56)

If we use calibration to measure accuracy, the weather forecaster's credence distribution becomes unstable. Such instability is a sign of irrationality. So from a calibration point of view, there's something rationally wrong with the forecaster's credences. But in reality there's nothing wrong with the forecaster's credences—they are a perfectly rational response to the evidence before her eyes! The problem lies with calibration as a measure of accuracy; calibration renders unstable credence distributions that are rationally permissible (if not rationally required!).

There are further ways in which calibration rewards agents for ignoring their evidence. Notice that any agent assigning credences over a partition of n propositions can secure a perfect calibration score by assigning each proposition a credence of $1/n$. For instance, if a six-sided die is to be rolled, an agent can guarantee herself perfect calibration (no matter how the roll comes out!) by assigning each possible outcome a credence of $1/6$. Depending on how you feel about the Principle of Indifference (Section 5.3), this might be a reasonable assignment when the agent has no evidence relevant to the members of the partition. But now suppose the agent gains highly reliable evidence that the die is biased in favor of coming up 6. Altering her credences to reflect that bias won't earn her a better calibration score than the uniform $1/6$ distribution, and might very well serve her worse.

One could make various attempts here to save calibration as a plausible measure of accuracy. For instance, calibration scores are less easily manipulable if we measure them only in the long-run. But this generates questions about the accuracy of credences in non-repeatable events, and soon we're assessing not actual long-run calibration but hypothetical calibration in the limit. Before long, we've made all the desperate moves used to prop up the frequency theory of "probability" (Section 5.1.1), and run into all the same problems.

The response here should be the same as it was with the frequency theory: Rather than deploy a notion that emerges only when events are situated in a larger collective, we find a notion (like propensity) that can be meaningfully applied to single cases considered one at a time. Looking back at Sam and Diane, our intuitive judgment that Diane is globally more accurate than Sam arises from local judgments that she was more accurate than him on each individual proposition. If you knew only the truth-value of X_1, you could still have said that Diane was more accurate than Sam on that one proposition. Our accuracy intuitions apply piece-wise; we assess credences one proposition at a time, then combine the results into a global accuracy measure.

10.2 The gradational accuracy argument for probabilism

10.2.1 The Brier score

We will now develop what's known as the "gradational accuracy" approach to evaluating credences. Our guiding idea will be that inaccuracy is distance from truth—a credence distribution gains accuracy by moving its values closer to the truth-values of propositions. Of course, credence values are real numbers,

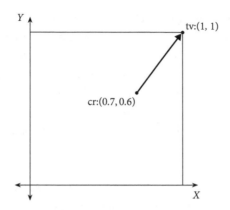

Figure 10.1 The Brier score

while truth-values are not. But it's natural to overcome that obstacle by letting 1 stand for truth and 0 stand for falsehood. Just as we have a distribution cr expressing the agent's credences in propositions, we'll have another distribution tv reflecting the truth-values of those propositions. Distribution tv assigns numerical values to the propositions in $\mathcal{L}$ such that $tv(X) = 1$ if X is true and $tv(X) = 0$ if X is false.[7]

Once we have distribution cr representing the agent's credences and distribution tv representing the truth, we want a scoring rule that measures how far apart these distributions are from each other. It's easiest to visualize the challenge on a diagram. To simplify matters, consider a credence distribution over only two propositions, X and Y. Our agent assigns $cr(X) = 0.7$ and $cr(Y) = 0.6$. I have depicted this credence assignment in Figure 10.1. In this diagram the horizontal axis represents the proposition X while the vertical axis represents Y. Any credence assignment to these two propositions can be represented as an ordered pair; I have placed a dot at the agent's cr-distribution of $(0.7, 0.6)$.

What about the values of tv? Let's suppose that propositions X and Y are both true. So $tv(X) = tv(Y) = 1$. I have marked $(1, 1)$—the location of tv on the diagram—with another dot. Now our question is how to measure the inaccuracy of the agent's credences; how should we gauge how far cr is from tv?

A natural suggestion is to use distance as the crow flies, indicated by the arrow in Figure 10.1. A quick calculation tells us that the length of the arrow is:

$$(1 - 0.7)^2 + (1 - 0.6)^2 = (0.3)^2 + (0.4)^2 = 0.25 \qquad (10.1)$$

Pythagorean Theorem aficionados will note the lack of a square-root in this distance expression (the arrow is actually 0.5 units long). For the time being, we're going to use inaccuracy measurements only for ordinal comparisons (which credence distribution is *farther* from the truth), so particular numerical values don't matter much—and neither does the square-root.

When generalized to a credence distribution over finitely many propositions $\{X_1, X_2, \ldots, X_n\}$, this distance measure of inaccuracy becomes

$$I_{BR}(cr, \omega) = (tv_\omega(X_1) - cr(X_1))^2 + (tv_\omega(X_2) - cr(X_2))^2 + \ldots$$
$$+ (tv_\omega(X_n) - cr(X_n))^2 \tag{10.2}$$

A few notes about this equation: First, what are the ωs doing in there? We usually want to evaluate the inaccuracy of your credence distribution relative to conditions in the actual world. But sometimes we'll wonder how inaccurate your credences would've been if you'd maintained your distribution but lived in a different possible world. For example, in Figure 10.1 we might wonder how inaccurate the credence distribution cr would have been had X and Y both been false. That is, we might want to calculate the distance between cr and the point $(0, 0)$. Equation (10.2) calculates the inaccuracy of credence distribution cr in an *arbitrary* possible world ω. $tv_\omega(X_i)$ represents the truth-value of proposition X_i in world ω; $I_{BR}(cr, \omega)$ then measures the inaccuracy of cr relative to conditions in that world. (So for the credence distribution $(0.7, 0.6)$ and the world $(0, 0)$, Equation (10.2) would yield an I_{BR}-value of $0.7^2 + 0.6^2 = 0.85$.)[8]

Second, Equation (10.2) tallies up inaccuracy one proposition at a time, then sums the results. For any credence distribution cr and particular proposition X_i, evaluating $(tv(X_i) - cr(X_i))^2$ is a way of gauging how far off distribution cr is *on that particular proposition*. Equation (10.2) makes that calculation for each individual proposition X_i, then adds up the results. In general, a scoring rule that sums the results of separate calculations made on individual propositions is called **separable**. Separable scoring rules track our intuition that accuracy assessments of an entire credence distribution can be built up piece-wise, considering the accuracy of one credence at a time; this was exactly the feature we found lacking in calibration's evaluation of Sam and Diane.[9]

The scoring rule described by Equation (10.2) is known as the Euclidean distance, the quadratic loss function, or most commonly the **Brier score**.[10] (This accounts for the "BR" subscript in I_{BR}.) The Brier score is hardly the only scoring rule available, but it is natural and widely used. So we will stick with it for the time being, until we examine other options in Section 10.3. At that point we'll find that even among the separable scoring rules, there may be ordinal non-equivalence—two separable scoring rules may disagree about which distribution is more accurate in a given world. Nevertheless, all the separable scoring rules have some features in common. For instance, while $I_{BR}(cr, \omega)$ is in some sense a global measure of the inaccuracy of cr in world ω, it doesn't take into account any wholistic or interactive features among the individual credences cr assigns. Separable scores can't, for example, take into account the sum or difference of $cr(X_i)$ and $cr(X_j)$ for $i \neq j$.[11]

Finally, I_{BR} and the other gradational measures we'll consider calculate the *in*accuracy of credence distributions in particular worlds. So an agent striving to be as accurate as possible will seek to *minimize* her score. Some authors prefer to work with credence distributions' **epistemic utility**, a numerical measure of epistemic value that rational agents *maximize*. Now there may be many aspects of a credence distribution that make it epistemically valuable or disvaluable besides its distance from the truth. But many authors work under the assumption that accuracy is the sole determiner of a distribution's epistemic value, in which case that value can be calculated directly from the distribution's inaccuracy. (The simplest way is to let the epistemic utility of distribution cr in world ω equal $1 - I_{BR}(cr, \omega)$.) If you find yourself reading elsewhere about accuracy arguments, be sure to notice whether the author asks agents to *minimize inaccuracy* or *maximize utility*. On either approach, the best credence is the one closest to the pin (the distribution tv). But with inaccuracy, as in golf, lowest score wins.

10.2.2 Joyce's accuracy argument for probabilism

In our discussion of calibration we said that it's rationally problematic for an agent's credence distribution to be "unstable"—for it to seem to the agent, by her own lights, that another credence distribution would be preferable to her own. We ultimately rejected assessing agents' credences using calibration, but now we have an alternative accuracy measure: the Brier score. If we could convince an agent that her credences are less accurate, as measured by the Brier score, than some other distribution over the same set of propositions, then it would seem irrational for her to maintain her credence distribution (as opposed to the other one).

How can we convince an agent that her credences are less accurate than some alternative? Inaccuracy is always measured *relative to a world*. Presumably the agent is interested how things stand in the *actual* world, but presumably she also has some uncertainty as to which propositions are true or false in the actual world. If she doesn't know the tv-values in the actual world, she won't be able to calculate her own Brier score in that world, much less the score of an alternative distribution.

But what if we could show her that there exists a single distribution that is more accurate than her own *in every logically possible world*? Then she wouldn't need to know which possible world was actual; she could determine on an a priori basis that however things stand in the actual world, she would be more accurate if she had that other distribution. In light of information like this, her present credences would look irrational. This line of thought is enshrined in the following principle:

Admissibles Not Dominated: If an agent's credence distribution is rationally permissible, and accuracy is measured with an acceptable scoring rule, then there does not exist another distribution that is more accurate than hers in every possible world.

Admissibles Not Dominated is a conditional. Contraposing it, we get that any credence distribution accuracy-dominated by another distribution on an acceptable scoring rule is rationally impermissible (or "inadmissible", in the accuracy literature's jargon).

Repurposing a theorem of de Finetti's (1974), and following on the work of Rosenkrantz (1981), Joyce (1998) demonstrated the

Gradational Accuracy Theorem: Given a credence distribution cr over a finite set of propositions $\{X_1, X_2, \ldots, X_n\}$, if we use the Brier score $I_{BR}(cr, \omega)$ to measure inaccuracy then:

- If cr does *not* satisfy the probability axioms, then there exists a probabilistic distribution cr' over the same propositions such that $I_{BR}(cr', \omega) < I_{BR}(cr, \omega)$ in every logically possible world ω; and
- If cr *does* satisfy the probability axioms, then there does not exist any cr' over those propositions such that $I_{BR}(cr', \omega) < I_{BR}(cr, \omega)$ in every logically possible world.

The Gradational Accuracy Theorem has two parts. The first part says that if an agent has a non-probabilistic credence distribution cr, we will be able to find a probabilistic distribution cr' defined over the same propositions as cr that

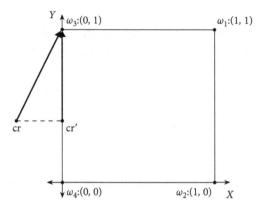

Figure 10.2 Violating Non-Negativity

accuracy-dominates cr. No matter what the world is like, distribution cr′ will be less inaccurate than cr. So the agent with distribution cr can be certain that, come what may, she is leaving a certain amount of accuracy on the table by assigning cr rather than cr′. There's a cost in accuracy, independent of what you think the world is like and therefore discernible a priori, to assigning a non-probabilistic credence distribution—much as there's a guaranteed accuracy cost to assigning logically inconsistent beliefs. On the other hand (and this is the second part of the theorem), if an agent's credence distribution *is* probabilistic, then no distribution (probabilistic or otherwise) is more accurate in every possible world. This seems a strong advantage of probabilistic credence distributions.[12]

Proving the second part of the theorem is difficult, but I will show how to prove the first part. There are three probability axioms—Non-Negativity, Normality, and Finite Additivity—so we need to show how violating each one leaves a distribution susceptible to accuracy domination. We'll take them one at a time, in order.

Suppose credence distribution cr violates Non-Negativity by assigning some proposition a negative credence. In Figure 10.2 I've imagined that cr assigns credences to two propositions, X and Y, bearing no special logical relations to each other. cr violates Non-Negativity by assigning cr(X) < 0. (The value of cr(Y) is irrelevant to the argument, but I've supposed it lies between 0 and 1.) We introduce probabilistic cr′ such that cr′(Y) = cr(Y) but cr′(X) = 0; cr′ is the closest point on the Y-axis to distribution cr.

We need to show that cr′ is less inaccurate than cr no matter which possible world is actual. Given our two propositions X and Y, there are four possible worlds.[13] I've marked them on the diagram as ω_1, ω_2, ω_3, and ω_4, determining

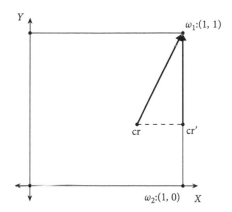

Figure 10.3 Violating Normality

the coordinates of each world by the truth-values it assigns to X and Y. (In ω_2, for instance, X is true and Y is false.) We now need to show that for each of these worlds, cr′ receives a lower Brier score than cr. In other words, we need to show that cr′ is closer to each world as the crow flies than cr is.

Clearly cr′ is closer to ω_2 and ω_1 than cr is, so cr′ is less inaccurate than cr relative to both ω_2 and ω_1. What about ω_3? I've indicated the distances from cr and cr′ to ω_3 with arrows. Because cr′ is the closest point on the Y-axis to cr, the points cr, cr′, and ω_3 form a right triangle. The arrow from cr to ω_3 is the hypotenuse of that triangle, while the arrow from cr′ to ω_3 is a leg. So the latter must be shorter, and cr′ is less inaccurate by the Brier score relative to ω_3. A parallel argument shows that cr′ is less inaccurate relative to ω_4. So cr′ is less inaccurate than cr relative to each possible world.

That takes care of Non-Negativity.[14] The accuracy argument against violating Normality is depicted in Figure 10.3. Suppose X is a tautology and cr assigns it some value other than 1. Since X is a tautology, there are no logically possible worlds in which it is false, so there are only the worlds ω_2 and ω_1 to consider. We construct cr′ such that cr′(Y) = cr(Y) and cr′(X) = 1. cr′ is closer than cr to ω_1 because the arrow from cr to ω_1 is the hypotenuse of a right triangle of which the arrow from cr′ to ω_1 is one leg. A similar argument shows that cr′ is closer than cr to ω_2, demonstrating that cr′ is less inaccurate than cr in every logically possible world.

Explaining how to accuracy-dominate a Finite Additivity violator requires a three-dimensional argument sufficiently complex that I will leave it for an endnote.[15] But we can show in two dimensions what happens if you violate one of the rules that follows from Finite Additivity, namely our Negation rule.

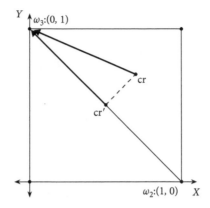

Figure 10.4 Violating Negation

Suppose your credence distribution assigns cr-values to two propositions X and Y such that X is the negation of Y. If you violate Negation, you'll have $cr(Y) \neq 1 - cr(X)$.

I've depicted only ω_2 and ω_3 in Figure 10.4 because only those two worlds are logically possible in this case (since X and Y must have opposite truth-values). The diagonal line connecting ω_2 and ω_3 has the equation $Y = 1 - X$; it contains all the credence distributions satisfying Negation. If cr violates Negation, it will fall somewhere off of this line. Then we can accuracy-dominate cr with the point closest to cr lying on the diagonal (call that point cr'). Once more, we've created a right triangle with cr, cr', and one of our possible worlds. The arrow representing the distance from cr to ω_3 is the hypotenuse of this triangle, while the arrow from cr' to ω_3 is its leg. So cr' has the shorter distance, and cr' is less inaccurate in ω_3 than cr according to the Brier score. A parallel argument applies to ω_2, so cr' is less inaccurate than cr in each of the two logically possible worlds.[16]

Joyce (1998, 2009) leverages the advantage of probabilistic credence distributions displayed by the Gradational Accuracy Theorem into an argument for probabilism:

Gradational Accuracy Argument for Probabilism

(Premise 1) A rationally permissible credence distribution cannot be accuracy-dominated on any acceptable scoring rule.

(Premise 2) The Brier score is an acceptable scoring rule.

(Theorem) If we use the Brier score, then any non-probabilistic credence distribution can be accuracy-dominated.

(Conclusion) All rationally permissible credence distributions satisfy the probability axioms.

The first premise of this argument is Admissibles Not Dominated. The theorem is the Gradational Accuracy Theorem. The conclusion of this argument is probabilism.

10.3 Objections to the accuracy argument for probabilism

Unlike Representation Theorem and Dutch Book Arguments, the Gradational Accuracy Argument for Probabilism has nothing to do with an agent's decision-theoretic preferences over practical acts. It clearly pertains to the *theoretical* rationality of credences assigned in pursuit of an *epistemic* goal: accuracy. (This is why Joyce's (1998) paper was titled "A Nonpragmatic Vindication of Probabilism".) This is a major advantage of the accuracy argument for probabilism. Of course, one has to be comfortable with the idea that belief-formation is a goal-directed activity—teleological, so to speak—and commentators have objected to that idea. (Examples appear in the Further Reading.)

But I want to focus on a more technical objection that has been with the gradational accuracy approach from its inception. Premise 2 of the Gradational Accuracy Argument states that the Brier score is an acceptable scoring rule. The Brier score is certainly not the only scoring rule possible; why do we think it's acceptable? And what does it even *mean* for a scoring rule to be acceptable in this context?

10.3.1 The absolute-value score

In his original (1998) presentation of the accuracy argument, Joyce selected the Brier score because it exhibits a number of appealing formal properties— what we might think of as adequacy conditions for an acceptable scoring rule. We've already seen that the Brier score is a separable rule. The Brier score also displays

Truth-Directedness: If a distribution cr is altered by moving at least one $cr(X_i)$ value closer to $tv_\omega(X_i)$, and no individual cr-values are moved farther away from tv_ω, then $I(cr, \omega)$ decreases.

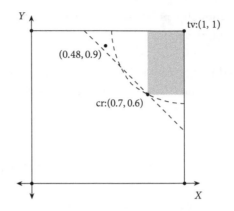

Figure 10.5 Truth-Directedness

The intuitive idea of Truth-Directedness is that if you change your credence distribution by moving some propositions closer to their truth-values, and leaving the rest alone, this should decrease inaccuracy. This condition is depicted in Figure 10.5. (Ignore the dashed elements in that diagram for now.) Assume once more that the agent assigns credences only to the propositions X and Y, and that both these propositions are true in the actual world. If the agent's credence distribution is $(0.7, 0.6)$, every point on or in the gray box (except for $(0.7, 0.6)$ itself) assigns an X-credence or a Y-credence closer to 1 than hers. On a truth-directed scoring rule, all of those distributions are more accurate than the agent's.

The Brier score isn't the only truth-directed scoring rule, or the only way of measuring distance on a diagram. Brier measures distance as the crow flies. But suppose you had to travel from the distribution $(0.7, 0.6)$ to the truth $(1, 1)$ by traversing a rectangular street grid, which permitted movement only parallel to the axes. The shortest distance between those two points measured in this fashion—what's sometimes called the "taxicab distance"—is

$$|1 - 0.7| + |1 - 0.6| = 0.3 + 0.4 = 0.7 \qquad (10.3)$$

I've illustrated this distance in Figure 10.6.

Generalizing the taxicab calculation to a distribution over finitely many propositions $\{X_1, X_2, \ldots, X_n\}$ yields

$$I_{\text{ABS}}(\text{cr}, \omega) = |\text{tv}_\omega(X_1) - \text{cr}(X_1)| + |\text{tv}_\omega(X_2) - \text{cr}(X_2)| + \ldots + |\text{tv}_\omega(X_n) - \text{cr}(X_n)| \qquad (10.4)$$

We'll call this the absolute-value scoring rule.

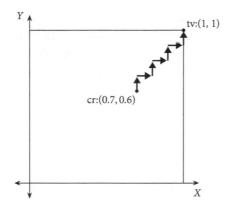

Figure 10.6 The absolute-value score

Both the absolute-value score and the Brier score satisfy Truth-Directedness. We can see this by attending to the dashed elements in Figure 10.5. The dashed line passing through $(0.7, 0.6)$ shows distributions that have the exact *same* inaccuracy as $(0.7, 0.6)$ if we measure inaccuracy by the absolute-value score.[17] Any point between that dashed line and $(1, 1)$ is *more* accurate than $(0.7, 0.6)$ by the absolute-value score. Notice that all the points in the gray box fall into that category, so the absolute-value score is truth-directed.

The dashed quarter-circle shows distributions that are just as inaccurate as $(0.7, 0.6)$ if we measure inaccuracy by the Brier score. Points between the dashed quarter-circle and $(1, 1)$ are less inaccurate than $(0.7, 0.6)$ according to the Brier score. Again, the gray box falls into that region, so the Brier score is truth-directed.

We can see in Figure 10.5 that the Brier score and the absolute-value score are ordinally non-equivalent measures of inaccuracy. To bring out the contrast, consider the distribution $(0.48, 0.9)$. Notice that Truth-Directedness doesn't settle whether this distribution is more or less accurate than $(0.7, 0.6)$—given that both X and Y have truth-values of 1, $(0.48, 0.9)$ does better than $(0.7, 0.6)$ with respect to Y but worse with respect to X. We have to decide whether the Y improvement is dramatic enough to merit the X sacrifice; Truth-Directedness offers no guidance concerning such tradeoffs. The Brier score and absolute-value score render opposite verdicts on this point. $(0.48, 0.9)$ lies inside the dashed line, so the absolute-value score evaluates this distribution as *less* inaccurate than $(0.7, 0.6)$. But $(0.48, 0.9)$ lies outside the quarter-circle, so the Brier score evaluates it as *more* inaccurate. Here we have a concrete case in which the absolute and Brier scores disagree in their accuracy rankings of two distributions.

Such disagreement is especially important when it comes to the Gradational Accuracy Argument. A Gradational Accuracy Theorem cannot be proven for the absolute-value score; in fact, replacing the Brier score with the absolute-value score in the statement of that theorem yields a falsehood. (We'll demonstrate this in the next section.) This makes the second premise of the Gradational Accuracy Argument crucial. The first premise says that rational credence distributions are not dominated on any acceptable scoring rule. If all the acceptable scoring rules were like the absolute-value score, then nonprobabilistic distributions would not be dominated and the argument could not go through. But if we can establish that the Brier score is acceptable, then we have an argument for probabilism.

10.3.2 Proper scoring rules

How do we decide whether the Brier score or the absolute-value score (or both, or neither) is an acceptable measure of inaccuracy? In his (1998), Joyce offered adequacy conditions beyond Truth-Directedness and separability that favored the Brier score over the absolute-value score. Maher (2002), however, argued that these properties were implausible as requirements on rationally acceptable scoring rules, and defended the absolute-value score. So we're left wondering how to select one over the other.

Historically, the Brier score was favored over the absolute-value score because Brier belongs to a broad class of scoring rules called the "proper" scoring rules. To understand this notion of propriety, we first need to understand *expected* inaccuracies.

Suppose I want to assess the inaccuracy of my friend Rita's credence distribution. We'll simplify matters by stipulating that Rita assigns only two credence values, $cr_R(X) = 0.7$ and $cr_R(Y) = 0.6$. Stipulate also that I am going to use the absolute-value score for inaccuracy measurement. We know from Equation (10.3) that if X and Y are both true, Rita's I_{ABS} score is 0.7. The trouble is, I'm not certain whether X or Y is true; I assign positive credence to each of the four truth-value assignments over X and Y. The table below shows my credence distribution (cr) over the four possibilities—which is distinct from Rita's:

	X	Y	cr	$I_{ABS}(cr_R, \cdot)$
ω_1	T	T	0.1	0.7
ω_2	T	F	0.2	0.9
ω_3	F	T	0.3	1.1
ω_4	F	F	0.4	1.6

The last column in this table shows the inaccuracy of Rita's distribution in each of the four possible worlds according to the absolute-value score. If X and Y are both true, her inaccuracy is 0.7; if X is true but Y is false, it's 0.9; etc.

The table tells me the inaccuracy of Rita's distribution in each possible world. I can't calculate her actual inaccuracy, because I'm not certain which possible world is actual. But I can calculate how inaccurate I *expect* Rita's distribution to be. The inaccuracy of a credence distribution is a numerical quantity, and just like any numerical quantity I may calculate my expectation for its value. My expectation for the I_{ABS} value of Rita's distribution cr_R is:

$$EI_{cr}(cr_R) = I_{ABS}(cr_R, \omega_1) \cdot cr(\omega_1) + I_{ABS}(cr_R, \omega_2) \cdot cr(\omega_2)$$
$$+ I_{ABS}(cr_R, \omega_3) \cdot cr(\omega_3) + I_{ABS}(cr_R, \omega_4) \cdot cr(\omega_4) \qquad (10.5)$$
$$= 0.7 \cdot 0.1 + 0.9 \cdot 0.2 + 1.1 \cdot 0.3 + 1.6 \cdot 0.4 = 1.22$$

For each world, I calculate how inaccurate cr_R would be in that world, then multiply by my credence cr that that world is actual.[18] Finally, I sum the results across all four worlds. Notice that because I'm more confident in, say, worlds ω_3 and ω_4 than I am in worlds ω_1 and ω_2, my expected inaccuracy for Rita's distribution falls near the higher end of the values in the fourth column of the table.

In general, if an agent employs the scoring rule I to measure inaccuracy, the agent's credence distribution is cr, and the finite set of worlds under consideration is $\{\omega_1, \omega_2, \ldots, \omega_n\}$, the agent's expected inaccuracy for any distribution cr' is:

$$EI_{cr}(cr') = I(cr', \omega_1) \cdot cr(\omega_1) + I(cr', \omega_2) \cdot cr(\omega_2) + \ldots + I(cr', \omega_n) \cdot cr(\omega_n)$$
$$(10.6)$$

This equation generalizes the expected inaccuracy calculation of Equation (10.5) above. The notation $EI_{cr}(cr')$ indicates that we are calculating the expected inaccuracy of credence distribution cr', as judged *from the point of view* of credence distribution cr.[19]

Equation (10.6) allows me to calculate my expected inaccuracy for any credence distribution, probabilistic or otherwise. If I wanted, I could even calculate my expected inaccuracy for my *own* credence distribution. That is, I could calculate $EI_{cr}(cr)$. But this calculation is fraught. When I calculate my expected inaccuracy for my own current credences and compare it to the inaccuracy I expect for someone else's credences, I might find that I expect that other distribution to be more accurate than my own. We will say that distribution cr' **defeats cr in expectation** if

$$EI_{cr}(cr') < EI_{cr}(cr) \qquad\qquad (10.7)$$

Your credence distribution defeats mine in expectation when, from the point of view of my own credence distribution, I expect yours to be less inaccurate than mine.

Being defeated in accuracy expectation is not quite as bad as being accuracy-dominated. Being defeated in expectation is kind of like having a twin sister who takes all the same classes as you but has a better GPA. (Being accuracy-dominated is like your twin's getting a better grade than you *in every single class*.) Still, being defeated in expectation is a rational flaw. Joyce writes:

> If, relative to a person's own credences, some alternative system of beliefs has a lower expected epistemic [inaccuracy], then, by her own estimation, that system is preferable from the epistemic perspective. This puts her in an untenable doxastic situation. She has a *prima facie* epistemic reason, grounded in her beliefs, to think that she should not be relying on those very beliefs. This is a probabilistic version of Moore's paradox. Just as a rational person cannot fully believe "X but I don't believe X," so a person cannot rationally hold a set of credences that require her to estimate that some other set has higher epistemic utility. [Such a] person is ... in this pathological position: her beliefs undermine themselves. (2009, p. 277)

The idea that rational agents avoid being defeated in expectation is related to our earlier weather-forecaster discussion of stability and credence elicitation. Lewis (1971) calls a distribution that assigns itself the highest expected accuracy **immodest**. ("When asked which method has the best estimated accuracy, the immodest method answers: 'I have'.") He then relates immodesty to an agent's epistemic goals:

> If you wish to maximize accuracy in choosing a [credence-assignment] method, and you have knowingly given your trust to any but an immodest method, how can you justify staying with the method you have chosen? If you really trust your method, and you really want to maximize accuracy, you should take your method's advice and maximize accuracy by switching to some other method that your original method recommends. If that method is also not immodest, and you trust it, and you still want to maximize accuracy, you should switch again; and so on, unless you happen to hit upon an immodest method. Immodesty is a condition of adequacy because it is a necessary condition for stable trust. (1971, p. 62)

These arguments from Joyce and Lewis support the following principle:

Admissibles Not Defeated: If an agent's credence distribution is rationally permissible, and she measures inaccuracy with an acceptable scoring rule, then there will not exist any distribution that she expects to be more accurate than her own.

Admissibles Not Defeated says that under an acceptable scoring rule, no credence distribution that is rationally permissible will take itself to be defeated in expectation by another distribution.[20]

Admissibles Not Defeated relates two elements: a credence distribution and a scoring rule. If we've already settled on an acceptable scoring rule, we can use Admissibles Not Defeated to test the rational permissibility of a credence distribution. But we can also argue in the other direction: If we know a particular credence distribution is rational, we can use Admissibles Not Defeated to argue that particular scoring rules are not acceptable.

For example, suppose I'm certain a fair die has just been rolled, but I know nothing about the outcome. I entertain six propositions, one for each possible outcome of the roll, and let's imagine that I assign each of those propositions a credence of 1/6. That is, my credence distribution cr assigns $\text{cr}(1) = \text{cr}(2) = \text{cr}(3) = \text{cr}(4) = \text{cr}(5) = \text{cr}(6) = 1/6$. This seems at least a rationally *permissible* distribution in my situation.

But now suppose that, in addition to having this perfectly permissible credence distribution, I also use the absolute-value scoring rule to assess accuracy. I entertain six possible worlds—call them ω_1 through ω_6, with the subscripts indicating how the roll comes out in a given world. In world ω_1, the roll comes out one, so $\text{tv}_{\omega_1}(1) = 1$ while the tv_{ω_1}-value of each of the other outcomes is zero. Thus we have

$$I_{\text{ABS}}(\text{cr}, \omega_1) = |1 - 1/6| + 5 \cdot |0 - 1/6| = 10/6 = 5/3 \qquad (10.8)$$

A bit of reflection will show that $I_{\text{ABS}}(\text{cr}, \omega_2)$ through $I_{\text{ABS}}(\text{cr}, \omega_6)$ also equal 5/3, for similar reasons. Recalling that I assign credence 1/6 to each of the six possible worlds, my expected inaccuracy for my own credence distribution is

$$\text{EI}_{\text{cr}}(\text{cr}) = 6 \cdot (5/3 \cdot 1/6) = 5/3 \qquad (10.9)$$

Next I consider my crazy friend Ned, who has the same evidence as me but assigns credence 0 to each of the six roll-outcome propositions. That is, Ned's distribution cr_N assigns $cr_N(1) = cr_N(2) = cr_N(3) = cr_N(4) = cr_N(5) = cr_N(6) = 0$. How inaccurate do I expect Ned to be? Again, in ω_1, $tv_{\omega_1}(1) = 1$ while the tv_{ω_1}-value of each other outcome is 0. So

$$I_{ABS}(cr_N, \omega_1) = |1 - 0| + 5 \cdot |0 - 0| = 1 \tag{10.10}$$

Similar calculations show that, as measured by the absolute-value score, in each possible world Ned's distribution will have an inaccuracy of 1. When I calculate my expected inaccuracy for Ned, I get

$$EI_{cr}(cr_N) = 6 \cdot (1 \cdot 1/6) = 1 \tag{10.11}$$

And now we run into a problem: 1 is less than 5/3. If I calculate inaccuracy using the absolute-value rule, I will expect Ned's distribution to be less inaccurate than my own; my credence distribution is defeated in expectation by Ned's. Yet Ned's distribution isn't *better* than mine in any epistemic sense—in fact, the Principal Principle would say that my distribution is rationally required while his is rationally forbidden! Something has gone wrong, and it isn't the credences I assigned. Instead, it's the scoring rule I used to compare my credences with Ned's.

We can use this example to construct an argument against the absolute-value score as an acceptable scoring rule. In the example, my credence distribution is rationally permissible. According to Admissibles Not Defeated, a rationally permissible distribution cannot be defeated in expectation on any acceptable scoring rule. On the absolute-value rule, my credence distribution *is* defeated in expectation (by Ned's). So the absolute-value scoring rule is not an acceptable inaccuracy measure. (This argument is similar to an argument we made against calibration as an accuracy measure, on the grounds that calibration made perfectly rational forecaster credences look unstable and therefore irrational.)

The Ned example cannot be used to make a similar argument against the Brier score. Exercise 10.4 shows that if I had used the Brier score, I would have expected my own credence distribution to be more accurate than Ned's. In fact, the Brier score is an example of a proper scoring rule:

Proper Scoring Rule: A scoring rule is proper just in case any agent with a probabilistic credence distribution who uses that rule takes herself to defeat in expectation every other distribution over the same set of propositions.

The absolute-value scoring rule is not proper. The Brier score is: a probabilistic agent who uses the Brier score will always expect herself to do *better* with respect to accuracy than any other distribution she considers.[21] The Brier score is not the only scoring rule with this feature. For the sake of illustration, here's another proper scoring rule:[22]

$$I_{LOG}(cr, \omega) = [-\log(1 - |tv_\omega(X_1) - cr(X_1)|)] + \ldots$$
$$+ [-\log(1 - |tv_\omega(X_n) - cr(X_n)|)] \tag{10.12}$$

Historically, the Brier score has been favored over the absolute-value score for inaccuracy measurement because Brier is a proper scoring rule.[23] Of course, propriety gives us no means of choosing between the Brier score and other proper scores such as the logarithmic rule of Equation (10.12). But it turns out we don't need to. Predd et al. (2009) showed that a Gradational Accuracy Theorem can be proven for *any* separable, proper scoring rule (not just the Brier score). So, for instance, on the logarithmic scoring rule any non-probabilistic credence distribution will be accuracy-dominated by some probabilistic distribution over the same propositions, but no probabilistic distribution will be dominated. The same is not true for the absolute-value score. In fact, if you look back to the Crazy Ned example, you'll find that Crazy Ned's non-probabilistic distribution accuracy-dominates my probabilistic distribution cr. In each of the six possible worlds, $I_{ABS}(cr_N, \omega) = 1$ while $I_{ABS}(cr, \omega) = 5/3$. On an improper scoring rule, a non-probabilistic distribution may accuracy-dominate a probabilistic one.

10.3.3 Are improper rules unacceptable?

We now have a clear argumentative path to probabilism. Suppose we establish that all acceptable scoring rules are proper. Then, regardless of any further distinctions we might make among the proper rules, it will turn out that all nonprobabilistic credence distributions can be accuracy-dominated, while no probabilistic distributions can be. Given Admissibles Not Dominated, we will be able to establish that credence distributions violating the probability axioms are irrational. So can we use propriety as an adequacy criterion for scoring rules?

A proper scoring rule is one on which probabilistic distributions always expect themselves to be more accurate than the alternatives. But why focus on what *probabilistic* distributions expect? Inaccuracy measurement has many applications, and in many of those applications (including one we'll see in Section 10.5) it is already assumed that probabilistic credence distributions are rational. In such situations we want an accuracy measure that interacts well with probabilistic distributions, so proper scoring rules are a natural fit, and it's traditional to apply the Brier score because of its propriety. But when an inaccuracy measure is used to *argue* for probabilism—as in the Gradational Accuracy Argument—it seems question-begging to privilege probabilistic distributions in selecting that scoring rule. For instance, our Crazy Ned argument against the absolute-value score *started* by assuming that my probabilistic distribution assigning credence 1/6 to each possible die-roll outcome was rationally permissible. We then criticized the absolute-value score on the grounds that it made that distribution look unstable and therefore irrational. Yet this criticism looks circular in the course of a debate about the rational status of credences satisfying the probability axioms.

In his (2009), Joyce moved from his old reasons for favoring the Brier score to a new approach that explicitly begins with the rational permissibility of probabilistic distributions. While I won't go into the specifics of that argument here, it takes as a premise that given any numerical distribution satisfying the probability axioms, there exists some situation in which it would be rationally permissible for an agent to assign those values as her credences. Admittedly, this premise—that probabilistic credences are rationally *permitted*—is weaker than the ultimate conclusion of Joyce's accuracy-dominance argument—that probabilistic credences are rationally *required*. Still, without any independent support for the premise, it feels like we're assuming the rationality of probabilistic credences in order to prove the rationality of probabilistic credences. It sounds like Linearity In, Linearity Out to me.[24]

Joyce does try to provide independent support for his premise. He argues that for any probabilistic distribution, we could imagine a situation in which an agent is rationally certain that those values reflect the objective chances of the propositions in question. By the Principal Principle, the agent would then be rationally required to assign the relevant values as her credences.

Yet recall our characters Mr. Prob, Mr. Bold, and Mr. Weak. Mr. Prob satisfies the probability axioms, while Mr. Bold violates Finite Additivity by having his credence in each proposition be the square-root of Mr. Prob's credence in that proposition. Mr. Bold happily assigns a higher credence to every uncertain proposition than Mr. Prob does. In arguing for probabilism, we look to establish that Mr. Bold's (and Mr. Weak's) credences are rationally forbidden. If we could establish that rational credences must match the numerical values of known frequencies or objective chances, then in many situations

Mr. Bold's distribution could be ruled out immediately, because frequencies and chances must each be additive.[25] But part of Mr. Bold's boldness is that even when he and Mr. Prob are both certain that a particular proposition has a particular nonextreme chance, he's willing to assign that proposition a higher credence than its chance value. Mr. Bold is willing to be more confident of a given experimental outcome than its numerical chance!

What if, when confronted with a fair die roll like the one in the Crazy Ned example, Mr. Bold maintains that it is rationally *impermissible* to assign a credence of 1/6 to each outcome? It's not that Mr. Bold disagrees with us about what the chances are; it's that he disagrees with us about whether rationally permissible credences equal the chances.[26] Faced with this position, our argument against the absolute-value score could not get off the ground, and would not favor the Brier score over absolute-value in constructing a Gradational Accuracy Argument. Similarly, Joyce's argument for his premise would go nowhere, because Mr. Bold clearly rejects the Principal Principle.[27] While we might intuitively feel like Mr. Bold's position is crazy, the accuracy-based *arguments* against it are question-begging.

10.4 Do we really need Finite Additivity?

Let's step back and take a broader view of the arguments discussed so far in this chapter. Some authors don't think accuracy considerations are central to assessing doxastic attitudes for rationality. But among those who embrace an accuracy-based approach, a few principles are uncontroversial. Everyone accepts Admissibles Not Dominated, and most authors seem okay with Admissibles Not Defeated. Everyone thinks accuracy measures should be truth-directed, and most are on board with separability. Controversy arises when we try to put more substantive constraints on the set of acceptable scoring rules. In order to run a gradational accuracy argument for probabilism, we need to narrow the acceptable scoring rules to the set of proper scores (or one of the other restricted sets Joyce considers in his 1998 and 1999). But arguments for such a restricted set often turn out to be question-begging.

What if we didn't try to narrow the set so far—what if we worked only with constraints on scoring rules that are entirely uncontroversial? In Exercise 10.3, you'll show that as long as one's scoring rule is truth-directed, Admissibles Not Dominated endorses Normality and Non-Negativity as rational constraints on credence. As usual, Finite Additivity is the most difficult Kolmogorov axiom to establish. But an excellent (1982) paper by Dennis Lindley shows how close we can get to full probabilism without strong constraints on our scoring rules.

Lindley assumes Admissibles Not Dominated, then lays down some very minimal constraints on acceptable scoring rules. I won't work through the details, but besides separability and Truth-Directedness he assumes (for instance) that an acceptable scoring rule must be smooth—your score doesn't suddenly jump when you slightly increase or decrease your credence in a proposition. Lindley shows that these thin constraints on scoring rules suffice to narrow down the class of rationally permissible credence distributions, and narrow it down more than just Normality and Non-Negativity would. In fact, every rationally permissible credence distribution is either probabilistic (it satisfies all three Kolmogorov axioms) or can be altered by a simple transformation into a probabilistic distribution. The permissible credence distributions stand to the probabilistic ones in something like the relation Mr. Bold and Mr. Weak stand to Mr. Prob. Mr. Prob satisfies Finite Additivity; Mr. Bold and Mr. Weak don't; but their credences can be converted into Mr. Prob's by a simple mathematical operation (squaring for Mr. Bold; square-rooting for Mr. Weak).[28]

How should we interpret this result? One reading would be that, as long as we rely exclusively on accuracy arguments, we will have to grant that Mr. Bold and Mr. Weak, despite not satisfying Finite Additivity, have doxastic attitudes just as rational as Mr. Prob's. Rational requirements on credences are stronger than just Normality and Non-Negativity—Lindley's result isn't anything goes, and he describes some distributions that satisfy those two constraints but are *not* transformable in the relevant manner into probabilities. But the requirements we get from Lindley are not as strong as Finite Additivity, and do not rule out Mr. Weak or Mr. Bold. So perhaps those characters are perfectly rational, and Finite Additivity is not a rational requirement.

A second reading, however, would be that Mr. Prob, Mr. Bold, and Mr. Weak aren't really distinguishable characters; they don't actually have differing doxastic attitudes in any significant sense. We have stipulated that upon hearing a fair coin was flipped, Mr. Prob assigns credence 1/2 to heads, while Mr. Bold assigns $1/\sqrt{2} \approx .707$. But do they have importantly different outlooks on the world? Their distributions are ordinally equivalent—Mr. Prob is more confident of X than Y just in case Mr. Bold is as well. And both of them satisfy certain structural constraints, such as Normality, Non-Negativity, and our credal Entailment rule. In real life these characters would think and act in many of the same ways; a functionalist might argue that their doxastic attitudes are identical.

Perhaps Mr. Bold stands to Mr. Prob in much the same relation that Fahrenheit measurements of temperature stand to Celsius. Representing temperatures as numbers requires us to introduce a measurement regime, which necessitates some arbitrary choice: what numbers should we assign to water's freezing point, to its boiling point, etc.? Whichever choices we make, the same underlying kinetic phenomena are portrayed—they may contain different *numbers*, but measurements of 0°C and 32°F describe the same state of the world. Perhaps, instead of there being two different characters Mr. Prob and Mr. Bold, the numerical credence distributions we've associated with these characters are just two different representations of the same underlying attitudes, utilizing two different measurement regimes. The relation between measurement schemes wouldn't be quite as straightforward as that between Celsius and Fahrenheit; something more than an affine transformation is involved. But from a mathematical point of view, getting from Mr. Prob's distribution to Mr. Bold's is a simple affair. Lindley identifies a whole family of numerical distributions that are simply transformable into probabilities, and Mr. Bold's is among them.

On this reading, Lindley's result establishes Finite Additivity as a rational requirement in the only way that could possibly matter. A rational agent's credences may be depicted by any one of a number of interrelated numerical distributions, depending on the measurement conventions of the person doing the depicting. To say that Finite Additivity is rationally required is to say that *at least one* of these distributions satisfies it; it's to say that a rational agent's attitudes are *representable* as additive, even if non-additive representations are available as well. Lindley shows that, given minimal conditions on an acceptable accuracy score, every admissible credence distribution either satisfies Finite Additivity or can be transformed into a distribution that does. And there's nothing more substantive than this to the claim that rationality requires Finite Additivity.[29]

To argue against this reading, one would have to argue that there can be significant, cognitive differences between an individual with Mr. Prob's credences and one with Mr. Bold's. If that were the case, then the difference between particular probabilistic and non-probabilistic distributions would not come down to just a choice among measurement schemes. We would be able to find individuals in the world who revealed through their thought, talk, or action that they were like Mr. Bold but not Mr. Prob, and we could have a meaningful conversation about whether their violation of Finite Additivity revealed them to be irrational.

In Chapter 1, I motivated the move from comparative to quantitative confidence models by noting that agents with ordinally equivalent opinions may nevertheless disagree on the relative sizes of confidence *gaps*. Given a tautology, a contradiction, and the proposition that a fair coin came up heads, Mr. Prob and Mr. Bold will rank these three propositions in the same order with respect to confidence. But if we asked an agent like Mr. Prob, he might say that he is more confident in heads than in the contradiction *by the same amount* that he is more confident in the tautology than in heads. A Bold-type wouldn't say that. (Mr. Bold has a larger gap between heads and the contradiction than he has between heads and the tautology.) These sorts of conversations do happen in the real world; perhaps they establish the doxastic differences we seek. Yet I worry about basing our case on agents' self-reports of their psychological states, which are notoriously unreliable. And I worry especially about relying upon conversations we've observed in a culture like ours, which teaches people to measure confidence on something like a linear percentage scale from a very young age.

In Part III of this book, I suggested we assess Bayesian epistemology by considering its applications; I focused especially on applications to confirmation and decision theory. Differences in confidence gaps between ordinally equivalent credence distributions may be highly significant when it comes to decision theory. If I am offered a gamble that yields a small profit on P but a major loss on $\sim P$, my decision will depend not only on whether I find P more likely than $\sim P$, but also on *how much more likely* I find it. So practical rationality may make confidence gap sizes observable in behavior.

Yet we saw in Chapter 8 that the differences between Mr. Prob's and Mr. Bold's credence distributions can be practically neutralized if those agents apply different valuation functions. If Mr. Prob combines his credences and utilities to generate preferences by maximizing expected value, and Mr. Bold combines his credences and identical utilities to generate preferences using a different function, Mr. Prob and Mr. Bold will wind up with the same preferences among acts. In that case, the numerical differences—including confidence-gap differences—between Mr. Prob's and Mr. Bold's credences will make no difference to how they behave. Moreover, Mr. Prob and Mr. Bold will both satisfy the preference axioms that make decision theory's account of practical rationality appealing. So decision theory seems perfectly compatible with reading Mr. Prob and Mr. Bold as just different representations of the same acting individual.

The situation seems to me much more open-ended when it comes to confirmation theory. As with decision theory, confirmation results depend not just on confidence rankings but also on quantitative relations among numerical credence values. In Section 6.4.2 we investigated credence distributions relative to which observing a black raven more strongly confirms the hypothesis that all ravens are black than does observing a non-black, non-raven. The Bayesian solution to the Ravens Paradox presented there describes two conditions on such distributions (Equations (6.10) and (6.11)). The second of those conditions is about the sizes of gaps—it asks whether learning a particular hypothesis would change how much more confident you were in one proposition than another. Despite their ordinal agreements, characters like Mr. Prob and Mr. Bold have different ratios between their credences in particular propositions. So Equation (6.11) might be satisfied by one of them but not by the other. This means that if Mr. Prob and Mr. Bold apply traditional Bayesian confirmation measures, they may disagree on whether the ravens hypothesis is more strongly confirmed by a black raven or by a red herring, which seems like a genuine difference in attitudes.[30] Confirmation is one of many non-decision-theoretic applications of Bayesian epistemology (coherence of a belief set, measuring information content, etc.) where it seems like confidence-gap sizes might make a real difference.

Perhaps in each of those applications we could play a trick similar to the one we used in decision theory. In decision theory we compensated for Mr. Bold's non-additive credence distribution by having him use a non-standard valuation function; the combination yielded act preferences identical to Mr. Prob's. What happens if Mr. Bold also uses a non-traditional confirmation measure? Perhaps there's an odd-looking confirmation measure Mr. Bold could apply which, despite Mr. Bold's credence differences with Mr. Prob, would leave the two agents with identical judgments about confirmational matters.[31] It's unclear, though, how such a non-traditional measure would stand up to the arguments, intuitive considerations, and adequacy conditions that have been deployed in the debate over confirmation measures. I know of no literature on this subject.

Where does that leave Finite Additivity as a rational constraint? As it stands, I think that applications of Bayesianism to theoretical rationality (how we infer, how we reason, how we determine what supports what) have a better chance of drawing real contrasts between Mr. Prob and Mr. Bold than practical applications do. It's also worth noting that Chapter 6's confirmation-theoretic results rely heavily on credence distributions' actually satisfying Finite Additivity. So it may turn out that an appealing account of agents' theoretical judgments will assess those judgments as rational only if the agent's attitudes are genuinely probabilistic. But that is pure speculation on my part.

10.5   An accuracy argument for Conditionalization

Up to this point we've considered accuracy-based arguments for only *synchronic* Bayesian norms. We've found that establishing probabilism on non-circular grounds is somewhat difficult. But if you've already accepted probabilism, a remarkable accuracy-based argument for updating by Conditionalization becomes available. The relevant result was proven by Hilary Greaves and David Wallace (2006).[32] We begin by restricting our attention to proper scoring rules. Doing so is non-circular in this context, because we imagine that we've already accepted probabilism as rationally required. This allows us to appeal to the fact that proper scores are credence-eliciting for probabilistic credences as a reason to prefer them.

Greaves and Wallace think of Conditionalization as a *plan* one could adopt for how to change one's credences in response to one's future evidence. Imagine we have an agent at time t_i with probabilistic credence distribution cr_i, who is certain she will gain some evidence before t_j. Imagine also that there's a finite partition of propositions $\{E_1, E_2, \ldots, E_n\}$ in $\mathcal{L}$ such that the agent is certain the evidence gained will be a member of that partition. The agent can then form a plan for how she intends to update—she says to herself, "If I get evidence E_1, I'll update my credences to such-and-such"; "If I get evidence E_2, I'll update my credences to so-and-so"; etc. In other words, an updating plan is a function from members of the evidence partition to cr_j distributions she would assign were she to receive that evidence. Conditionalization is the plan that sets $cr_j(\cdot) = cr_i(\cdot \mid E_m)$ in response to learning E_m between t_i and t_j.

Next, Greaves and Wallace show how, given a particular updating plan, the agent can calculate from her point of view at t_i an expectation for how inaccurate that plan will be.[33] Roughly, the idea is to figure out what credence distribution the plan would generate in each possible world, measure how inaccurate that distribution would be in that world, multiply by the agent's t_i confidence in that possible world, then sum the results. More precisely, the expectation calculation proceeds in six steps:

1. Pick a possible world ω to which the agent assigns nonzero credence at t_i.
2. Figure out which member of the partition $\{E_1, E_2, \ldots, E_n\}$ the agent will receive as evidence between t_i and t_j if ω turns out to be the actual world. (Because possible worlds are maximally specified, there will always be a unique answer to this question.) We'll call that piece of evidence E.
3. Take the updating plan being evaluated and figure out what credence distribution it recommends to the agent if she receives evidence E

between t_i and t_j. This is the credence distribution the agent will assign at t_j if ω is the actual world and she follows the plan in question. We'll call that distribution cr_j.

4. Whichever scoring rule we've chosen (among the proper scoring rules), use it to determine the inaccuracy of cr_j if ω is the actual world. (In other words, calculate $I(cr_j, \omega)$.)

5. Multiply that inaccuracy value by the agent's t_i credence that ω is the actual world. (In other words, calculate $I(cr_j, \omega) \cdot cr_i(\omega)$.)

6. Repeat this process for each world to which the agent assigns positive credence at t_i, then sum the results.

This calculation has the t_i agent evaluate an updating plan by determining what cr_j distribution that plan would recommend in each possible world. She assesses the recommended distibution's accuracy in that world, weighting the result by her confidence that the world in question will obtain. Repeating this process for each possible world and summing the results, she develops an overall expectation of how accurate her t_j credences will be if she implements the plan.

Greaves and Wallace go on to prove the following theorem:

Accuracy Updating Theorem: For any proper scoring rule, probabilistic distribution cr_i, and evidential partition in $\mathcal{L}$, a t_i agent who calculates expected inaccuracies as described above will find Conditionalization more accurate than any updating plan that diverges from it.

The Accuracy Updating Theorem demonstrates that from her vantage point at t_i, an agent with probabilistic credences and a proper scoring rule will expect to be most accurate at t_j if she updates by Conditionalization. Given a principle something like Admissibles Not Defeated for updating plans, we can use this result to argue that no updating plan deviating from Conditionalization is rationally acceptable.

Does this argument show that the agent is rationally required to update by Conditionalization between t_i and t_j? If she's interested in minimizing expected inaccuracy, then at t_i she should certainly *plan* to update by conditionalizing—of all the updating plans available to the agent at t_i, she expects Conditionalization to be most accurate. Yet being required to make a plan is different from being required to implement it. Even if the agent remembers at t_j what she planned at t_i, why should the t_j agent do what her

t_i self thought best? Among other things, the t_j agent has more evidence than her t_i self did.

This is the same problem we identified in Chapter 9 for diachronic Dutch Strategy arguments. The Accuracy Updating Theorem establishes a *synchronic* point about which policy a t_i agent concerned with accuracy will hope her t_j self applies.[34] But absent a substantive premise that agents are rationally required later on to honor their earlier plans, we cannot move from this *synchronic* point to a genuinely *diachronic* norm like Conditionalization.

10.6 Exercises

Problem 10.1. 🌙 On each of ten consecutive mornings, a weather forecaster reports her credence that it will rain that day. Below is a record of the credences she reported and whether it rained that day:

Day	1	2	3	4	5	6	7	8	9	10
cr(rain)	1/2	1/4	1/3	1/3	1/2	1/4	1/3	1	1/2	1/4
Rain?	Y	N	N	N	Y	Y	N	Y	N	N

Unfortunately, the forecaster's reports turned out not to be perfectly calibrated over this ten-day span. But now imagine she is given the opportunity to go back and change two of the credences she reported over those ten days.[35] What two changes should she make so that her reports over the span become perfectly calibrated? (Assume that changing her credence report does not change whether it rains on a given day.)

Problem 10.2. 🌙🌙 Throughout this problem, assume the Brier score is used to measure inaccuracy.

(a) Suppose we have an agent who assigns credences to two propositions, X and Y, and those credences are between 0 and 1 (inclusive). Draw a box diagram (like those in Figures 10.2, 10.3, and 10.4) illustrating the possible distributions she might assign over these two propositions. Then shade in the parts of the box in which $cr(X) \geq cr(Y)$.

(b) Now suppose that $Y \vDash X$. Use your diagram from part (a) to show that if an agent's credence distribution violates the Entailment rule by assigning $cr(Y) > cr(X)$, there will exist a distribution distinct from hers that is more accurate than hers in every logically possible world. (Hint: When $Y \vDash X$, only three of the four corners of your box represent logically possible worlds.)

(c) In Exercise 9.2 we encountered Roxanne, who assigns the following credences (among others) at a given time:

$$cr(A \& B) = 0.5 \qquad\qquad cr(A) = 0.1$$

Construct an alternate credence distribution over these two propositions that is more accurate than Roxanne's in every logically possible world. (<u>Hint</u>: Let $A \& B$ play the role of proposition Y, and A play the role of X.) To demonstrate that you've succeeded, calculate Roxanne's inaccuracy and the alternate distribution's inaccuracy in each of the three available possible worlds.

Problem 10.3. 🐦🐦 Assuming only that our inaccuracy scoring rule is truth-directed, argue for each of the following from Admissibles Not Dominated:
(a) Non-Negativity
(b) Normality

Problem 10.4. 🐦 Return to the Crazy Ned example of Section 10.3.2, in which you assign 1/6 credence to each of the six possible die roll outcomes while Ned assigns each a credence of 0. This time we'll use the Brier score (rather than the absolute-value score) to measure inaccuracy in this example.
(a) Calculate the inaccuracy of your credence distribution in a world in which the die comes up one. Then calculate Ned's inaccuracy in that world.
(b) Calculate your expected inaccuracy for your own distribution, then calculate *your* expected inaccuracy for *Ned's* distribution.
(c) How do your results illustrate the fact that the Brier score is a proper scoring rule?

Problem 10.5. 🐦 Suppose that at t_i an agent assigns credences to exactly four propositions, as follows:

proposition	cr_i
$P \& Q$	0.1
$P \& \sim Q$	0.2
$\sim P \& Q$	0.3
$\sim P \& \sim Q$	0.4

The agent is certain that between t_i and t_j, she will learn whether Q is true or false.

(a) Imagine the agent has a very bizarre updating plan: No matter what she learns between t_i and t_j, she will assign the exact same credences to the four propositions at t_j that she did at t_i. Using the six-step process described in Section 10.5, and the Brier score to measure inaccuracy, calculate the agent's expected inaccuracy for this updating plan from her point of view at t_i. (Hint: You only need to consider four possible worlds, one for each of the four possible truth-value assignments to the propositions P and Q.)

(b) Now imagine instead that the agent's updating plan is to generate her t_j credences by conditionalizing her t_i credences on the information she learns between the two times. Calculate the agent's t_i expected inaccuracy for *this* updating plan (using the Brier score to measure inaccuracy once more).

(c) How do your results illustrate Greaves and Wallace's Accuracy Updating Theorem?

Problem 10.6. 🌙🌙🌙 In this exercise you will prove a limited version of Greaves and Wallace's Accuracy Updating Theorem. Suppose we have an agent who assigns credences to exactly four propositions, as follows:

proposition	cr_i	cr_j
$P \& Q$	s	w
$P \& \sim Q$	t	x
$\sim P \& Q$	u	y
$\sim P \& \sim Q$	v	z

where cr_i is probabilistic and regular. Suppose also that the agent is certain at t_i that between then and t_j she will learn the truth about whether Q obtains. Finally, assume the agent uses the Brier score to measure inaccuracy.

(a) If the agent updates by Conditionalization and learns Q between t_i and t_j, what will be the values of w, x, y, and z (expressed in terms of s, t, u, and v)?

(b) We will now systematically consider updating plans that diverge from Conditionalization, and show that for each such plan, there exists an alternative plan that the agent expects at t_i to have lower inaccuracy.

To begin, suppose that the agent has an updating plan on which she assigns a nonzero value to either x or z in the event that she learns Q is true. Show that if we calculate expected inaccuracy using the six-step process described in Section 10.5, she will expect this plan to have a higher inaccuracy than the plan that assigns the same w through z values in the event that she learns $\sim Q$, the same w and y values if she learns Q, but assigns 0 to both x and z if she learns Q.

(A similar argument can be made to show that the agent should assign $w = y = 0$ if she learns $\sim Q$.)

(c) Your work in part (b) allows us to restrict our attention to updating plans that assign $x = z = 0$ when Q is learned. Use the Gradational Accuracy Theorem to argue that among such plans, for any plan that has the agent assign a non-probabilistic t_j distribution after learning Q, there exists another plan that has her assign a probabilistic distribution at t_j after learning Q and that she expects to have a lower inaccuracy from her point of view at t_i.

(A similar argument can be made for the agent's learning $\sim Q$.)

(d) Given your results in parts (b) and (c), we may now confine our attention to updating plans that respond to learning Q by assigning a probabilistic t_j distribution with $x = z = 0$. Suppose we hold fixed what such a plan assigns when the agent learns $\sim Q$, and test different possible assignments to w and y when the agent learns Q. Find the values of w and y that minimize the agent's t_i expected inaccuracy for her updating plan.

(Useful algebra fact: A quadratic equation of the form $f(k) = ak^2 + bk + c$ with positive a attains its minimum when $k = \frac{-b}{2a}$.)

(e) How do the results of parts (a) and (d) confirm Greaves and Wallace's point that updating by Conditionalization minimizes expected inaccuracy? (Notice that an argument similar to that of part (d) could be made for a plan that disagrees with Conditionalization on what to do when the agent learns $\sim Q$.)

Problem 10.7. ✐ Of the three kinds of arguments for probabilism we've considered in this part of the book—Representation Theorem arguments, Dutch Book arguments, and accuracy-based arguments—do you think any of them succeeds in establishing requirements of rationality? Which type of argument do you find most convincing? Explain your answers.

10.7 Further reading

INTRODUCTIONS AND OVERVIEWS

Richard Pettigrew (2013b). Epistemic Utility and Norms for Credences. *Philosophy Compass* 8, pp. 897–908

Eminently readable introduction to accuracy-based arguments for Bayesian norms and particular arguments for probabilism and Conditionalization.

Richard Pettigrew (2016). *Accuracy and the Laws of Credence*. Oxford: Oxford University Press

Book-length presentation of the entire accuracy program.

Classic Texts

Bas C. van Fraassen (1983). Calibration: A Frequency Justification for Personal Probability. In: *Physics Philosophy and Psychoanalysis*. Ed. by R. Cohen and L. Laudan. Dordrecht: Reidel, pp. 295–319

Abner Shimony (1988). An Adamite Derivation of the Calculus of Probability. In: *Probability and Causality*. Ed. by J.H. Fetzer. Dordrecht: Reidel, pp. 151–61

Classic arguments for probabilism on calibration grounds.

Bruno de Finetti (1974). *Theory of Probability*. Vol. 1. New York: Wiley

Contains de Finetti's proof of the mathematical result underlying Joyce's Gradational Accuracy Theorem.

James M. Joyce (1998). A Nonpragmatic Vindication of Probabilism. *Philosophy of Science* 65, pp. 575–603

Foundational article that first made the accuracy-dominance argument for probabilism.

Hilary Greaves and David Wallace (2006). Justifying Conditionalization: Conditionalization Maximizes Expected Epistemic Utility. *Mind* 115, pp. 607–32

Presents the minimizing-expected-inaccuracy argument for updating by Conditionalization.

E_XTENDED_ D_ISCUSSION_

James M. Joyce (2009). Accuracy and Coherence: Prospects for an Alethic Epistemology of Partial Belief. In: *Degrees of Belief*. Ed. by Franz Huber and Christoph Schmidt-Petri. Vol. 342. Synthese Library. Springer, pp. 263–97

Joyce further discusses the arguments in his earlier accuracy article and various conditions yielding privileged classes of accuracy scores.

Dennis V. Lindley (1982). Scoring Rules and the Inevitability of Probability. *International Statistical Review* 50, pp. 1–26

Paper discussed in Section 10.4 in which Lindley shows that even with very minimal conditions on acceptable accuracy scores, every rationally permissible credence distribution is either probabilistic or can be converted to a probabilistic distribution via a simple transformation.

Hannes Leitgeb and Richard Pettigrew (2010a). An Objective Justification of Bayesianism I: Measuring Inaccuracy. *Philosophy of Science* 77, 201–35
Hannes Leitgeb and Richard Pettigrew (2010b). An Objective Justification of Bayesianism II: The Consequences of Minimizing Inaccuracy. *Philosophy of Science* 77, pp. 236–72

Presents alternative accuracy-based arguments for synchronic and diachronic Bayesian norms.

Kenny Easwaran (2013). Expected Accuracy Supports Conditionalization— and Conglomerability and Reflection. *Philosophy of Science* 80, 119–42

Shows how expected inaccuracy minimization can be extended in the infinite case to support such controversial norms as Reflection and Conglomerability.

Hilary Greaves (2013). Epistemic Decision Theory. *Mind* 122, pp. 915–52
Jennifer Carr (2017). Epistemic Utility Theory and the Aim of Belief. *Philosophy and Phenomenological Research* 95, pp. 511–34
Selim Berker (2013). Epistemic Teleology and the Separateness of Propositions. *Philosophical Review* 122, pp. 337–93

These papers criticize the teleological epistemology of accuracy-based arguments for rational constraints.

Notes

1. In Chapter 9 I suggested that rational appraisals concern how things look from the agent's own point of view. (It's important that an agent be able to tell *for herself* that her credences leave her susceptible to a Dutch Book.) An agent is often unable to assess the accuracy of her own beliefs, since she lacks access to the truth-values of the relevant propositions. This makes the a priori aspect of the argument for Belief Consistency crucial—an agent with inconsistent beliefs can see *from her own standpoint* that at least some of those beliefs are false, regardless of what contingent facts she may or may not have at her disposal.

2. Small technical note: In the definition of calibration, we ignore values of x that the distribution doesn't assign to any propositions. Shimin Zhao also pointed out to me that, while we define calibration for any *real x*, an agent who assigns credences over a finite set of propositions can be perfectly calibrated only if all of her credence values are rational numbers!

3. Like so many notions in Bayesian epistemology, the idea of accuracy as calibration was hinted at in Ramsey. In the latter half of his (1931), Ramsey asks what it would be for credences "to be consistent not merely with one another but also with the facts" (p. 93). He later writes, "Granting that [an agent] is going to think always in the same way about all yellow toadstools, we can ask what degree of confidence it would be best for him to have that they are unwholesome. And the answer is that it will in general be best for his degree of belief that a yellow toadstool is unwholesome to be equal to the proportion of yellow toadstools which are in fact unwholesome" (p. 97).

4. There's also been some interesting empirical research on how well-calibrated agents' credences are in the real world. A robust finding is that everyday people tend to be overconfident in their opinions—only, say, 70% of the propositions to which they assign credence 0.9 turn out to be true. (For a literature survey see Lichtenstein, Fischoff, and Phillips 1982.) On the other hand, Murphy and Winkler (1977) found weather forecasters' precipitation predictions to be fairly well calibrated—even before the introduction of computer, satellite, and radar improvements made since the 1970s!

5. This example is taken from Joyce (1998).

6. If you're a Regularity devotee (Section 4.2), you may think the forecaster shouldn't assign absolute certainty to snow—what she sees out the window could be clever Hollywood staging! Setting the forecaster's credence in snow to 1 makes the numbers in this example simpler, but the same point could be made using an example with regular credences.

7. Compare the practice in statistics of treating a proposition as a dichotomous random variable with value 1 if true and 0 if false.

8. Notice that we're keeping the numerical values of the distribution cr constant as we measure inaccuracy relative to different possible worlds. $I_{BR}(cr, \omega)$ doesn't measure the

inaccuracy in world ω of the credence distribution the agent *would* have in that world. Instead, given a particular credence distribution cr of interest to us, we will use $I_{BR}(\text{cr}, \omega)$ to measure how inaccurate *that particular numerical distribution* is relative to each of a number of distinct possible worlds.

9. In this chapter we will apply scoring rules only to credence distributions over finitely many propositions. If you're wondering what happens when infinitely many propositions get involved, see (Kelley ms) for some important results and useful references.

10. Named after George Brier—another meteorologist!—who discussed it in his (1950).

11. Notice also that each X_i contributes equally to the sum $I_{BR}(\text{cr}, \omega)$. Thus I_{BR} treats each proposition to which the agent assigns a credence in some sense equally. If you thought it was more important to be accurate about some X_j than others, you might want to insert constants into the sum weighting the $(\text{tv}(X_i) - \text{cr}(X_i))^2$ terms differently. The main mathematical results of this chapter would go through even with such weightings; this follows from a lemma called "Stability" at Greaves and Wallace (2006, p. 627).

12. The second part of the Gradational Accuracy Theorem stands to the first part much as the Converse Dutch Book Theorem stands to the Dutch Book Theorem (Chapter 9).

13. Strictly speaking there are four world-*types* here, a world being assigned to a type according to the truth-values it gives X and Y. But since all the worlds of a particular type will enter into accuracy calculations in the same way, I will simplify discussion by pretending there is exactly one world in each type.

14. Notice that a similar argument could be made for any cr lying outside the square defined by ω_4, ω_2, ω_3, and ω_1. So this argument also shows how to accuracy-dominate a distribution that violates our Maximum rule.

   One might wonder why we *need* an argument that credence-values below 0 or above 1 are irrational—didn't we stipulate our scale for measuring degrees of belief such that no value could fall outside that range? On some ways of understanding credence, arguments for Non-Negativity are indeed superfluous. But one might define credences purely in terms of their role in generating preferences (as discussed in Chapter 8) or sanctioning bets (as discussed in Chapter 9). In that case, there would be no immediate reason why a credence couldn't take on a negative value.

15. Suppose you assign credences to three propositions X, Y, and Z such that X and Y are mutually exclusive and $Z \vDash\!\!\dashv X \vee Y$. We establish X-, Y-, and Z-axes, then notice that only three points in this space represent logically possible worlds: $(0, 0, 0)$, $(1, 0, 1)$, and $(0, 1, 1)$. The distributions in this space satisfying Finite Additivity all lie on the plane passing through those three points. If your credence distribution cr violates Finite Additivity, it will not lie on that plane. We can accuracy-dominate it with distribution cr' that is the closest point to cr lying on the plane. If you pick any one of the three logically possible worlds (call it ω), it will form a right triangle with cr and cr', with the segment from cr to ω as the hypotenuse and the segment from cr' to ω as a leg. That makes cr' closer than cr to ω.

16. To give the reader a sense of how the second part of the Gradational Accuracy Theorem is proven, I will now argue that no point lying inside the box in Figure 10.4 and on the illustrated diagonal may be accuracy-dominated with respect to worlds ω_2 and ω_3. In other words, I'll show how satisfying Negation wards off accuracy domination (assuming one measures inaccuracy by the Brier score).

Start with distribution cr′ in Figure 10.4, which lies on the diagonal and therefore satisfies Negation. Imagine drawing two circles through cr′, one centered on ω_2 and the other centered on ω_3. To improve upon the accuracy of cr′ in ω_2, one would have to choose a distribution closer to ω_2 than cr′—in other words, a distribution lying inside the circle centered on ω_2. To improve upon the accuracy of cr′ in ω_3, one would have to choose a distribution lying inside the circle centered on ω_3. But since cr′ lies on the line connecting ω_2 and ω_3, those circles are tangent to each other at cr′, so there is no point lying inside *both* circles. Thus no distribution is more accurate than cr′ in both ω_2 and ω_3.

17. The dashed line is like a contour line on a topographical map. There, every point on a given contour line lies at the same altitude. Here, every point on the dashed line has the same level of inaccuracy.

18. Here I'm employing a convention that "$cr(\omega_1)$" is the value cr assigns to the proposition that X and Y have the truth-values they possess in world ω_1. In other words, $cr(\omega_1)$ is the cr-value on the first line of the probability table.

19. Readers familiar with decision theory (perhaps from Chapter 7) may notice that the expected-inaccuracy calculation of Equation (10.6) strongly resembles Savage's formula for calculating expected utilities. Here a "state" is a possible world ω_i that might be actual, an "act" is assigning a particular credence distribution cr′, and an "outcome" is the inaccuracy that results if ω_i is actual and one assigns cr′. Savage's expected utility formula was abandoned by Jeffrey because it yielded implausible results when states and acts were not independent. Might we have a similar concern about Equation (10.6)? What if the act of assigning a particular credence distribution is not independent of the state that a particular one of the ω_i obtains? Should we move to a Jeffrey-style expected inaccuracy calculation, and perhaps from there to some analogue of Causal Decision Theory? As of this writing, this question is only just beginning to be explored in the accuracy literature, in articles such as Greaves (2013) and Konek and Levinstein (2019).

20. Notice that Admissibles Not Defeated entails our earlier principle Admissibles Not Dominated. If distribution cr′ accuracy-dominates distribution cr, it will also have a lower expected inaccuracy than cr from cr's point of view (because it will have a lower inaccuracy in every possible world). So being accuracy-dominated is a particularly bad way of being defeated in expectation. (As in sports, it's bad enough to get defeated, but even worse to get *dominated*.) Admissibles Not Defeated says that permissible credence distributions are never defeated in expectation; this entails that they are also never dominated.

21. On a proper scoring rule, a probabilistic agent will always expect her own accuracy to be better than that of any other distribution. On the absolute-value rule, a probabilistic agent will sometimes expect other distributions to be better than her own. Some scoring rules fall in the middle: on such rules, a probabilistic agent will never expect anyone else to do *better* than herself, but she may find other distributions whose expected accuracy is *tied* with her own. To highlight this case, some authors distinguish "strictly proper" scoring rules from just "proper" ones. On a strictly proper scoring rule a probabilistic agent will never find any other distribution that ties hers for accuracy expectation; a merely proper rule allows such ties. I am using the term "proper" the way these authors use "strictly proper". For an assessment of how the distinction between propriety and

strict propriety interacts with the results of this chapter and with varying notions of accuracy dominance (such as "strong" vs. "weak" accuracy domination), see Schervish, Seidenfeld, and Kadane (2009). For an argument that one's commitments to propriety and strict propriety should stand or fall together, see Campbell-Moore and Levinstein (2021).

22. This rule is intended to be applied only to cr-values between 0 and 1 (inclusive).

23. To better understand the Brier score, we visualize it as the Euclidean distance between two points in space. Strictly speaking, though, Euclidean distance is the square-root of the Brier score. As long as we make only ordinal comparisons (whether one distribution is more accurate than, or just as accurate as, another distribution in a given world), that square-root doesn't matter. So all the arguments in previous sections (including the arguments that non-probabilistic distributions can be dominated) go through either way. But square-roots can make a difference to expectation calculations. It turns out that while the Brier score is a proper scoring rule, its square-root (the Euclidean distance) is not.

24. From a Linearity-In, Linearity-Out point of view, Joyce's (2009) argument does have one advantage over attempts to favor the Brier score using propriety considerations. If you're truly worried about making linearity assumptions in the process of establishing probabilism, you might be concerned that Admissibles Not Defeated centers around linear *expectations* of inaccuracy. Joyce's (2009) argument runs from his premise to probabilism using only Admissibles Not Dominated along the way, without invoking Admissibles Not Defeated at all.

25. See note 4 in Chapter 5.

26. Compare Fine (1973, Sect. IIID).

27. See Hájek (2009a) for a very different kind of objection to Joyce's argument.

28. It's worth comparing Lindley's result to Cox's Theorem (though the latter does not invoke considerations of accuracy). Richard Cox (1946, 1961) laid down a set of minimal constraints on an agent's credence distribution, such as: the agent assigns equal credences to logically equivalent propositions; the agent's credence in ~P is a function of her credence in P; her credence in P & Q is a function of her credence in Q and her credence in P given Q; the latter function is twice differentiable; etc. He then showed that any credence distribution satisfying these constraints is isomorphic to a probabilistic distribution. For discussion of the mathematics, and of various philosophical concerns about Cox's conditions, see Paris (1994), Halpern (1999), Van Horn (2003), and Colyvan (2004).

29. I'm inclined to read Lindley's own interpretation of his result along these lines. For one thing, Lindley titles his paper "Scoring Rules and the Inevitability of Probability". For another, after noting on page 8 that Admissibles Not Defeated is a kind of Pareto optimality rule, he writes that an agent who chooses any of the distributions permitted by that rule and a minimally acceptable scoring rule is thereby "effectively introducing probabilities".

30. The same goes for Bayesian results mentioned in Chapter 6, note 39 showing that a red herring cannot confirm the ravens hypothesis to anything more than an exceedingly weak degree. These results depend on particular credal differences and ratios being "minute" in absolute terms, so they might go through for Mr. Prob but not for Mr. Bold (or vice versa).

31. Since Mr. Bold's credences are the square-root of Mr. Prob's, an obvious move would be to take whatever confirmation measure Mr. Prob uses and replace all of its credal expressions with their squares.

32. As we'll see, the Greaves and Wallace result focuses on minimizing *expected* inaccuracy. For Conditionalization arguments based on accuracy-domination, see Briggs and Pettigrew (2020) and Williams (ms). For an alternative expected-accuracy approach to updating, see Leitgeb and Pettigrew (2010a,b).

33. It's important that Greaves and Wallace restrict their attention to what they call "available" updating plans. Available plans guide an agent's credal response to her total evidence (including the evidence she imagines she'll receive); they do not allow an agent to set her credences based on further factors not in evidence. For instance, consider the updating plan according to which an agent magically assigns credence 1 to each proposition just in case it's true and credence 0 just in case it's false—even if her evidence isn't fine-grained enough to indicate the truth-values of all the relevant propositions. This would be an excellent plan in terms of minimizing inaccuracy, but it isn't a feasible updating strategy for an agent going forward. This updating plan does not count as "available" in Greaves and Wallace's sense, and so does not compete with Conditionalization for the most accurate updating plan.

34. Like Reflection, the resulting norm is a *synchronic* requirement on an agent's attitudes toward propositions about *diachronic* events.

35. Perhaps via time-machine?

PART V

CHALLENGES AND OBJECTIONS

Part V takes up prominent challenges and objections to Bayesian epistemology.

Chapters 11 and 12 concern phenomena that it feels like the Bayesian formalism ought to be able to handle, but fails to manage in its traditional presentation. In Chapter 11, the challenge is to model updates in which an agent forgets information, or struggles to locate herself in time or space. In Chapter 12, the problem is to account for the current confirmational significance of evidence previously acquired, or to allow that rational agents might be uncertain of logical truths. In each case, we will consider modifications of the core Bayesian rules that might allow the framework to address these shortcomings.

Chapters 13 and 14, on the other hand, contemplate replacing traditional Bayesianism with something else entirely. In Chapter 13, we consider whether the Problem of the Priors motivates an entirely different statistical approach to inductive inference, such as frequentism or likelihoodism. In Chapter 14, we ask whether levels of confidence might best be modeled by some other mathematical structure than a real-valued distribution over a language of propositions. (The alternative structures we consider in that chapter still have a role for the Kolmogorov axioms, and so strike me as part of the extended Bayesian family.)

In each case, our first order of business will be to make precise exactly what objection is being raised. We will often find multiple, interlocked challenges to Bayesianism that require disambiguation. Then we will take up various proposals to extend or replace the Bayesian formalism to address these problems. We will assess how well the alternatives solve problems for Bayesianism, and suggest new problems that arise in their wake.

These chapters are largely independent of each other, and may be read in any order. (Though understanding the Problem of the Priors—Section 13.1—will probably be helpful for Chapter 14.) While Chapter 13 is a fairly unified whole, a reader with specific interests should be able to dip into just one section of any of the other chapters and get something useful out.

11

Memory Loss and Self-locating Credences

Section 4.1.1 pointed out a couple curious features of Conditionalization, Bayesians' traditional diachronic updating norm. First, when an agent conditionalizes on a proposition that she learns, she becomes certain of that proposition. This means that any agent who learns a piece of contingent information and updates by conditionalization violates the Regularity Principle (Section 4.2). Second, once an agent becomes certain of a proposition, further conditionalizations will keep that certainty intact. If the agent always changes her credences by Conditionalization, she will never lose any certainties.

These consequences of Conditionalization prove especially counterintuitive in two types of cases we have not yet considered in this book. An agent who gains a piece of information at one time (and perhaps becomes certain of it) may lose that information later on, through a natural process of forgetting. Yet this evolution of her credences does not comport with Conditionalization. Or a fact that was once secure for an agent may slide out from under her: while I may be rationally certain right now that today is Sunday, this certainty will be less rational in twenty-four hours.

So we have two types of problematic cases for updating by Conditionalization: cases involving memory loss, and cases involving self-locating propositions (like "today is Sunday"). This chapter will consider each type of case in turn. First we'll consider whether memory loss is a rational phenomenon capable of generating genuine counterexamples to Conditionalization. Then we'll consider how Conditionalization could be amended to take forgetting into account. Next we'll turn to cases involving self-location, asking whether the problem there is created by Conditionalization or by some other feature of our Bayesian system. Then we'll once more consider alterations to Conditionalization to fix the problem. Throughout these discussions, we'll also assess whether other Bayesian tools (Jeffrey Conditionalization, Dutch Strategies, accuracy arguments, etc.) can help us address memory loss and self-location.

Fundamentals of Bayesian Epistemology 2: Arguments, Challenges, Alternatives. Michael G. Titelbaum, Oxford University Press. © Michael G. Titelbaum 2022. DOI: 10.1093/oso/9780192863140.003.0011

11.1 Memory loss

11.1.1 The problem

W.J. Talbott writes:

> Consider the following example: I can remember what I had for dinner one week ago, but I have no memory of what I had one month ago or one year ago. This is not usually taken to reflect on my rationality, for there can be little doubt that if philosophers held a special election to name the most epistemically rational human being alive, neither the winner of that election nor any of the top finishers would, except on certain special occasions, remember what they had eaten for dinner one year earlier either . . . .
>
> If we assume that I had spaghetti for dinner one year ago (on March 15, 1989), then where t_1 is 6:30pm on March 15, 1989; t_2 is 6:30pm on March 15, 1990; and S is the proposition that Talbott had spaghetti for dinner on March 15, 1989 . . . if I am assumed to satisfy Temporal Conditionalization at every time from t_1 to t_2, it must be the case that I be as certain of S at t_2 as I was at t_1 . . . . But, in fact (and I knew this at t_1), if at t_2 I am asked how probable it is that I had spaghetti for dinner on March 15, 1989, the best I can do is to calculate a probability based on the relative frequency of spaghetti dinners in my diet one year ago.   (1991, pp. 138–9)[1]

At some point on March 15, 1989, the proposition S that he has spaghetti for dinner that night enters into Talbott's evidence. He is thus certain of S at t_1 (while he is eating the dinner). If he updates exclusively by Conditionalization over the following year, Talbott will remain certain of S at t_2 (one year later). Yet as Talbott points out, it hardly seems a rational requirement that he remain certain of S on March 15, 1990. In fact, it seems most rational for 1990 Talbott to base his opinions about what he had for March 15, 1989 dinner on the frequency with which he was eating particular dinners around that time. (Talbott pegs his personal $cr_2(S)$ value around 0.1.) So we have a counterexample to Conditionalization.

When confronted with such counterexamples, Conditionalization defenders often respond that forgetting isn't rational. They often say that the norms of Bayesian epistemology describe the doxastic behavior of an **ideally rational agent**. Conditionalization is supposed to describe how an ideally rational agent would alter her credences over time. So if an ideally rational agent would never forget anything, there can be no counterexamples to Conditionalization in which the agent forgets. As Talbott articulates this response, "An ideally rational agent would not suffer impairments of her faculties and judgment, and an ideally rational agent would not be subject to the fallibility of human memory" (1991, p. 141).

While talk of ideally rational agents is common in epistemology, it often confuses crucial issues. We are supposed to generate our notion of an ideally rational agent by starting with a set of rational requirements, then imagining an agent who satisfies them all. Yet once the talk of ideal agents begins, we think we have some independent grasp on what such perfect agents would be like. The question of whether an ideally rational agent would suffer memory loss should just be the question of whether rationality forbids forgetting; the former question cannot be answered independently to adjudicate the latter. Whether memory loss is a failing in *some* sense—whether an agent would in some sense be more perfect were she to permanently retain information—is irrelevant if the sense in question is not the rational one.

So the position that memory loss is irrational is not so easily defended as it might seem. Yet even if one could defend that position, it would be insufficient to avoid memory loss-related counterexamples to Conditionalization. Consider the following ingenious example from Frank Arntzenius:

> There are two paths to Shangri La, the Path by the Mountains, and the Path by the Sea. A fair coin will be tossed by the guardians to determine which path you will take: if heads you go by the Mountains, if tails you go by the Sea. If you go by the Mountains, nothing strange will happen: while traveling you will see the glorious Mountains, and even after you enter Shangri La, you will forever maintain your memories of that Magnificent Journey. If you go by the Sea, you will revel in the Beauty of the Misty Ocean. But, just as you enter Shangri La, your memory of this Beauteous Journey will be erased and be replaced by a memory of the Journey by the Mountains.   (2003, p. 356)

Suppose that at t_1 the guardians of Shangri La have just explained this arrangement to you, but have not yet flipped their fateful coin. By the Principal Principle (Section 5.2.1), your credence that the coin comes up heads should be $cr_1(H) = 1/2$. Suppose that some time later, at t_2, you find yourself traveling the Path by the Sea. At that point you have $cr_2(H) = 0$, because only a tails result leads you to the Sea. Arriving in Shangri La at t_3, you nevertheless remember traveling the Path by the Mountains (due to the guardians' trickery). However you realize you would possess those memories[2] whichever path you traveled, so you take them as no evidence about the outcome of the coin flip. Presumably, then, it's rational to assign $cr_3(H) = 1/2$.

So far this example is no different from Talbott's spaghetti—an agent becomes certain of a proposition at an earlier time, then violates Conditionalization by losing that certainty at a later time due to memory malfunction. A theorist who holds that any loss of memory is a rational failing will see no counterexample to Conditionalization here.

But now consider the version of the case in which the coin comes up heads, and you travel the Path by the Mountains. While traveling that path you assign $cr_2(H) = 1$, but upon reaching Shangri La the same logic holds as above: your memories of the Mountains no longer provide any evidence about the outcome of the coin flip, so you revert to $cr_3(H) = 1/2$. Once more you violate Conditionalization, but your memory has not been altered or failed in any way. Arntzenius dramatizes the situation by examining your outlook at t_2, as you travel past the mountains and consider what your future credences will be:

> You know that your degree of belief in heads is going to go down from one to 1/2. You do not have the least inclination to trust those future degrees of belief.... Nonetheless, you think you will behave in a fully rational manner when you acquire those future degrees of belief. Moreover, you know that the development of your memories will be completely normal. It is only because something strange would have happened to your memories had the coin landed tails that you are compelled to change your degree of belief to 1/2 when that counterfactual possibility would have occurred.   (2003, p. 357)

In the Shangri La story it's rational to violate Conditionalization not only when you suffer memory loss, but also when you merely suspect that you *may* have forgotten something. Perhaps it's irrational to forget information. But is it also irrational to assign nonzero credence to the contingent possibility that one forgot something at some point in the past? To rule out all these Conditionalization violations, one must maintain not only that the ideally rational agent has a perfect memory but also that she is perfectly certain that her memory is perfect (despite possessing no evidence to that effect).

Shangri La also reveals the futility of another possible response to memory loss. We've already seen that even independent of memory loss concerns, many epistemologists follow Richard C. Jeffrey in rejecting Conditionalization's assumption that evidence arrives as certainty. They advocate rejecting the Conditionalization regime in two ways: first, we endorse the Regularity Principle (Section 4.2) that forbids certainty in logically contingent propositions; and second, we adopt Jeffrey Conditionalization (Section 5.5) as our Bayesian updating norm. While independently motivated, these maneuvers also seem to handle the memory loss cases. For example, it's perfectly consistent with Jeffrey Conditionalization for Talbott to assign a very high credence on March 15, 1989 that he has spaghetti for dinner that night, while assigning a much lower credence to that proposition one year later.

Yet moving to Jeffrey Conditionalization won't solve all of our problems either. In the Shangri La example, rationality requires you to set $cr_3(H) = 1/2$ upon arriving in Shangri La. We'd like our Bayesian updating rule to generate that requirement for us. In other words, if you assigned some other credence to heads at t_3, we'd like the rule to reveal a rational mistake.

Jeffrey Conditionalization doesn't seem able to do that for us. In a Jeffrey Conditionalization, an agent's experience affects her by redistributing her credences over some finite "originating" partition $\{B_1, B_2, \ldots, B_n\}$. The rule then tells us how this change over the Bs affects the agent's other credences. Between t_2 and t_3 in Shangri La, what's the partition over which experience changes your credences? One obvious candidate is the partition containing H and $\sim H$. But Jeffrey Conditionalization never tells an agent how to alter her credences over the originating partition; some other mechanism sets those credences, and Jeffrey Conditionalization works out the effects of that setting on other credences. So Jeffrey Conditionalization won't be able to tell us that rationality requires $cr_3(H) = 1/2$; it's consistent with any $cr_3(H)$ value at all.[3]

11.1.2 A possible solution

Section 4.3 discussed the fact that if an agent updates by Conditionalization throughout her life, two things will be true of her:

1. Any piece of evidence she possesses at a given time will be retained at all later times.
2. Her series of credences will be representable by a hypothetical prior.

These two conditions are independent. It's possible to describe an agent whose evidence builds like a snowball, but whose credences aren't representable by a hypothetical prior. It's also possible to describe an agent whose credence assignments over time are representable by a hypothetical prior, but who loses information over time. (You examined one such an agent in Exercise 4.7.)

Endorsing Conditionalization as a rational norm makes both conditions necessary for rationality. But we now have reason to abandon the first one. In cases involving memory loss or the threat thereof, an agent may lose information without rational defect.[4] So what happens to our theory of updating if we drop the first condition, but keep the second?

To be clear, the suggestion is that we set aside Conditionalization as our Bayesian diachronic norm, but keep the following condition on rational updating:

Hypothetical Representability: Given any finite series of credence distributions $\{cr_1, cr_2, \ldots, cr_n\}$ that the agent assigns over time, there exists at least one regular probability distribution Pr_H such that for all $1 \leq i \leq n$,

$$cr_i(\cdot) = Pr_H(\cdot \mid E_i)$$

(where E_i is the conjunction of the agent's total evidence at t_i).

What does Hypothetical Representability demand of an agent's credences over time? First, Hypothetical Representability is strictly weaker than Conditionalization. Section 4.3's Hypothetical Priors Theorem told us that any agent who conditionalizes is representable by a hypothetical prior. But the converse of that theorem is false: An agent can be representable by a hypothetical prior even if she sometimes fails to conditionalize. This occurs when the agent loses information from time to time (as with Jane, the agent in Exercise 4.7). Nevertheless, if we take an agent like that and focus on an episode in which she strictly gains evidence (or keeps her evidence constant), Hypothetical Representability will require her to update by conditionalizing over the course of that episode. So in cases where Conditionalization is plausible—that is, cases in which an agent loses no information—Hypothetical Representability yields the same results as Conditionalization.

What about other cases? One interesting case is strict information loss: when an agent's later evidence is a proper subset of her earlier evidence. Here, Hypothetical Representability recommends reverse-temporal conditionalization.[5] In a normal conditionalization, an agent's credence in a proposition at the *later* time equals her credence in that proposition at the *earlier* time conditional on the evidence she *gains* between the two times. But in a reverse-temporal conditionalization, an agent's credence at the *earlier* time equals her credence at the *later* time conditional on the evidence she *loses* between those two times. In other words, if an agent's evidence between t_i and t_j changes only by losing proposition E, then for any H in $\mathcal{L}$,

$$cr_i(H) = cr_j(H \mid E) \tag{11.1}$$

Notice that if this is right, then were the agent at some time t_k after t_j to re-gain the information she lost between t_i and t_j, normal conditionalization on E would set cr_k identical to her earlier cr_i. So, for instance, between 1989 and 1990 Talbott forgets what he ate for dinner on March 15, 1989. But if you supplied him on March 15, 1990 with the information he had lost, his spaghetti credences that night would presumably look just like they did after dinner on March 15, 1989.

Reverse-temporal conditionalization also occurs in the Shangri La example. Initially—before the guardians flip their coin—you assign $cr_1(H) = 1/2$. By t_2 you've conditionalized on travelling the Path by the Mountains, so you assign $cr_2(H) = 1$. But upon reaching Shangri La you lose that information, and now assign $cr_3(H) = 1/2$. If you satisfy Hypothetical Representability, the relationship between cr_1 and cr_2 is exactly the same as the relationship between cr_3 and cr_2; it's just that the temporal order has been reversed.[6]

Hypothetical Representability generates intuitively plausible results for both of our memory loss cases. It obviously succeeds better than traditional Conditionalization, but also succeeds better than Jeffrey Conditionalization, by actually constraining the agent's t_3 credences in Shangri La. What about more complicated cases, in which an agent both gains and loses information between two times? Here's one way to think about how Hypothetical Representability applies to those cases. Suppose that between times t_j and t_k, an agent gains a body of evidence representable by the conjunction E, but loses evidence representable by the conjunction D. (More precisely, E_j & $E = \models E_k$ & D; think of D as the agent's evidential *deficit* at the later time.) Hypothetical Representability will require that for any H in $\mathcal{L}$,

$$cr_k(H \mid D) = cr_j(H \mid E) \qquad (11.2)$$

To wrap one's mind around this, it helps to attach names to the agent's time-slices at the two times. So let's call the t_j edition of the agent "Jen", and the t_k edition "Ken". According to Hypothetical Representability, if you gave Jen all the information that Ken has but she doesn't, and gave Ken all the information Jen has but he doesn't, the two of them would then reach identical conclusions about any proposition H.[7]

Hopefully I've now made the case that satisfying Hypothetical Representability will lead an agent to update in an intuitively reasonable fashion. But are there any deeper, more theoretical arguments available for Hypothetical Representability as a rational constraint? In Chapters 9 and 10 we saw Dutch Strategy- and accuracy-based arguments for Conditionalization (respectively). Could arguments on these grounds be given for Hypothetical Representability? Well, in one sense they can, because Conditionalization entails Hypothetical Representability, so any argument for the former is also an argument for the latter. But really what we want to know is: Once we've allowed cases in which a rational agent loses information over time, can Dutch Strategies or accuracy arguments support Hypothetical Representability's application to such cases?

I don't know of any extensive investigation of such topics, but the initial prognosis does not look good. Here's why: Consider Dutch Strategies first. In Section 9.3.2 I argued that a Dutch Strategy targets an agent's plans for updating in the future. So suppose that as you travel the Path by the Mountains, you plan to set your credences once you reach Shangri La according to reverse-temporal conditionalization (as Hypothetical Representability requires). You know that once you've done so, you will assign $cr_3(H) = 1/2$, so you will be willing to accept $0.50 for a betting ticket on which you pay out $1 on heads. But at t_2 you're certain that the coin came up heads—otherwise you wouldn't be traveling the Path by the Mountains—so you're certain that that ticket will generate a net loss for you of $0.50. At t_2 there exists a bet that you plan to accept at t_3, but that generates a sure loss relative to all the doxastic possibilities you entertain at t_2. This looks like a Dutch Strategy against planning to be representable by a hypothetical prior.

A similar point can be made using diachronic accuracy arguments. The Greaves and Wallace argument we discussed in Section 10.5 assesses the updating plan an agent entertains at a given time by calculating that plan's expected accuracy given what she knows at that time. At t_2 in Shangri La you're certain the coin comes up heads, so you will assess the accuracy of updating plans relative only to worlds in which H is true. This means that any plan assigning a high $cr_3(H)$ value will look more accurate than plans assigning lower values. So being representable by a hypothetical prior—and therefore assigning $cr_3(H) = 1/2$—will not be the plan with the best expected accuracy.[8]

What should we conclude from all this? If you're very committed to Dutch Strategy and/or diachronic accuracy arguments, you might conclude that we now have decisive arguments against assigning $cr_3(H) = 1/2$ in the Shangri La case. In fact, it looks like we have arguments against assigning anything lower than $cr_3(H) = 1$. (Similar points could be made for Talbott's spaghetti example.)[9] So you might conclude that decreasing one's credences in light of memory loss (or the threat thereof) is indeed some sort of rational failure. On the other hand, one might apply a *modus tollens* here rather than a *modus ponens*, and conclude that we don't yet have a good enough general understanding of Dutch Strategy and diachronic accuracy arguments to apply them correctly to cases involving information loss. If that's right, then these argument-types are (at least for now) incapable of supplying us with guidance about rational updates in the face of information loss.

So in the next section I'll try another approach. I'll offer a different kind of argument that rationality requires being representable by a hypothetical prior.

11.1.3 Suppositional Consistency

Being representable by a hypothetical prior implies two things about an agent's credences over time. First, those credences consistently align with the *same* function from bodies of total evidence to sets of attitudes. In Section 4.3, we described this function as representing the agent's ultimate epistemic standards. Given any body of total evidence, an agent's epistemic standards tell her what credences to adopt toward any proposition. If an agent satisfies Hypothetical Representability, she keeps her ultimate epistemic standards constant over time. The easiest way to see this is that if an agent's credences are representable by a hypothetical prior, then at any two times at which her total evidence is the same, her credences must be the same as well.

Why might rationality require an agent to maintain the same epistemic standards over time? If you're an Objective Bayesian in the normative sense (Section 5.1.2), then you believe in the Uniqueness Thesis, which implies that only one set of epistemic standards is rationally permissible. Since rationality requires an agent to apply this unique set of epistemic standards at all times, in order to be rational she must apply the same set of epistemic standards at each time (namely, the uniquely rational set).

But what if you're a Subjective Bayesian in the normative sense, and so think that multiple distinct epistemic standards are rationally permissible? If an agent applies one set of standards at a particular time, does rationality require her to apply the same set of standards later on? This is currently a matter of much philosophical controversy. Some Subjective Bayesians have supported the diachronic consistency of epistemic standards using Dutch Strategy arguments. But this approach has two flaws: First, while Dutch Strategy arguments work particularly well for cases in which an agent's evidence remains constant across times (and perhaps for cases in which that evidence increases), we just saw that they go awry for cases of information loss. But second and more importantly, I argued in Section 9.3.2 that while a Dutch Strategy argument may establish that it's rational to *plan* to update in a particular fashion, it can't establish the diachronic principle that an agent is rationally required at a later time to do what she planned earlier on. For similar reasons, Dutch Strategies cannot establish that an agent is required to keep her epistemic standards constant over time.

So the Subjective Bayesian must turn to other forms of argument for diachronic standard-consistency, or perhaps to appeals to intuition. It does feel like there's something rationally off about an agent who reports one doxastic attitude toward a proposition at a given time, then a different attitude toward the same proposition a bit later, despite the fact that her evidence hasn't changed. But I won't wade any farther into that dialectic here.[10]

Instead, I want to focus on a second implication of Hypothetical Representability. Applying a constant set of epistemic standards over time is necessary but not sufficient for representability by a hypothetical prior. Hypothetical Representability requires those standards to display a particular type of mathematical structure. To see why, notice that to say that a set of epistemic standards is a *function* from bodies of total evidence to credence distributions is just to say that it generates the same credal outputs whenever it's given the *exact same* evidential inputs. Put into practice, this means that at any two times when the agent has the same total evidence, she must assign the same credence values. But this doesn't say anything about the function's outputs when *different* bodies of total evidence are supplied. Merely requiring fidelity to *some* function from evidence to credences doesn't say anything about how the agent's responses to different evidence sets should be related. Yet Hypothetical Representability puts substantive cross-evidential constraints in place.

To understand those constaints, let's introduce a bit of new notation. Given a function representing an agent's epistemic standards, let cr_E represent the credence distribution that function outputs for an input of total evidence E. So, for instance, $cr_X(Y)$ will be the credence value prescribed for proposition Y when the agent has total evidence X. Also, for the sake of easy readability, define $cr_X(Y \mid Z)$ as a shorthand for the ratio $cr_X(Y \& Z)/cr_X(Z)$. Using this notation, Hypothetical Representability requires an agent's epistemic standards to satisfy the following constraint:[11]

Suppositional Consistency: Given any propositions A, B, and C in $\mathcal{L}$,

$$cr_A(B \mid C) = cr_{A \& C}(B)$$

Intuitively, Suppositional Consistency says that moving a proposition back and forth between what an agent supposes and what her evidence includes should make no difference to the credences she assigns. If an agent's epistemic standards satisfy Suppositional Consistency, then the credence she assigns to proposition B when her total evidence is A & C will equal the credence she assigns to B when her total evidence is A and she further *supposes* that C.

This idea came up when we first motivated updating by Conditionalization. A fair six-sided die was rolled, and I asked for your credence that it came up six conditional on the supposition that it came up even. When I then told you that the roll actually did come up even, your new unconditional credence in six was the same as your old conditional credence. Initially you evaluated the prospects for a six outcome relative to the *supposition* of even; later you evaluated those prospects relative to the *fact* of even. But either way, you were evaluating the prospects of a six relative to the same set of conditions. It's just that in one case you took those conditions to actually obtain, while in the other case you considered what it would be like *if* they obtained.

Suppositional Consistency generalizes this idea across your epistemic standards, applying it to cases in which you gain evidence, lose evidence, or both. If you're suppositionally consistent over time, then whenever the conjunction of your evidence and your suppositions comes to the same thing, you'll assign the same credence to a given proposition. (Notice that this is stronger than just the principle that if your total evidence doesn't change, your credences shouldn't either.)

Suppositional Consistency captures the cross-evidential constraints enforced by Hypothetical Representability. If rationality requires agents to keep their ultimate epistemic standards constant, and those epistemic standards are also required to be suppositionally consistent, then Hypothetical Representability is a requirement of rationality. And this, in turn, requires agents to conditionalize when no information is lost, and reverse-temporally conditionalize when no information is gained.

To summarize: The traditional Bayesian Conditionalization norm makes rational agents representable by a hypothetical prior, but also forbids them to lose information. Yet it's possible to satify Hypothetical Representability over a period of time even if evidence is sometimes lost during that period. To do so, an agent must keep the same epistemic standards throughout, and those epistemic standards must satisfy Suppositional Consistency. Personally, I think it's exceedingly plausible that all rational epistemic standards are suppositionally consistent, treating supposed and real evidence symmetrically.[12] (Perhaps this presumption is buoyed by the application of Dutch Strategies and accuracy arguments to the no-information-loss case.) If we can then establish either Objective Bayesianism, or the Subjective Bayesian position that rationality requires constant epistemic standards over time, we will establish Hypothetical Representability as well. But the jury is still out on all of these questions.[13]

11.2 Self-locating credences

The contemporary discussion of self-locating credences was jump-started when Adam Elga introduced philosophers[14] to the Sleeping Beauty Problem:

> Some researchers are going to put you to sleep. During the two days that your sleep will last, they will briefly wake you up either once or twice, depending on the toss of a fair coin (Heads: once; Tails: twice). After each waking, they will put you back to sleep with a drug that makes you forget that waking.
>
> When you are first awakened, to what degree ought you believe that the outcome of the coin toss is Heads?
>
> *First answer*: 1/2, of course! Initially you were certain that the coin was fair, and so initially your credence in the coin's landing Heads was 1/2. Upon being awakened, you receive no new information (you knew all along that you would be awakened). So your credence in the coin's landing Heads ought to remain 1/2.
>
> *Second answer*: 1/3, of course! Imagine the experiment repeated many times. Then in the long run, about 1/3 of the wakings would be Heads-wakings—wakings that happen on trials in which the coin lands Heads. So on any particular waking, you should have credence 1/3 that that waking is a Heads-waking, and hence have credence 1/3 in the coin's landing Heads on that trial. This consideration remains in force in the present circumstance, in which the experiment is performed just once.   (2000, 143–4)

Elga's two answers to the problem reveal an unexpected tension: In Section 5.2.1 we suggested that a rational agent aligns her credence that a particular event-type will produce a particular outcome-type with the long-run frequency with which that outcome-type occurs. We also suggested that a rational agent aligns her credence with the objective chance of a particular outcome (as codified by the Principal Principle). These two suggestions seemed to be in harmony, since we expect a repeated event to produce long-run frequencies in line with its associated chances. Yet Elga's first answer to the Sleeping Beauty Problem applies chance information to generate a heads credence of 1/2, while his second answer applies long-run frequencies to recommend 1/3. Two types of information that should line up don't—so which should we listen to?

This problem seems like it should be easy to solve with the standard Bayesian machinery. Let's follow the literature in calling the agent in the story "Beauty", and let's suppose that Beauty is first put to sleep on Sunday night, after which the coin is flipped. Before Beauty goes to sleep on Sunday, but after the experiment is explained to her and she knows the coin flip will be fair, it's uncontroversial how her credences should look. (For instance, her credence that the coin will come up heads should be 1/2.) So why not just take that Sunday-night credence distribution, conditionalize it on whatever she learns between Sunday night and when she's awakened, and settle once and for all her rational credence in heads upon awakening?

Well, what *does* Beauty learn when she awakens? Let's suppose that her first awakening happens Monday morning, and if there is a second awakening it occurs on Tuesday. Then when Beauty awakens, she becomes certain that it's either Monday or Tuesday. This is an example of **self-locating information**, a category that includes information about what time it is ("It's five o'clock"), where the agent is located ("I'm in Margaritaville"), and who he is ("I'm Jimmy Buffett"). Self-locating information creates all sorts of problems for Conditionalization.

11.2.1 The problem

To represent an agent's credences concerning her identity or present spatio-temporal location, we're going to revise our understanding of propositions. To this point we've associated propositions with sets of possible worlds. We've assumed that for any proposition, there are some possible worlds in which the proposition is true, and others in which it is false.

But some propositions don't divide possible worlds quite that cleanly. For example, consider the proposition that it's now 1 p.m. That proposition is false at the moment at which I'm currently writing, but was true approximately 40 minutes ago. So should we count that proposition as true or false in my world (the actual world)?

Traditional possible worlds don't provide enough information to settle the truth-values of propositions like *it's now 1 p.m.* It's as if I asked you to help me practice my scene in a play, handed you the entire script, then waited patiently for you to start feeding me cues. The script describes everything in the play, but it doesn't give you *all* the information you need—you need to know which scene we're on, and which character I'm playing. Similarly, a possible world is maximally specified in the sense that for any possible world and any particular event, that event either does or does not occur in that world. But even if you had a complete list of all the events that would ever occur in the actual world, that wouldn't be enough for you to know *who* you are in that world, *where* you are, or *what time* it is. It wouldn't be enough to tell you whether it's now 1 p.m.

The most common response to this problem, following in the tradition of David Lewis (1979), is to augment traditional possible worlds by adding **centers** to them. Given a traditional possible world, a center picks out a particular individual at a particular time and place within that world. An ordered pair of a world and a center in it is then called a **centered possible world**. (The traditional possible worlds without centers are called **uncentered worlds**.)

With this structure in place, we may now consider **centered propositions**, which are associated with sets of centered possible worlds. The propositions we dealt with in previous chapters—now called **uncentered propositions**—are a proper subset of the centered propositions. Uncentered propositions don't discriminate among the centers associated with a given uncentered world; if two centered worlds are indexed to the same uncentered world, and one of them belongs to the set associated with a particular uncentered proposition, then the other one does too. Centered propositions, on the other hand, may discriminate among centered possible worlds indexed to the same uncentered world. The proposition that it's now 1 p.m. is a centered proposition but not an uncentered one; it's true at some centers in the actual world but not others. The proposition that Shaquille O'Neal and Michelangelo share the same birthday (March 6th) is an uncentered proposition; its truth doesn't vary depending on who or where you are within the world. An uncentered proposition tells you something about which world you're in, but doesn't tell you anything about where you are in that world. A centered proposition can provide self-locating information.[15]

Everything we said about (uncentered) propositions in Chapters 2 and 3 transfers easily to centered propositions. Centered propositions enter into logical relations, defined in terms of centered possible worlds. (Two *centered* propositions are consistent if they are both true in at least one *centered* world; two *centered* propositions are equivalent if they are associated with the same set of *centered* possible worlds; etc.) We can understand a rational credence distribution as a probability distribution over a language of centered propositions. The Kolmogorov axioms and Ratio Formula operate just as before. So, for instance, Finite Additivity requires my credence in the disjunction that it's either 1 p.m. or 2 p.m. to be the sum of my credence that it's 1 p.m. and my credence that it's 2 p.m. (since those disjuncts are mutually exclusive).

The trouble comes when we get to updating. Suppose that at some point in my life I gain a piece of self-locating evidence, and update on it by Conditionalization. Conditionalization generates certainties, so I become certain of a centered proposition. But Conditionalization also retains certainties, so if I keep conditionalizing, I will remain certain of that centered proposition. Yet this may not be rational. Suppose that 40 minutes ago I was staring at a perfectly reliable clock, and so became certain that it was 1 p.m. Despite having had that experience (back then), it is presently perfectly rational for me to be less than certain that it's now 1 p.m. Rational updating in centered propositions does not always go by Conditionalization.

Perhaps we should switch to a regime that combines Regularity with Jeffrey Conditionalization (Section 5.5)? Such an updating approach never sends contingent propositions to certainty, so doesn't have to worry about their sticking there; centered propositions about the time would be free to float up and down in confidence across successive updates. Yet Namjoong Kim (2009) argues that Jeffrey Conditionalization doesn't interact well with self-location either.

In a Jeffrey update, experience directly influences an agent's unconditional credences over a particular partition of propositions. Unconditional credences in other propositions in the agent's language are then set using conditional credences whose conditions are propositions within the originating partition. So suppose I look out my window, and experience directly influences my credences over the partition containing the centered propositions *it's raining now* and *it's not raining now*. I want to set my unconditional credence in the proposition *that it rains on Sunday* (for some particular Sunday that I've got in mind). Jeffrey Conditionalization would have me set that unconditional credence using such conditional credences as cr(it rains on Sunday | it's raining now). Yet Jeffrey Conditionalization is appropriate only if those conditional credences remain constant over time. (This was the Rigidity property we saw in Section 5.5.) And there seem to be moments when it would be quite rational for me to change my value for that conditional credence—for instance, as the calendar flips over from the Sunday in question to Monday morning.

Ultimately Conditionalization and Jeffrey Conditionalization suffer from the same affliction. An agent's evidence relates her to the world; it helps her discern the truth-values of various propositions. Conditionalization and Jeffrey Conditionalization assume that as time goes on, changes occur on only one side of that relation: An agent's evidence may change, but the truth-values of the target propositions remain fixed. Yet centered propositions present the agent with moving targets; as she gains (or loses) information, their truth-values may change as well. Traditional updating rules aren't up to this more complicated task.[16]

At this point it's tempting to object that I've misdiagnosed the problem—the flaw isn't in the updating rules; it's in the approach to content. There's an intuitive sense in which, when I ask myself at 1 p.m. "Is it now 1 p.m.?", I'm asking something different than when I ask 40 minutes later "Is it now 1 p.m.?" The two questions are asked about different moments; I'm asking for information about two different *nows*. The Lewisian approach to content holds that when I assign a credence of 1 at 1 p.m. to the proposition that it's now 1 p.m., and when I assign a lower credence 40 minutes later that it's now 1 p.m., those two credences are assigned to the same centered proposition. (So I've gone from certainty to less-than-certainty in one and the same proposition, in violation of Conditionalization.) Perhaps we should adopt a theory of content that reads those two credences as attitudes toward distinct propositions; this would remove the counterexamples to Conditionalization.

There are many theories of self-locating content in the philosophy of language literature. Some of them break with Lewis and maintain that if I wonder at two different times whether it's 1 p.m., the propositions I entertain at those distinct times have distinct contents. It's possible to build a Bayesian updating scheme on top of these non-Lewisian approaches to self-locating content. Yet I can report that none of the approaches anyone has tried leaves Conditionalization unscathed. Even with an alternate approach to content, Conditionalization needs to be either modified or supplemented (or both) to yield a plausible updating scheme.[17] So the problem for Bayesian updating raised by self-location can't be solved entirely by altering our theory of content.

In the next two sections I'll survey a couple of changes to Conditionalization that have been proposed to help it deal with self-location. In doing so I'll continue to work with a Lewisian approach to content, not because I personally think it's the best,[18] but because it's the most common and easily manageable, and because most of the important contrasts between proposals can still be seen in that context.

11.2.2 The HTM approach

The first formal, systematic Bayesian approach to updating self-locating credences was proposed in the artifical intelligence literature by Joseph Y. Halpern (2004, 2005), who offered it as an extension of his joint work with Mark Tuttle. Christopher J.G. Meacham (2008) later offered an extensionally equivalent formalism couched in a possible-worlds framework more familiar to philosophers. I will describe Meacham's version here, but honor its origins by calling it the **HTM approach**.

The HTM approach starts with the thought that for the many decades during which Bayesians focused exclusively on uncentered propositions, Conditionalization worked pretty well for them. So let's keep Conditionalization for uncentered propositions, then add a mechanism for handling the rest. Put formally, the HTM approach offers the following two-step updating process:

1. Suppose that between times t_i and t_j, an agent gains evidence E. Her first updating step is to redistribute her credence over uncentered worlds by conditionalizing. To do so, she takes all the uncentered worlds she entertained at t_i, assigns credence 0 to those ruled out by E, then renormalizes her credences over the rest. (That is, she multiplies the credences assigned to all the remaining uncentered worlds by the same constant, so that the sum of those credences is 1.)

2. The agent now has to distribute her credences over the centered worlds associated with each of the remaining uncentered worlds. To do so, she works through the uncentered worlds one at a time. Taking a particular uncentered world (call it u), she considers all the centers associated with that world she entertained as live possibilities at t_i. She assigns credence 0 to any centered world incompatible with E. Then she takes the credence assigned to u in Step 1 above, and distributes it among the remaining centered worlds associated with that uncentered world.[19]

A couple of comments about this updating process: The first step ensures that for any uncentered proposition U, the agent will assign $cr_j(U) = cr_i(U \mid E)$. So the HTM approach updates uncentered propositions by Conditionalization, exactly as intended.[20] Meanwhile, you may have noticed that the second step is underspecified—while it says that the credence associated with a given uncentered world u should be distributed over a particular set of centered worlds associated with u, it doesn't say exactly how that distribution should go. Halpern suggests that Indifference Principle reasoning might make it natural to distribute u's credence equally over the centers associated with u; Meacham suggests we maintain ratios between credences in u's centers from before the update occurred. Which proposal we go with will be immaterial in the discussion that follows.

To see an example of the HTM approach in action, consider the following case: At t_1, you have no idea what time it is. You're wearing a watch that you're

60% confident is running reliably. The rest of your credence is divided equally between the possibility that the watch is entirely stopped, and the possibility that it's running but is off by exactly an hour (perhaps due to daylight savings time). You then glance at your watch, notice that it's running (the second hand is moving), and that it currently reads 1 p.m. If you apply the HTM approach, your credences will evolve as follows:

	cr_1	Step 1	Step 2	
Stopped	0.20	0	12 p.m.	0
			1 p.m.	0
			2 p.m.	0
Reliable	0.60	0.75	12 p.m.	0
			1 p.m.	0.75
			2 p.m.	0
Running, but off an hour	0.20	0.25	12 p.m.	0–0.25
			1 p.m.	0
			2 p.m.	0–0.25

You begin at t_1 with credences of 0.20, 0.60, and 0.20 respectively in the uncentered possibilities that your watch is stopped, reliable, and running but off by an hour. After t_1 you receive evidence E that the watch is running and reads 1 p.m. In Step 1 of the HTM process, you rule out as incompatible with E the uncentered world in which the watch is stopped. So your credence in that world goes to 0, along with all the centered worlds associated with that world. The other two uncentered possibilities are consistent with E, so you renormalize your credences over those, multiplying each by a normalization factor of 5/4 so that they sum to 1.

Now we proceed to Step 2, and consider various centers associated with each uncentered world. In the world where your watch is reliable, E rules out every time possibility except 1 p.m., so the full 0.75 credence assigned to that uncentered world goes to the associated 1 p.m. centered world. In the uncentered world where your watch is running but off by an hour, E rules out the 1 p.m. center. But it leaves open both the 12 p.m. and 2 p.m. centers, because your evidence doesn't reveal whether the watch is running fast or slow. So the 0.25 credence in that uncentered world must be distributed between those two centers by some strategy that Step 2 leaves unspecified. I've therefore indicated above that your credences in the 12 p.m. and 2 p.m. centers must be between 0 and 0.25, but I haven't assigned them specific values. (We'll come back to this underspecification in Section 11.2.3 below.)

Once we have the values in the table, we can use them to calculate further credences. For example, what is your credence after learning E that it's currently before 2 p.m.? That will be 0.75 (from the centered possibility that your watch is reliable and it's 1 p.m.) plus whatever credence is assigned to the centered possibility that your watch is off by an hour and it's 12 p.m.

Now that we know how the HTM approach works, how can we tell if it's a rational procedure for updating self-locating credences? You'll show in Exercise 11.6 that applying the HTM approach to the Sleeping Beauty Problem yields a 1/2 credence in heads upon awakening. Is that really what rationality requires of Beauty?

As in our discussion of updating rules for memory loss, it's tempting to support this new updating approach with a Dutch Strategy or an accuracy-based argument. But once more, such arguments let us down in the present context. Hitchcock (2004) describes two different Dutch Strategies that can be constructed against Beauty, one that's guaranteed to take her money if she assigns any credence other than 1/2 upon awakening, and another that takes her money if she assigns anything other than 1/3. The difference between these strategies comes in how many bets the bookie is required to place. In the first strategy, the bookie places a particular bet with Beauty on Monday but not on Tuesday. In the second strategy, the bookie places a bet with Beauty every time she awakens during the experiment—once if heads, twice if tails. The first strategy seems objectionable because it requires the bookie to know something Beauty doesn't. (To decide whether to place the bet on a given day, he has to know whether that day is Monday or Tuesday, but that's exactly what Beauty *doesn't* know on either of those days.) Yet the second strategy seems objectionable because the number of bets placed varies with the truth-value of the very proposition on which the bet is being placed (that the coin came up heads).

While a debate has ensued in the literature, no consensus has emerged as to which of these Dutch Strategies (if either) is dispositive. Meanwhile, on the accuracy front, Kierland and Monton (2005) identify two different approaches to tallying up accuracy, and note that on one approach thirding minimizes expected inaccuracy, while on the other approach halfing fares best. To complicate matters further, Briggs (2010) argues that which Dutch Strategy and which accuracy argument you accept for Sleeping Beauty should depend on whether you endorse Causal or Evidential Decision Theory (Section 7.3)!

So I don't think Dutch Strategy or accuracy-based arguments are going to adjudicate the Sleeping Beauty Problem any time soon. Is there another way we could assess the HTM approach? Here it's worth noting that the HTM approach to updating has the following consequence:

Relevance-Limiting Thesis: If an update doesn't eliminate any uncentered possible worlds, then it does not change the agent's credence distribution over uncentered propositions.

The intuitive idea of the Relevance-Limiting Thesis is that purely centered evidence—evidence that tells an agent only about who she is or where she's located in the actual world, without telling her anything about what the actual world is like—is irrelevant to uncentered propositions. A batch of self-locating evidence may redistribute an agent's credences across centers, but if it doesn't *rule out* any uncentered worlds as actual, then it shouldn't change the agent's uncentered credences.

While the Relevance-Limiting Thesis may be intuitively plausible, it is subject to counterexample. All of the counterexamples are fairly artificial, because it's difficult to find a real-life situation in which an agent gains information without *any* uncentered implications. So here's one such artificial example:

> Ten people are arranged in a circle in a cylindrical room. One of them is you. You have no qualitative way of distinguishing yourself from the others in the room—you are all dressed alike, you all look identical, and you have no memories of anything that happened before this moment. (You don't even know your own name!) Your only way of picking yourself out uniquely among the members of the room is with indexical expressions ("me", "the person currently having *this* thought", etc.).
>
> A fair coin is flipped to determine the contents of a bag. If the coin comes up heads, the bag will contain nine black balls and one white ball. If the coin comes up tails, the bag will contain one black ball and nine white balls. You do not receive any direct information about the outcome of the coin flip, but once the bag is filled it is passed around the room. Each person draws out one ball and passes the bag until the bag is entirely empty. You cannot see the ball anyone else has drawn, but your ball is black.[21]

The science-fiction setup of the example is designed to prevent your new evidence when you pick the black ball from eliminating any uncentered possible worlds. If you were certain, say, that you were Shakira, then you could eliminate any uncentered worlds in which Shakira picks a white ball. But given your inability to self-identify non-indexically, the new information you gain is best expressed by the centered proposition "I pick a black ball". This information doesn't eliminate either of the uncentered worlds you entertain; it's compatible with both the heads-world and the tails-world that your ball would be black. So according to the Relevance-Limiting Thesis, picking a black ball should not change your credence in heads versus tails. Yet intuitively, observing a black ball should dramatically increase your confidence in heads. Consider the relevant Bayes factor: your chance of picking a black ball given heads is nine times your chance of picking a black ball given tails. So the mystery bag scenario is a counterexample to the Relevance-Limiting Thesis, and to the HTM approach.[22]

11.2.3 Going forward

Since Halpern and Meacham took their first stab at a Bayesian self-locating updating scheme, over a dozen alternative approaches have been proposed in the literature. I'll discuss one of them here, then refer you to the Further Reading to learn about more.

Where and how does the HTM approach go wrong? In many cases in which an agent has self-locating uncertainty, that uncertainty is mirrored by uncertainty about some uncentered proposition. Return to our example of the potentially unreliable watch. There we weren't sure how much of your t_2 credence associated with the uncentered possibility "off by an hour" should be distributed to the temporal location of 12 p.m. and how much of it should be distributed to 2 p.m. But notice that since the watch reads 1 p.m. at that point, you're certain of two biconditionals at t_2: it's noon just in case the watch is fast by an hour, and it's 2 p.m. just in case the watch is slow by an hour. So your t_2 credence in noon should equal your t_2 credence in the uncentered proposition that the watch is fast, and your credence in 2 p.m. should equal your credence that it's slow. (Recalling from Exercise 2.7 that if you're certain of a biconditional between two propositions, you should be equally confident in each.)

When an agent can match centered propositions with uncentered propositions in this fashion, any centered information she gains will come with uncentered information as well. In that case, we can update as the HTM approach requires: manage the core substance of the update by conditionalizing on uncentered information, then distribute credence from uncentered to centered propositions using the biconditionals. But this strategy runs into trouble when an agent learns a crucial piece of centered information for which she has no uncentered equivalent. In the mystery bag example, the centered "I pick a black ball" is relevant to whether the coin came up heads, but no uncentered proposition is available with which it can be put into a biconditional. So the first step of HTM updating misses the significance of "I pick a black ball" entirely, and the HTM approach founders on this example.

A simple fix is to restrict the applicability of the HTM approach. We apply HTM only when the following condition is met: At each time under consideration, for each centered proposition, the agent has an uncentered proposition which she is certain forms a true biconditional with the centered one. In the mystery bag example, this condition fails for the centered proposition "I pick a black ball." So we avoid the HTM approach's incorrect verdicts for mystery bag by restricting it from being applied to that example at all.

Of course, this raises the question of how we're going to get updating verdicts for the mystery bag case at all. A similar question could be asked about the Sleeping Beauty Problem. When Beauty awakens on Monday morning, she doesn't know what day it is, and information about the current date could be of crucial significance to her credences about the coin. (Since she's awake, if it's currently Tuesday then the coin came up tails!) But Beauty lacks any uncentered proposition that she's certain has the same truth-value as "Today is Tuesday." So once more, the restricted HTM approach doesn't yield any direct guidance for Beauty's credences upon awakening.

While the details are too complex to go into here, workarounds are available that *indirectly* apply the HTM approach to these examples. The trick is to focus on a narrow set of times for which the centered propositions don't provide any difficulties, or to add a feature to the example that gives the agent uncentered propositions for biconditionals without distorting the credences of interest. Exercise 11.6 below gives a suggestion how the latter might be accomplished; full details are available in (Titelbaum 2013a, Ch. 9). (In case you're wondering, the restricted HTM approach yields a 1/3 answer to the Sleeping Beauty Problem.)

It's also worth thinking about how to integrate the two halves of this chapter: Can we get a single Bayesian updating scheme that applies to memory-loss cases, cases with self-locating information, and even cases that combine the two? After all, it looks like the forgetting drug is a pretty key feature of the Sleeping Beauty Problem.[23] And in Arntzenius's Shangri La example, the agent needs to manage credences in such centered propositions as "I can recall traveling the Path by the Mountains" and "I'm now in Shangri La." (See Exercise 11.7.)

There hasn't been much formal work in the literature on Bayesian approaches to memory loss, so there's been even less on integrating memory loss with self-location. Nevertheless, both Meacham (2008) and Sarah Moss (2012) mention in passing how their self-locating updating schemes could be extended to handle forgetting cases, with Meacham proposing a condition much like Hypothetical Representability. A single updating system that combines Hypothetical Representability and something like the restricted version of HTM I've been discussing is carefully constructed, explored, and applied in (Titelbaum 2013a).

11.3 Exercises

Unless otherwise noted, you should assume when completing these exercises that the credence distributions under discussion satisfy the probability axioms and Ratio Formula. You may also assume that whenever a conditional credence expression occurs or a proposition is conditionalized upon, the needed proposition has nonzero unconditional credence so that conditional credences are well defined.

Problem 11.1. 🎵 A six-sided die (not necessarily fair!) has been rolled. At t_1 it's part of your evidence that the roll came up odd. At that time you assign credence 1/6 that it came up one and credence 1/3 that it came up three. Between t_1 and t_2 you learn that the roll came up prime. But between t_2 and t_3 you forget that the roll came up odd; at t_3 you assign credence 1/2 that it came up even (while retaining the evidence that it came up prime). If you satisfy Hypothetical Representability, what is your t_3 credence that the roll came up three?

Problem 11.2. 🎵🎵 Suppose we have two times t_j and t_k, with E_j representing the conjunction of the agent's total evidence at t_j and E_k playing a similar role with respect to t_k. Prove that the statements below all follow from Hypothetical Representability:
  (a) For any propositions H, D, and E in $\mathcal{L}$ such that $E_j \ \& \ E \ =\!\vDash\ E_k \ \& \ D$, $\mathrm{cr}_k(H \,|\, D) = \mathrm{cr}_j(H \,|\, E)$.
  (b) If the agent *loses* no evidence from t_j to t_k, then she updates by Conditionalization between those two times.
  (c) If the agent *gains* no evidence from t_j to t_k, then she updates by reverse-temporal conditionalization between those two times. (Compare Equation (11.1).)

Problem 11.3. 🎵🎵 Prove that an agent updates at all times by Conditionalization if and only if the following two conditions are both met:

1. Any piece of evidence she possesses at a given time is retained at all later times.
2. Hypothetical Representability.

Problem 11.4. 𝄞 Explain why any agent who updates using the HTM approach will satisfy the Relevance-Limiting Thesis.

Problem 11.5. 𝄞
(a) Make a probability table for the mystery bag example using "Coin lands heads" and "I pick a black ball" as your atomic propositions. Suppose that t_1 is the time after the coin has been flipped but before you've picked a ball out of the bag. Fill in t_1 credences that would be rational if you set your credences equal to the objective chances. Then fill in the t_2 credences that would be rational if you simply conditionalized on the proposition "I pick a black ball." If you updated that way, what would be your t_2 credence in heads?
(b) Now suppose you start with the same t_1 credences but update by the HTM approach. Then what are your t_2 credences? What is your t_2 credence in heads? Explain.

Problem 11.6. 𝄞𝄞
(a) Show that on the HTM approach, when Beauty awakens on Monday morning in the Sleeping Beauty Problem not knowing what day it is, her credence in heads should be 1/2.
(b) Now suppose the researchers add one small detail to the protocol of their Sleeping Beauty experiment. After Beauty goes to sleep on Sunday night, they will not only flip their fateful coin, but also roll a fair die. If the die roll comes up odd, they will paint the walls of Beauty's room red on Monday and blue on Tuesday. If the die roll comes up even, they will paint the walls of Beauty's room blue on Monday and red on Tuesday. The procedures for waking her up and putting her back to sleep are the same; on any day when Beauty is awake, she will be awake long enough to notice the color on the walls.[24]

On Sunday night, Beauty is informed about all the details of the experiment, including the painting plan. She goes to sleep, then finds herself awake sometime later, certain that it's either Monday or Tuesday, not certain which day it is, but also certain that the walls of her room are red. According to the HTM approach, what should Beauty's credence in heads be at that point?

(c) Compare your answers to parts (a) and (b). Does it seem to you that the HTM approach is getting things right here? Should the painting procedure make any difference to Beauty's credence in heads upon awakening?

(d) For each of the following centered propositions, provide an uncentered proposition such that when Beauty awakens in the red room, she is certain that the centered proposition and the uncentered proposition have the same truth-value:

  (i) Today is Monday.

  (ii) I am currently in a red room.

 (iii) Today is Monday or Tuesday.

Problem 11.7. 🎵🎵 You are now going to build a Bayesian model of the Shangri La problem that incorporates both its self-locating and its memory loss components.

(a) Start by building a probability table on the three atomic propositions "The coin comes up heads," "I can recall the Path by the Mountains," and "I am now in Shangri La." Next, make three columns for cr_1, cr_2, and cr_3. Finally, use the details of the story on p. 383 to fill out what you think is the rational credence distribution for each of those three times. Assume that in the version of the story you're representing, the agent travels the Path by the Mountains, and while on that path (at t_2) is certain that she is traveling it and can recall doing so.

(b) For each of the following three updating schemes, explain why it can't produce the pattern of credences in your probability table: updating by Conditionalization; the HTM approach; and the restricted HTM approach described on page 402.

Problem 11.8. 🎵🎵🎵 Wandering through the State Fair, you encounter a very strange carnival game. The rules are posted, and each customer reads them before she plays:

At the start of today we filled our pond with 100 fish, each fish red or blue. We flipped a fair coin to determine whether the pond would contain seventy-five red and twenty-five blue fish, or vice versa.

Every customer who decides to play our game will first be told how many people played today before they did. Then the customer will fish one fish at random out of the pond and keep it. (Their fish will be replaced in the pond by a fish of the same color.)

The lucky seventh player of the day will get a very special treat. Once they fish out their fish, we will find all the customers from earlier in the day who drew the *opposite* color fish—if there are any—and select one of those customers at random to introduce to our lucky seventh player. (Each customer with the opposite-colored fish will have the same chance of being selected.) We hope this is the beginning of a beautiful friendship!

You decide to play the game, and are told that you are the lucky seventh player of the day! You then proceed to the pond and fish out a fish at random, finding that it's blue. The organizers go away for a while, and come back with a woman named Linda, who's got a red fish. The two of you chat amiably for a while, and do indeed begin to strike up a friendship. . . .

(a) Let's start by focusing on your credences. After you read the rules for the game, but before you fish out your fish, how confident should you be that the pond contains predominantly blue fish?

(b) Once you fish out your fish, and update by Conditionalizing on the fact that it's blue, how confident should you be that the pond is predominantly blue?

(c) When you're introduced to Linda, you learn that at least one of the six people who fished before you got a red fish. If the pond is indeed predominantly blue, what's the chance that at least one of the first six people gets a red fish? (<u>Hint</u>: Try first calculating the chance that all six of them get blue fish.)

(d) So, taking your confidence from part (b) and conditionalizing on the fact that at least one of the six people before you received a red fish, how confident should you be after meeting Linda that the pond is predominantly blue?

(e) Now let's focus on *Linda's* credences. After she reads the rules, but before she fishes out her fish, how confident should she be that the pond is predominantly blue?

(f) After Linda randomly fishes out a fish that turns out to be red, how confident should she be that the pond is predominantly blue?

(g) Sometime later, Linda is selected as the person to be introduced to the lucky seventh player. She is introduced to you, and learns that the seventh fish drawn was blue. Conditionalizing on this information, how confident should she now be that the pond is predominantly blue?

(h) Did the credences you calculated in parts (d) and (g) turn out to be different? If so, why? Did you or Linda start the game with importantly different priors? Is there some difference in your total evidence at the end of the game that would justify a difference in your credences? Or did we leave some piece of relevant evidence out in our calculations? Whichever answer you give, state as precisely as you can what you think is going on.[25]

11.4 Further reading

INTRODUCTIONS AND OVERVIEWS

Andy Egan and Michael G. Titelbaum (2022). Self-Locating Belief. Forthcoming in *The Stanford Encyclopedia of Philosophy*.

Background information on available philosophical theories of self-locating content.

Michael G. Titelbaum (2016). Self-Locating Credences. In: *The Oxford Handbook of Probability and Philosophy*. Ed. by Alan Hájek and Christopher R. Hitchcock. Oxford: Oxford University Press, pp. 666–80

Surveys existing schemes for updating self-locating beliefs, grouped by broad strategic approach.

Michael G. Titelbaum (2013b). Ten Reasons to Care about the Sleeping Beauty Problem. *Philosophy Compass* 8, pp. 1003–17

Survey of literature on the Sleeping Beauty Problem, including its connections to such diverse areas as decision theory, philosophy of language, and the philosophy of quantum mechanics.

CLASSIC TEXTS

W. J. Talbott (1991). Two Principles of Bayesian Epistemology. *Philosophical Studies* 62, pp. 135–50

Classic discussion of the rationality of memory loss within a Bayesian framework.

Adam Elga (2000). Self-locating Belief and the Sleeping Beauty Problem. *Analysis* 60, pp. 143–7

David Lewis (2001). Sleeping Beauty: Reply to Elga. *Analysis* 61, 171–6

Exchange that drew philosophers' attention to the Sleeping Beauty Problem.

Frank Arntzenius (2003). Some Problems for Conditionalization and Reflection. *The Journal of Philosophy* 100, pp. 356–70

Examines a number of problem cases for Conditionalization and Reflection involving self-location and/or memory loss. Introduces the Shangri La example.

Extended Discussion

Isaac Levi (1980). *The Enterprise of Knowledge.* Boston: The MIT Press

While Levi's terminology is fairly different from ours, Chapter 4 works through all of the mathematics behind our discussion of memory loss and hypothetical priors. Also discusses various historically important Bayesians' positions on hypothetical priors.

Michael G. Titelbaum (2013a). *Quitting Certainties: A Bayesian Framework Modeling Degrees of Belief.* Oxford: Oxford University Press

Covers the topics in this chapter at a much greater level of philosophical and technical depth. Chapters 6 and 7 discuss memory loss and Suppositional Consistency. Chapters 8 through 11 propose an updating rule for situations involving context-sensitive information, compare that rule to rival approaches, and apply it to a variety of situations (including Everettian interpretations of quantum mechanics).

Notes

1. I have changed Talbott's notation for moments in time to square with our usage in this book. Also, Talbott is concerned with the spaghetti example not only as a counterexample to Conditionalization (what he calls "Temporal Conditionalization") but also to van Fraassen's (1984) Reflection Principle. As we saw in Section 5.2.2, Conditionalization and Reflection are intimately related; with respect to this chapter's concerns, they stand or fall together.

2. Some authors like to use the words "remember" and "memories" factively. Replacing my talk of "memories" with "quasi-memories" or some such would make no difference to what's rationally required in the example.

3. You might think it's not Jeffrey Conditionalization's job to set a specific $cr_3(H)$ value. As an updating norm, Jeffrey Conditionalization is concerned with diachronic consistency. Yet $cr_3(H) = 1/2$ seems required not by diachronic considerations, but instead by your t_3 evidence about the fairness of the coin (and lack of trustworthy evidence about its outcome). So perhaps Jeffrey Conditionalization can be liberal about $cr_3(H)$, while the Principal Principle restricts that value to 1/2.

   But now consider a version of Shangri La in which you don't know anything at t_1 about the coin's bias, or lack thereof. The evidence you possess at t_1 doesn't mandate any particular $cr_1(H)$ value on its own; you set your $cr_1(H)$ value according to your personal epistemic standards. (Perhaps that value isn't even 1/2.) I submit that at t_3, you should assign the same credence to heads that you did at t_1. This value cannot be required by the Principal Principle, nor by any other purely synchronic evidential norm, since we stipulated that such norms were insufficient to fix your credence in heads at t_1. Rather, it is required by diachronic considerations, in particular considerations of keeping your epistemic standards consistent over time. The next section presents an updating rule that's responsive to such considerations, even in cases involving memory loss.

4. Schervish, Seidenfeld, and Kadane respond to Arntzenius by noting that it's "already assumed as familiar in problems of stochastic prediction . . . that the information the agent has at t_2 includes all the information that she or he had at time t_1. This is expressed mathematically by requiring that the collection of information sets at all times through the future form what is called a *filtration*" (2004, p. 316). Their point is that even before Shangri La was introduced, statisticians were well aware that Conditionalization applies only when no information is lost over time.

5. Compare the discussion at Levi (1987, p.198).

6. We can see how this follows from Hypothetical Representability by noting that (with respect to the propositions of interest in this problem) E_1 is identical to E_3. So conditionalizing a hypothetical prior on E_3 will yield exactly the same distribution as conditionalizing that prior on E_1, making cr_3 identical to cr_1.

7. Equation (11.2) thus represents an intertemporal version of Elga's interpersonal "guru-flection" principle (Elga 2007).

   I should also note that Equation (11.2) assumes that E_j and E_k are logically consistent. (If they weren't, some of the relevant conditional credence expressions would have credence-0 conditions.) Throughout our discussion of memory loss, I will assume that the agent's total evidence sets at any two times are consistent with each other. (So, for example, an agent can't lose a piece of information at one time, then learn its negation at another.) This won't be a problem if we assume that evidence is factive, because truths are logically consistent with each other. At least, it won't be a problem until Section 11.2, when we consider self-location and centered propositions. . .

8. In Chapter 10, note 33, I pointed out that Greaves and Wallace restrict their attention to "available" updating plans. One might argue that relative to an intuitive notion of availability (not necessarily Greaves and Wallace's notion) the memory loss (or threat thereof) that you experience at t_3 in Shangri La keeps plans that assign $cr_3(H) > 1/2$

from being available to you at that time. If so, being representable by a hypothetical prior might be the *available* plan providing the best expected accuracy relative to your attitudes at t_2. But now it looks like most (all?) of the argument for assigning $cr_3(H) = 1/2$ is based on considerations of availability, with accuracy argumentation no longer doing the heavy lifting.

9. Compare also the Dutch Strategy we constructed against an agent who loses a certainty in Exercise 9.3.

10. For an investigation into the intuitions, a survey of some of the philosophical discussion, and a modest attempt to justify those intuitions, see Titelbaum (2015a).

11. Keep in mind that our discussion in this chapter is confined to *diachronic* rationality constraints. We assume throughout that credence distributions dictated by rationally permissible epistemic standards satisfy traditional synchronic Bayesian norms, in particular the probability axioms and the Ratio Formula.

12. For what it's worth, all the "logical probability" distributions Carnap explored (Section 6.2) satisfied Suppositional Consistency. Isaac Levi also extensively examined Suppositional Consistency (which he called "confirmational conditionalization") in a series of publications beginning with Levi (1974).

13. I should mention one further problem down the Subjective Bayesian path. Suppose we have a Subjective Bayesian who holds that losing evidence because of memory loss (or the threat thereof) is no sign of irrationality, yet requires a rational agent to maintain her epistemic standards over time. What happens if an agent forgets her epistemic standards? It seems inconsistent to forgive an agent's forgetting her evidence but not her standards. Yet once an agent has forgotten her earlier epistemic standards, how can we require her to maintain them going forward? Or are epistemic standards so deeply embedded in our identity as epistemic agents that we cannot possibly forget them—or cannot forget them while remaining the same agent?

14. The Sleeping Beauty Problem is an adaptation of the absent-minded driver paradox from Piccione and Rubinstein (1997). Elga attributes the "Sleeping Beauty" name to Robert Stalnaker, who Elga says "first learned of examples of this kind in unpublished work by Arnold Zuboff".

15. Some authors call centered propositions *de se* **propositions**, and uncentered propositions *de dicto*. There's also a bit of variation in the definitions: I've followed one group of authors in defining centered propositions as a superset of the uncentered; other authors define centered propositions to be mutually exclusive with uncentered. (For them, a proposition counts as centered only if it somewhere discriminates between two centers within the same possible world.) This terminological issue has little ultimate significance.

16. In his influential (1979) paper on self-locating content, Lewis wrote, "It is interesting to ask what happens to decision theory if we take all attitudes as *de se*. Answer: very little. We replace the space of worlds by the space of centered worlds, or by the space of all inhabitants of worlds. All else is just as before" (p. 534). Some commentators have taken this to mean that Lewis thought Conditionalization would work just fine with self-locating contents. But Schwarz (2010) notes that: (1) Lewis mentions *decision theory* rather than updating in that quote, and it's not clear that updating rules are part of decision theory; and (2) Lewis makes clear in a footnote to his (1996) that he's aware of

the self-location problem for both Conditionalization and Jeffrey Conditionalization: "Before I turned out the light, I saw that it was just minutes before midnight. In the course of a long and sleepless night, I undergo a redistribution of credence from the proposition that it is now before midnight to the proposition that it is now after midnight. It is far from obvious that this revision goes by probability kinematics [Jeffrey Conditionalization], let alone by conditionalizing" (p. 309, n. 6). So whatever Lewis thought in the late 1970s, he was clearly suspicious of conditionalizing over self-locating contents well before the Sleeping Beauty Problem brought the issue to most philosophers' attention in the 2000s.

17. The best-known attempt to solve Conditionalization's problems with self-location solely by altering the underlying theory of content appeared in Stalnaker (2008). Stalnaker began by building an alternative to Lewis's theory of content, then used it to construct a model of the Sleeping Beauty Problem. With those pieces in place he wrote, "The strategy for determining exactly what Sleeping Beauty's degrees of belief should be, when she wakes up, is to start by determining how her degrees of belief should be apportioned *on Sunday*.... Then her degrees of belief on Monday (and/or Tuesday) will be determined simply by conditionalizing on the information that she receives on waking up." (pp. 63–4, emphasis in original). Yet Brian Weatherson convincingly argued in his (2011) that Stalnaker's semantics would require Beauty to assign Monday credences to propositions she couldn't even *entertain* on Sunday. Since Conditionalization can generate credences for a proposition at a later time only if the agent assigned that proposition a credence at an earlier time, the credences Stalnaker needed couldn't be generated by Conditionalization alone. Stalnaker ultimately (2011) conceded that even with his non-Lewisian approach to content, Conditionalization would require supplementation to properly model the Sleeping Beauty Problem.

18. One of my concerns is that categorizing propositions according to whether they are self-locating or not—dividing centered propositions up into those that are uncentered and those that aren't—may draw the line in the wrong place for epistemological work. If the underlying problem for the traditional Bayesian machinery is that some propositions change their truth-values over time, we need to separate propositions into those whose truth-values are changeable and those whose truth-values aren't. The former group, which I would follow MacFarlane (2005) in labeling "context-sensitive" propositions, includes self-locating propositions, but may also include propositions involving knowledge, epistemic modals, predicates of taste, etc. Moreover, the relevant issue may not be which propositions are *actually* capable of changing their truth-values across contexts, but which propositions it's rational for a given agent to suspect *might* change truth-values. Thus the proper category to focus on may be *epistemically context-sensitive* propositions. For more on this designation, Titelbaum (2013a, pp. 177ff.).

19. This presentation of the HTM approach simplifies matters by assuming the agent entertains only finitely many centered worlds.

20. Here's a way to see that the HTM approach updates uncentered propositions by Conditionalization: Step 1 treats uncentered worlds like the state-descriptions in a probability table, updating them via exactly the process we described for updating probability tables on page 93. That updating process conditionalizes any proposition that can be stated as a disjunction of the state-descriptions involved. Since every uncentered proposition is

412 MEMORY LOSS AND SELF-LOCATING CREDENCES

a disjunction of uncentered worlds, Step 1 has the effect of applying Conditionalization to all uncentered propositions. Step 2 merely distributes uncentered credence among centered possibilities; it doesn't redistribute credences across uncentered worlds. So Step 2 keeps the conditionalization of uncentered propositions intact.

21. This example, and much of the surrounding discussion, is adapted from Titelbaum (2013a, §10.1). The first such counterexample to the Relevance-Limiting Thesis I'm aware of appears in Bradley (2011, §9); Bradley attributes the example to Matt Kotzen. The Relevance-Limiting Thesis first appeared in Titelbaum (2008).

22. We'll consider another problem for the HTM approach in Exercise 11.6. Meacham eventually (2010a) cited that problem as one reason he abandoned the HTM approach in favor of another self-locating updating scheme.

23. Though Kierland and Monton (2005) propose a problem structurally analogous to Sleeping Beauty that substitutes duplication for memory loss: No matter how the flip comes out, Beauty awakens only once, on Monday morning. But if the coin comes up tails, the researchers create a duplicate of Beauty who shares all of her memories. So when Beauty awakens on Monday morning, she is certain what day it is but uncertain just who she is. Again, the question is how confident Beauty should be that the coin came up heads; but this Döppelganger Beauty problem features neither memory erasure nor the threat thereof.

24. This specific variant of the Sleeping Beauty Problem was first proposed in Titelbaum (2008). Similar ideas appear in examples due to Kierland and Monton (2005), Neal (2006), Rosenthal (2009), and Meacham (2010a).

25. Thanks to Teddy Seidenfeld for extensive discussion of the solution to this problem.

12

Old Evidence and Logical Omniscience

In Chapter 11, we noted two features of everyday life (memory loss and self-location) that traditional Bayesian updating doesn't seem well designed to model. Still, recent authors have offered extensions of or minor modifications to Conditionalization that seem to address these concerns while staying faithful to the basic spirit of the Bayesian approach.

This chapter takes up two problems—the Problem of Old Evidence and the Problem of Logical Omniscience—on which the Bayesian literature has been focused for longer. The problems are connected because some authors have proposed that a solution to the Problem of Logical Omniscience would offer a path to solving the Problem of Old Evidence. These problems also seem to strike closer to the heart of Bayesianism and its five core normative rules.

The Problem of Old Evidence arises because once a proposition goes to credence 1, it is no longer positively relevant to anything, and so cannot count as confirming anything. This initially looks like yet another troublesome side effect of the way Conditionalization treats evidence. But ultimately it's a deep problem about the Bayesian conception of confirmation, a problem that may require us to rethink that conception and what it's good for.

The Problem of Logical Omniscience also initially presents as a side effect, this time of basing the Bayesian norms on probability mathematics. Kolmogorov's Normality axiom assigns probability 1 to all logical truths, which generates the Bayesian requirement that rational agents be certain of logic at all times. It feels like we should be able to tweak the axioms and loosen this requirement, so as to reflect more reasonable requirements on agents' logical abilities. Yet every simple tweak undermines the Bayesian formalism's ability to fully carry out the applications we considered in Chapters 6 and 7. Thus, while many authors are still trying to formalize logical non-omniscience, others have begun to reconsider whether an omniscience requirement is such a bad idea after all.

Fundamentals of Bayesian Epistemology 2: Arguments, Challenges, Alternatives. Michael G. Titelbaum,
Oxford University Press. © Michael G. Titelbaum 2022. DOI: 10.1093/oso/9780192863140.003.0012

12.1 Old evidence

As we discussed in Chapter 6, the Bayesian account of induction (and of scientific inference more generally) revolves around the notion of "confirmation". A piece of evidence supports a hypothesis when it confirms that hypothesis; we may also say that the evidence provides some justification for the hypothesis, or supplies reason for it. Many Bayesians have found the following principle about confirmation appealing:

Confirmation and Learning: Evidence E confirms hypothesis H for a rational agent if and only if learning E would increase the agent's confidence in H.

Chapter 6 also discussed the key Bayesian insight that confirmation can be analyzed in terms of probability, using the following principle:

Confirmation and Probability: Evidence E confirms hypothesis H relative to probability distribution Pr if and only if $Pr(H \mid E) > Pr(H)$.

In slogan form, confirmation is positive probabilistic relevance.

How are these two principles related? At first blush they seem to be about different things: Confirmation and Learning discusses confirmation relative to an agent, while Confirmation and Probability concerns confirmation relative to a probability distribution. But by this point you know enough Bayesian lore to connect them. Suppose we understand confirmation for an agent as confirmation relative to that agent's credence distribution cr. And suppose rationality requires agents to update by Conditionalization. Then for any rational agent, learning E increases her confidence in H just in case $cr(H \mid E) > cr(H)$. So if we're assessing confirmation relative to the credence distribution cr of a rational agent, Confirmation and Learning and Confirmation and Probability give us interchangeable methods of identifying confirmatory evidence.

12.1.1 The problem

In his chapter "Why I Am Not a Bayesian", Clark Glymour introduced what has since become a famous problem for Bayesian confirmation theory:

> Scientists commonly argue for their theories from evidence known long before the theories were introduced.... The argument that Einstein gave

in 1915 for his gravitational field equations was that they explained the anomalous advance of the perihelion of Mercury, established more than half a century earlier. Other physicists found the argument enormously forceful, and it is a fair conjecture that without it the British would not have mounted the famous eclipse expedition of 1919. Old evidence can in fact confirm new theory, but according to Bayesian kinematics, it cannot. (1980, pp. 306–7)

In this example, let the hypothesis H be Einstein's General Theory of Relativity (GTR), represented by the field equations Glymour mentions. Let the evidence E be the fact that Mercury's orbit displays an anomalous perihelion advance of 43 seconds of arc per century.[1] While this E was difficult to square with the classical Newtonian theory of gravitation, Einstein proved in 1915 that E was predicted by GTR. Howson and Urbach write, "this prediction arguably did more to establish that theory and displace the classical theory of gravitation than either of its other two dramatic contemporary predictions, namely the bending of light close to the sun and the gravitational red-shift" (2006, p. 298).

Why is this a problem? Let's start with Confirmation and Learning. As Earman notes, E was common knowledge in the physics community well before 1915: "the nature of Mercury's perihelion had been the subject of inten-sive study by Le Verrier, Newcomb, and other astronomers" (1992, p. 119). Intuitively, an agent who already *knows* something can't subsequently *learn* it. So when we set out to explain how E confirmed H for Einstein in 1915, it looks like we can't say that learning E would've increased his confidence in H at that time. In other words, the left-hand side of Confirmation and Learning's biconditional seems true for Einstein in 1915, but the right-hand side looks false. (Later we'll see a counterfactual strategy that tries to make the right-hand side true.)

Now consider Glymour's challenge to Confirmation and Probability. Since Einstein knew E prior to 1915, we should count it as part of his background corpus K in 1915. In Bayesian confirmation theory, a proposition X counts as part of an agent's K at a given time just in case $cr(X) = 1$ (Section 6.2.1). So for Einstein's 1915 credence distribution we have $cr(E) = 1$. But in that case E can't have been relevant to anything for Einstein at that time; the probability axioms and Ratio Formula imply that if $cr(E) = 1$, then for any $H \in \mathcal{L}$, $cr(H \mid E) = cr(H)$. (See Exercise 12.1.) So relative to Einstein's credence distribution in November of 1915—when he first showed that GTR explains Mercury's perihelion advance—that piece of evidence E wasn't positively relevant to anything. If we accept Confirmation and Probability, then the perihelion of Mercury couldn't confirm *anything* for Einstein in 1915, much less his General Theory of Relativity. And the same goes for any other physicist antecedently aware of the astronomical facts.

Glymour used this case to illustrate the **Problem of Old Evidence** for Bayesian epistemology.[2] According to Glymour, the problem is that when we try to explain how the perihelion of Mercury could be evidence for GTR, we find that "none of the Bayesian mechanisms apply, and if we are strictly limited to them, we have the absurdity that old evidence cannot confirm new theory." (1980, p. 307).

Since Glymour posed this challenge, a number of Bayesians (Christensen 1999; Eells 1985; Garber 1983; Zynda 1995) have tried to disambiguate exactly what the problem is supposed to be. As we'll see below, there are actually a number of challenges to Bayesian epistemology in the vicinity, but we'll start with a rough cut into two. Though the distinction isn't original to Christensen, I like his labels for the two problems: the *diachronic* problem of old evidence and the *synchronic* problem of old evidence.

The diachronic problem starts from the fact that over the course of 1915, Einstein went through a transition when it came to his attitudes towards GTR. At the end of the year he was highly confident of GTR; at the beginning of that year he hadn't been. What prompted this transition in Einstein's attitudes, and made the transition rational? Einstein may have cited E (Mercury's anomalous perihelion advance) in his arguments for GTR, but the transition can't have been prompted by his learning E—because he already knew E well prior to 1915! The diachronic problem is to explain the rational transition in Einstein's attitudes in a manner compatible with Bayesian theory.

The synchronic problem arises after the transition in Einstein's attitudes has already occurred. Suppose that sometime just after 1915, once Einstein was already highly confident of GTR, we went to him and asked what justified his high confidence in the theory. He might (quite reasonably) cite the anomalous perihelion advance of Mercury. In other words, he might say that the proposition we've been calling E confirms the proposition we've been calling H. Yet this can't mean that $cr(H \mid E) > cr(H)$ for Einstein's credence distribution, because at the time we're asking, he already has $cr(E) = 1$. The synchronic problem is to give a Bayesian explanation of the sense in which E supports Einstein's standing confidence in H, even though E is already part of Einstein's evidence.

As we discuss these two problems, we will follow Glymour in largely focusing on the Einstein example, and assuming that Einstein's attitudes developed in a rational fashion. Perhaps if one examined the historical record closely enough, one could discover various ways in which Einstein actually proceeded irrationally in his development of GTR. But that would be beside the point. The example (and various further details of it we will develop along the way) is meant to stand in for a general set of reasonable scientific practices that one might want Bayesianism to explain. However things actually went with Einstein, there will remain contexts in which it's rational for scientists to cite evidence that's been known for a long time in explaining transitions in their attitudes, or in defending the attitudes they currently possess.

12.1.2 Solutions to the diachronic problem

Let's start with a couple of proposals addressing the diachronic problem of old evidence. Both of these proposals grant that it was rational for Einstein at the end of 1915 to have a high credence in GTR that he didn't have before. But they deny that the change was due to Einstein's learning about the perihelion of Mercury—after all, he'd known about the perihelion for quite some time. These proposals focus on different potential causes of his attitude change.

New theories. One way to assign a new, high credence to a proposition is not to have assigned it any credence before. Perhaps the crucial transition that happened in 1915 was that Einstein created a theory, GTR, that he hadn't conceived of before. Seeing that this new theory explained a fact of which he was already well aware—the anomalous perihelion of Mercury—he assigned it an appropriately high credence.

This may not be a historically accurate recapitulation of what actually happened to Einstein during 1915. (Perhaps this proposal is a more realistic account of what happened to other physicists who first learned of GTR during 1915.) But it raises an important challenge for Bayesian epistemology: How do we account for shifts in an agent's credence distribution when she comes to recognize new possibilities not previously articulated in her doxastic space? Throughout this book we've represented the doxastic possibilities an agent entertains using a finite set of possible worlds, each described by a state-description composed from a finite list of atomic propositions. When an agent updates by Conditionalization, her numerical credence distribution over that set of possible worlds may change, but the underlying language across which the distribution is assigned remains fixed.

So what happens when an agent comes to entertain a new proposition that she hadn't considered before? Perhaps she learns of a new theory; perhaps she gains a new concept; perhaps she becomes acquainted with something to which she couldn't refer before. This isn't the kind of learning modeled by Conditionalization, in which the agent's new set of doxastic possibilities is a proper subset of those she previously entertained. And it isn't forgetting (Chapter 11), in which the agent entertains once more doxastic possibilities she had previously ruled out. When a new proposition is added to an agent's language, her doxastic space becomes more fine-grained. At least one possible world she had previously entertained is now split into two: one in which the new proposition is true, and another in which the new proposition is false.

How should the agent's credences evolve in such a situation? Perhaps the credence assigned to the old (coarse-grained) world should simply be divided between the two new (finer-grained) worlds. But the response to new theories may be more complex than that. For example, consider a nineteenth-century astronomer who entertains only two possible explanations for the anomalous orbit of Mercury: Newtonian physics is false, or Newtonian physics is true but some previously undetected planet is warping Mercury's orbit.[3] If we introduced this astronomer to Einstein's theory, that would partition the astronomer's "Newtonian physics is false" possibility into two new possible worlds (one in which GTR is true, the other in which Newtonian physics is false but GTR is as well). Plausibly, the two new worlds wouldn't just divvy up the credence previously assigned to "Newton was wrong"; recognizing the new possibility might also drain some of the astronomer's confidence away from the undetected planet hypothesis.

Clearly the question of how an agent should alter her credences upon recognizing new possibilities is subtle and rich. But just as clearly, the answer can't just straightforwardly be "through traditional Conditionalization". The **problem of new theories** is an important challenge for Bayesian updating schemes.[4]

Logical learning. By the end of 1915, Einstein had a newfound confidence in GTR. This wasn't because he learned of Mercury's perihelion advance during that year; he'd known about it long before. One could debate whether 1915 was the first year in which Einstein truly had a grasp of GTR, and thus was capable of assigning a credence to it. But that would require scouring the historical record, and making some difficult judgments about exactly how well an agent must grasp a theory or proposition in order to assign it a doxastic attitude.

Yet there's something important that clearly did happen during late 1915, and which obviously increased Einstein's confidence in GTR. Einstein worked out the *connection* between GTR and the perihelion of Mercury; he demonstrated that GTR predicts the perihelion effect to a high degree of precision with few adjustable parameters. Plausibly, it was learning about this connection between GTR and the perihelion—not learning about the perihelion itself—that led Einstein to increase his confidence in GTR in late 1915.

But if that's right, the Bayesian now has a different problem. From our current vantage point of scientific sophistication, we can see that various background facts known by Einstein and other physicists prior to 1915 *entail* that GTR predicts the perihelion advance of Mercury. It's just that no one recognized this entailment before Einstein. So the fact Einstein learned in late 1915—and that increased his confidence in GTR—was a logical truth. This creates a problem, because traditional Bayesianism can't model the learning of logical truths. An agent who satisfies the probability axioms will assign credence 1 to every logical truth in her language at all times. So there's no way a probabilistic agent can go from a credence less than 1 to certainty in such a proposition. This is the problem of logical learning; we will consider it and the related Problem of Logical Omniscience in Section 12.2.

12.1.3 Solutions to the synchronic problem

Perhaps Einstein first considered GTR during 1915; perhaps he considered it before, but became more confident in the theory once he saw its connection to Mercury's perihelion. At best, these proposals will explain the diachronic transition that brought Einstein to a substantial confidence in GTR by the end of 1915. But they won't address the synchronic problem of old evidence: how can a Bayesian account for the fact that *after* his transition to a high confidence in GTR, Einstein (and others) cited the perihelion advance of Mercury as a key piece of evidence confirming the theory?

Historical backtracking. On a probabilistic approach to confirmation, we want to say that E confirms H for an agent when she assigns cr(H | E) > cr(H). But given the probability calculus, this inequality fails when cr(E) = 1. (If Einstein is already certain of Mercury's anomalous perihelion advance, that fact can't be positively relevant to anything on his credence distribution.) A straightforward solution is to find some other credence distribution associated with the agent that doesn't assign 1 to E already, and use it to gauge E's confirmation of H.

It's natural to seek such a distribution in the agent's past. After all, if we take an agent who's certain of some E, and ask her, "In what sense does E support H for you?", she might answer, "E was one of the things that made me confident of H in the first place." This explains a support fact in her present using a confidence increase from her past. Perhaps it's sufficient for E to confirm H right now that there was some time in the past when the agent learned E and increased her confidence in H as a result.

While this proposal may cover some of the desired cases, there are plenty of old evidence examples for which it just won't suffice. First, it won't work for cases in which the agent came to consider H only once she was certain of E. In Einstein's case, he knew about the perihelion of Mercury long before he formulated GTR. As we've seen, an extension of traditional Bayesian updating is needed to cover the introduction of GTR into Einstein's conceptual space. But setting that point aside, it's clear there wasn't any time when Einstein learned of Mercury's perihelion and this new fact increased his confidence in GTR.

Second, even in cases in which the agent has been aware of H all along, the proposal requires other facts to have been learned in a very specific order. Suppose that from the agent's current point of view, E confirms H only relative to some other fact in her background corpus K. If the agent learned K after she learned E, there may have been no moment at which learning E increased the agent's confidence in H. For example, suppose I am playing cards and learn the proposition E that my opponent has a 7 in her hand. By itself, this doesn't confirm the proposition H that she possesses a card that's higher than all of mine. But then I subsequently view my own hand, and learn proposition K that my highest card is a 5. With all this information available, I might say that E confirms H for me. But there was never a time when learning E made me more confident of H.

Third, we need to address overdetermination cases. Sometimes an agent has multiple pieces of evidence that independently confirm the same hypothesis; we don't want one piece of evidence's supporting (or even entailing) H to obscure the other's confirmatory role. Suppose that after getting all the card information above, I then learn E', that my opponent also has a 9 in her hand. I might say that E' confirms H for me just as much as E does. Yet learning E' didn't increase my confidence in H, because by the time I learned E' I already had $cr(H) = 1$.

Counterfactual backtracking. Given these problems (and others) for the historical backtracking approach, some authors have suggested we look for a credence distribution on which E is positively relevant to H not in the agent's past, but in a close possible world. Recall the problem that Glymour's Einstein example created for the Confirmation and Learning principle. If an agent already knows E, then there's no way the agent can learn E, so it's difficult to read the right-hand side of that principle's biconditional as true. We can avoid this problem by reading the right-hand side as a counterfactual: *were* it possible for the agent to learn E, doing so *would* increase her confidence in H. In other words, we test for confirmation by asking, "If the agent didn't already know E, would learning it make her more confident in H?"

Howson and Urbach are prominent proponents of this counterfactual approach; as they see it, "to regard any evidence, once known, as confirming one has to go counterfactual" (2006, p. 299). How do we put the approach into practice? First, we need a subjunctive reading of Confirmation and Probability, something like: E confirms H just in case *were* the agent to assign $cr(E) < 1$, she would also assign $cr(H \mid E) > cr(H)$. Then we need to take the credence distribution that assigns $cr(E) = 1$ in the actual world, and determine what it would look like were that certainty removed. Howson and Urbach write, "the old-evidence 'problem' really is not a problem, merely an implicit reminder that if E is in K then it should first be deleted, as far as that can be done, before assessing its evidential weight" (2006, p. 299). Deleting E from the agent's background corpus K does not mean just taking E out and leaving the rest of K intact. We'll want to remove other propositions from K along with E, such as propositions logically equivalent to E or sets of propositions that jointly entail E. Luckily, this problem of "contracting" a consistent corpus to excise a particular proposition has been addressed by a formal theory commonly known as AGM. (Alchourrón, Gärdenfors, and Makinson 1985) Howson and Urbach apply a probabilistic descendant of AGM to work out a counterfactual credence distribution for the perihelion case (2006, pp. 300–1).[5]

Yet it's not clear that this type of counterfactual will always generate the confirmatory judgments we want. Patrick Maher (1996, p. 156) offers the example of an author who knows his work is popular (call this proposition P), and who judges that its popularity confirms that it's important (call this I). Maher suggests that if the author didn't know his work was popular, he might think popularity was no evidence of importance. So when we go to the counterfactual world where the author doesn't already know P, we find that his credence distribution doesn't render P positively relevant to I. The counterfactual version of Confirmation and Probability fails in this case.

In a related vein, Earman notes:

It is relevant that in 1907 Einstein wrote, "I am busy on a relativistic theory of the gravitational law with which I hope to account for the still unexplained secular change of the perihelion motion of Mercury. So far I have not managed to succeed." (Seelig 1956, p. 76) Thus it is not beyond the pale of plausibility that if Einstein hadn't known about the perihelion phenomenon, he wouldn't have formulated GTR. And if someone else had formulated the theory, Einstein might not have understood it well enough to assign it any degree of belief at all.   (1992, p. 123)

This is a particular example of a quite general scientific phenomenon, in which theories are formulated specifically to explain (or at least to accommodate) certain pieces of data. In such cases, the scientist's attitudes toward H may look quite unrecognizable when we go to worlds in which E isn't known. As often transpires when we suggest a counterfactual account of an actual-world condition,[6] the possible world realizing the antecedent of the subjunctive conditional may not yield exactly what we hoped for or expected.

Relativizing beyond credence. Confirmation and Probability relativizes E's confirmation of H to some probability distribution Pr. We have been trying to read that Pr as the credence distribution of some agent, either actual or counterfactual. Perhaps the Problem of Old Evidence requires us to back off from that commitment.

Objective Bayesians (in the normative sense) want facts about confirmation to be independent of how any individual subject sees the world. On their position, the question of whether the perihelion of Mercury confirms General Relativity isn't relative to any particular agent's credence distribution, or personal history. They will want that question answered relative to a probability distribution representing the unique rational epistemic standard—what Carnap (Section 6.2.2) called distribution $\mathbf{m}$.

But we can't determine whether E confirms H simply by asking whether $\mathbf{m}(H \mid E) > \mathbf{m}(H)$. The unique rational standard is represented by a regular distribution emptied of all empirical evidence. If we want to recover post-1915 Einstein's judgment that Mercury's perihelion confirmed GTR, we need to recognize that judgment's dependence on a vast corpus of physical background knowledge (about the masses of stars and planets, their locations and velocities, etc.). We should identify the appropriate corpus K, then ask whether $\mathbf{m}(H \mid E \, \& \, K) > \mathbf{m}(H \mid K)$.

Maher develops an objectivist view on which "E confirms H" states "that E confirms H relative to background evidence $[K]$, but $[K]$ is usually taken to be fixed to sufficient accuracy by the context and so is normally not explicitly indicated." (1996, p. 166) This proposal keeps rational epistemic standards fixed, but relativizes confirmation to some background context—whether it be the context against which Einstein was developing physics with his compatriots, the context in which *we* discuss whether Mercury's perihelion confirmed GTR, etc. Crucially, K need not be Einstein's actual knowledge base at any given time, and it will not include E itself.

But once we've introduced the move of relativizing confirmation to a background corpus distinct from any particular agent's body of total evidence, there's no reason Subjective Bayesians can't make this move as well. If we can make good sense of what I called an agent's "ultimate epistemic standards" in Section 4.3—her general tendencies for assessing evidence independent of the body of evidence she happens to possess—then we can replace the objectivist unique rational prior with a distribution Pr_H representing the agent's idiosyncratic standards. E would confirm H relative to a particular agent *and* a particular context just in case $\text{Pr}_H(H \mid E \,\&\, K) > \text{Pr}_H(H \mid K)$, where Pr_H is the hypothetical prior representing that agent's standards and K is the contextually salient background corpus.[7]

Or we could let context do more work than just providing a background corpus. For example, suppose a group of scientists is studying some thermodynamic effect in a sample of gas. Given the settings of various parameters (pressure, volume, etc.), statistical thermodynamics specifies a probability distribution over all the possible ways the molecules in that sample might be arranged. Even if individuals in the group have different personal credences about the sample's current status, they might agree to rely on this contextually salient distribution in determining the evidential significance of experimental results. That distribution would furnish a Pr against which confirmation of theories by evidence could be gauged.

The general idea is that even for a Subjective Bayesian, the probabilities used to establish confirmation relations need not come from any particular individual's credence distribution. Context may make salient a distribution appropriate for the job. Yet it's unclear how far we can take this approach. Even if we grant that context may specify—or at least constrain—probabilities against which confirmation can be measured, will these values include not just likelihoods but also priors? Perhaps a small group of scientists working on a common project will share enough assumptions about the relevant priors to allow for useful assessments of confirmation. But when we turn to Einstein's case, what contextual distribution assigned a meaningful prior probability to the General Theory of Relativity, such that we can assess whether facts about Mercury's perihelion would boost that probability?

12.1.4 More radical solutions

<u>Regularity and Jeffrey Conditionalization.</u> Up to this point, our discussion of old evidence has assumed that a rational agent moves through life by collecting evidence in the form of propositions, then conditionalizing on that evidence to raise it to certainty. Our basic problem has been that once this occurs for some E, the agent's certainty in E prevents it from being positively relevant to any particular H.

But we've already questioned this model of updating and evidence collection a number of times. In the historical case, Earman notes that "the literature of [Einstein's] period contained everything from 41 to 45 [seconds] of arc per century as the value of the anomalous advance of Mercury's perihelion, and even the weaker proposition that the true value lies somewhere in this range was challenged by some astronomers and physicists" (1992, p. 121). Einstein probably wasn't *certain* of the proposition E that Mercury's perihelion has a 43 arc-second anomalous advance, and probably shouldn't have been. In Section 4.2 we discussed the Regularity Principle, which bans sending contingent propositions to certainty. This principle is often paired with Jeffrey Conditionalization (Section 5.5) as an updating rule, which allows experience to impinge on an agent's opinions not by sending any proposition to certainty, but by changing her unconditional credence distribution across a partition.

The Regularity/Jeffrey Conditionalization regime looks promising as a solution to the Problem of Old Evidence. On this regime, the Confirmation and Probability test for confirmation will always be probative relative to a rational agent's credence distribution, since that distribution will not accumulate certainties over time. (Or at least, the test will be probative for contingent propositions. We might still have questions—both synchronic and diachronic—about how logical truths confirm other propositions, since Normality will require rational agents to assign logical truths credence 1.)[8]

Yet there are deep questions about how to understand confirmation if Conditionalization is set aside. Suppose Confirmation and Learning expresses your fundamental intuition about evidential support: E is evidence for H just in case learning E makes one more confident in H. On a Jeffrey Conditionalization approach, it's no longer clear what it means to "learn" E (since experience no longer raises contingent propositions' credences to 1). One might try to weaken Confirmation and Learning to something like:

Confirmation and Increase: Evidence E confirms hypothesis H for a rational agent if and only if any experience that increased her confidence in E would also increase her confidence in H.

But this principle doesn't match Confirmation and Probability. It's possible to create a situation in which every Jeffrey Conditionalization across a particular partition that increases confidence in E increases confidence in H, yet E is negatively relevant to H. There are also situations in which $\Pr(H \mid E) > \Pr(H)$, yet a Jeffrey Conditionalization that increases confidence in E decreases confidence in H. (See Exercise 12.2 for examples of each.) So the condition for confirmation specified in Confirmation and Increase is neither necessary nor sufficient for satisfying Confirmation and Probability's relevance condition for confirmation.

Thus on a Jeffrey Conditionalization approach, we may have to choose between equating confirmation with positive probabilistic relevance and tying it to an intuitive principle about learning. Here one might argue for the positive relevance approach by noting that Confirmation and Probability is well motivated by considerations independent of learning; as we showed in Chapter 6, probabilistic relevance satisfies many adequacy conditions for a proper explication of confirmation. So if we embrace Regularity, model rational updates using Jeffrey Conditionalization, and use Confirmation and Probability as our test for confirmation relative to an agent's credences, it looks like we'll get a plausible theory of confirmation that avoids old evidence concerns.

Or at least, we'll get a plausible theory so long as we stick to qualitative confirmation judgments. Things get trickier when we quantitatively measure *degrees* of confirmation. The fundamental concern behind the Problem of Old Evidence was that changing one's attitude toward E shouldn't change how one views E's support for H. Christensen (1999), therefore, suggests that on a proper solution to the Problem of Old Evidence, updates that change an agent's unconditional credence in E shouldn't change how strongly E confirms H for her. He demonstrates that if we measure the confirmation of H by E using the measure we called s in Section 6.4.1, then E's confirmation of H will remain unchanged through any Jeffrey Conditionalization that directly changes the agent's unconditional distribution across the $E/\sim E$ partition. (Exercise 12.3.)

Yet as we noted in Section 6.4.1, s fails to display other properties that seem to be adequacy conditions for a proper measure of evidential support. And Christensen himself offers intuitive counterexamples to measure s (1999, p. 456ff.) So it's unclear whether Regularity and Jeffrey Conditionalization can together provide a subjectivist account of confirmation that handles old evidence and is plausible both qualitatively *and* quantitatively.

Limitations of the theory. It may be time now to ask some hard questions about the purposes of Bayesian confirmation theory, and about that theory's limitations.

Let's say that an agent, a piece of evidence E, and a hypothesis H together pass the "learning test" just in case learning E would increase that agent's confidence in H. Confirmation and Learning says that passing the learning test is both necessary and sufficient for E to confirm H for the agent. The right-to-left component of this principle—according to which passing the learning test suffices for confirmation—serves an important Bayesian purpose. We want to know when it's rational for an agent to increase her confidence in a hypothesis upon gaining some evidence. Confirmation and Learning tells us that a confidence increase is rational only when the evidence confirms that hypothesis, which in turn (given Confirmation and Probability) occurs only when the evidence is positively probabilistically relevant. This gives the working scientist a criterion for when an experimental result ought to make her more confident of a particular theory.

But this purpose would be served just as well by a weakened version of the principle:

Confirmation and Learning (weakened): Evidence E confirms hypothesis H for a rational agent if learning E would increase the agent's confidence in H.

Here I've replaced the "if and only if" in the original principle with a mere "if". In other words, passing the learning test is now a sufficient but not a necessary condition for confirmation. Why should we insist on a necessary *and* sufficient condition for confirmation? Only if we're hoping that the Confirmation and Learning principle will serve a further purpose pursued by many early Bayesians: providing a philosophical *analysis* of the scientific notion of confirmation.

In response to the question, "How can we tell when evidence confirms a hypothesis for an agent?", Bayesians had an excellent initial thought. If learning that evidence would increase the agent's (rational) confidence in the hypothesis, that's a good sign of a confirmatory relationship. Yet Confirmation and Learning—in its strong, biconditional form—takes this helpful indicator and elevates it to a necessary and sufficient condition for confirmation. Perhaps this is an overreach; perhaps there are other ways evidence can confirm a hypothesis, other tests that reveal confirmation when the learning test fails to do so.

To acknowledge that possibility, let's work with the weakened version of the principle for a while. Among other things, the weakened principle allows us to distinguish true *counterexamples* to Bayesian confirmation theory from cases that merely reveal the *limitations* of that approach. A counterexample shows that a theory renders false verdicts; a limitation occurs when you'd like the theory to render a verdict but it won't. For Bayesian confirmation theory, a counterexample would be a case in which learning E would indeed increase a rational agent's confidence in H, yet E doesn't confirm H for that agent. If such a case existed, even the weakened Confirmation and Learning principle would be *false*. But the cases in our old evidence discussion work the other way around: intuitively, E confirms H for an agent, yet it's not possible for her to rationally increase her confidence in H by learning E. Such cases reveal the limits of what Confirmation and Learning can tell us about confirmation—the principle is going to help us detect some cases of confirmation, but not others.

Can we say more about which cases might be missed? It helps to draw a distinction between confirmation as a *process* and confirmation as a *standing relation*. Suppose I tell you I just confirmed a suspicion I've had for quite some time. That kind of confirmation is a process, occurring over a span of time in response to particular stimuli. (The "confirmation" of judicial nominees is a temporally extended process as well.) The diachronic problem of old evidence seems most concerned with this notion of confirmation: what was it over the course of 1915 that confirmed GTR for Einstein? And the learning test (perhaps augmented by one of the solutions to the diachronic problem discussed in Section 12.1.2 above) seems best paired with this notion of confirmation. Positive probabilistic relevance relative to a rational agent's credence distribution reveals which bits of evidence would confirm a hypothesis for the agent in the process sense.

But many old evidence cases consider confirmation as a standing relation; they take an agent's confidence in a hypothesis at a particular moment and ask what the agent takes to support the hypothesis *at that time*—what the agent thinks justifies her current confidence. This standing justificatory relation is a perfectly good sense of the term "confirmation", and is what the synchronic problem of old evidence concerns. Yet it may not be the notion that Bayesian confirmation theory was designed to detect. Honestly, we should've already been worried when we noticed that positive probabilistic relevance is symmetric. Many epistemologists take justification to be an asymmetric relation. (The proposition that it's raining right now is justified for me by the proposition that I can see raindrops falling outside my window—not vice versa.) If so, it can't be fully captured by positive relevance.[9] Williamson (2000, p. 187) also points out that whenever a probabilistic agent assigns $0 < \mathrm{cr}(E) < 1$, she will also assign $\mathrm{Pr}(E \mid E) = 1$; so every uncertain proposition is positively relevant to itself. Yet standing confirmation isn't generally reflexive; a rational agent takes very few propositions to be evidence for themselves.

Now the fact that learning one proposition would increase the agent's confidence in a second proposition—or the fact that learning the first increased her confidence in the second at some point in the past—might be a good *indicator* of the presence of some standing justificatory relation. But even if one of the strategies from Section 12.1.2 takes care of Bayesian confirmation theory's diachronic problem, the synchronic problem of old evidence reveals the learning test's limitations as an indicator of standing confirmation.

Proponents of a new theory often burnish its credentials by showing off all of the problems they think it can solve, accompanied by maximalist claims that it will succeed in every case. Once the theory is accepted, we develop a more mature understanding of its capabilities and limitations. Perhaps Bayesian confirmation theory simply isn't the best approach for detecting and analyzing standing support relations. Or perhaps it has a place as one component of a broader, unified approach that analyzes confirmation as both a relation and a process.[10]

12.2 Logical omniscience

Savage (1967) asks us to consider the plight of "a person required to risk money on a remote digit of π" (p. 308). For example, suppose I offer to sell Andre the following ticket:

> This ticket entitles the bearer
> to $1 if the trillionth digit of π is
> a 2, and nothing otherwise.

If he's rational, what is Andre's fair price for that ticket—how much would he be willing to pay to possess it? From our discussions of decision theory (Chapter 7) and Dutch Books (Chapter 9), we conclude that if Andre's rational, his price for this ticket will equal his credence in the proposition that the trillionth digit of π is a 2.

But I happen to know (thanks to Talbott 2005) that the trillionth digit of π *is* a 2. And since mathematical truths are true in every possible world, the Normality axiom requires rational agents to have a credence of 1 in that proposition. So if Andre's rational, he should be willing to pay any amount up to $1 for the ticket above.

Does that seem reasonable? Perhaps Andre doesn't know what I know about the trillionth digit of π. In that case, is it really rational for him to pay, say, $0.90 for the ticket? (And on the flip side, would it be *irrational* for him to purchase a ticket that paid off on any digit other than 2?) Perhaps, while Andre doesn't know the true identity of the trillionth digit, he does know about conjectures that each of the digits 0 through 9 occurs equally often within the decimal expansion of π. In that case, it might seem rational for him to set a fair betting price of $0.10.

But the Normality axiom doesn't care what Andre knows or doesn't know. It requires him to be certain of all tautologies, no matter his evidence. This requirement—known as the **logical omniscience requirement**—has struck many authors as unreasonable, especially because there's no Bayesian requirement that a rational agent be certain of *all* truths. As Daniel Garber memorably put it:

> Asymmetry in the treatment of logical and empirical knowledge is, on the face of it, absurd. It should be no more irrational to fail to know the least prime number greater than one million than it is to fail to know the number of volumes in the Library of Congress.   (1983, p. 105)

Does rationality really require certainty in all logical truths? Can we build a Bayesian formalism that loosens this requirement? We will first disambiguate this general concern into distinct, specific problems, then consider how one might respond to each.[11]

12.2.1 Clutter avoidance and partial distributions

One problem with Normality as we've presented it is that in any propositional language $\mathcal{L}$ closed under the traditional connectives, there will be infinitely many tautologies. Normality demands that the rational agent assign each and every one of those tautologies a credence of 1; thus it requires the rational agent to possess infinitely many attitudes. But that may be impossible for finite beings such as ourselves. And even if possible, ferreting out all of the tautologies and assigning them certainty may be a waste of our cognitive capacity, energy, and attention. Gilbert Harman (1986, p. 12) famously defended the following principle:

Clutter Avoidance: One should not clutter one's mind with trivialities.

But if cognitive clutter is the problem, Normality isn't the sole culprit. As we've stated it, Non-Negativity requires assigning a non-negative credence to *every* proposition in $\mathcal{L}$; Finite Additivity requires particular values for each disjunction; and the Ratio Formula requires a value for each conditional credence with a nonzero condition. Harman also complained about updating by Conditionalization:

> One can use conditionalization to get a new probability for P only if one has already assigned a prior probability not only to E but to P & E. If one is to be prepared for various possible conditionalizations, then for every proposition P one wants to update, one must already have assigned probabilities to various conjunctions of P together with one or more of the possible evidence propositions and/or their denials .... To be prepared for coming to accept or reject any of ten evidence propositions, one would have to record probabilities of over a thousand such conjunctions for each proposition one is interested in updating. To be prepared for twenty evidence propositions, one must record a million probabilities. For thirty evidence propositions, a billion probabilities are needed, and so forth.   (1986, p. 26)

To me, Harman's Clutter Avoidance complaint indicates that we haven't been careful enough in formulating our Bayesian principles. The damage can be repaired in three steps.

First, we should acknowledge that nothing in the traditional arguments for Bayesianism requires that credences be distributed over a *full* propositional language. One cannot be Dutch Booked, or accuracy-dominated, by virtue of failing to adopt an attitude toward a particular proposition. (If I don't assign a credence to a proposition, then I will have no fair betting price for tickets involving that proposition, and I simply won't buy or sell any bets a bookie offers me on it.) Dutch Book and accuracy arguments show only that if one *does* assign credences over particular sorts of propositions, and those credences fail to meet certain conditions, then one's distribution will have various undesirable features. Thus these arguments do nothing to favor a complete distribution—a distribution that assigns real numbers to every proposition in a language $\mathcal{L}$ closed under connectives—over a **partial distribution**—a distribution that assigns reals to only some of the propositions in such a language. Meanwhile, if one thinks that the best arguments for Bayesianism come from its applications, then while a partial credence distribution may yield fewer applications to particular cases (fewer confirmational judgments, fewer determinate decisions, etc.), it will not break either Bayesian confirmation or decision theory.

Second, we should understand our Bayesian rules as evaluative norms of rational consistency, not as prescriptive recipes for reasoning. It's easy to read Finite Additivity as demanding that your credence in a disjunction be *calculated from* your credences in its mutually exclusive disjuncts, and to read Conditionalization as saying that you must *determine* your unconditional credences at a given time from conditional credences you assigned earlier on. But the rules don't have to be read that way. Finite Additivity might simply say that if you find an agent who assigns credences to P, Q, and $P \lor Q$ (with P and Q mutually exclusive), then that agent's distribution is rationally flawed if the third credence isn't the sum of the first two. Similarly, we might find an agent who assigns a credence to P at t_j, but assigned no credence at t_i to that P conditional on the E she learned between those two times—perhaps P is a proposition she never considered until t_j. Harman is right that "one can use conditionalization to get a new probability for P only if one has already assigned a prior probability not only to P but to $P \& E$." But perhaps we shouldn't think of Conditionalization as a tool to be *used* for crafting new credences from old; instead, we can think of it as a check on diachronic attitudinal consistency. *If* an agent assigns both $cr_i(P \mid E)$ and $cr_j(P)$, *then* those two credences should be equal. But Conditionalization confers no rational demerits if the agent simply fails to assign one of those values.[12]

Third, however, putting these two points into practice requires a bit of technical tweaking. Here's why: We want to say that it's rational, for instance, for an agent to assign credences only to the contingent propositions P and $\sim P$, and to nothing else. But we also want to rationally criticize that agent if she assigns $cr(P) = cr(\sim P) = 0.7$. Strictly speaking, an agent who assigns only those two unconditional credences hasn't violated the Bayesian rules for partial distributions we've just been considering. She satisfies Non-Negativity; every proposition to which she assigns an unconditional credence receives a non-negative value. She trivially satisfies Normality, because she doesn't adopt an attitude toward any tautology, and we are now reading that rule as saying only, "*If* you assign a credence to a tautology, its value must be 1". Similarly, she trivially satisfies Finite Additivity, by not adopting attitudes toward any disjunctions. Yet intuitively, this agent's distribution conflicts with the probability axioms!

Bayesians who allow rational partial credence distributions have traditionally addressed this problem as follows: We require that an agent's partial credence distribution be **extendable** to a complete distribution that satisfies the probability axioms. That is, it must be possible to construct a complete probabilistic distribution that assigns the same value as the agent does to every proposition about which she has an opinion. This is clearly impossible for the partial distribution assigning $cr(P) = cr(\sim P) = 0.7$. Suppose we had a complete distribution that included those values. By virtue of its completeness, it would also assign some real number to $P \vee \sim P$. If it assigned this tautology a credence of 1, the distribution would violate Finite Additivity, because the value assigned to the disjunction $P \vee \sim P$ wouldn't be the sum of the values assigned to its mutually exclusive disjuncts. But if the distribution satisfied Finite Additivity by assigning that tautologous disjunction a value of 1.4, then it would violate Normality. Thus the partial distribution isn't extendable to a complete probabilistic distribution, which reveals its irrationality.

A similar move addresses Harman's clutter complaint about Conditionalization. We simply require that an agent's t_i and t_j credence distributions be *extendable* to complete distributions that satisfy the probability axioms (and Ratio Formula) and relate to each other as Conditionalization requires. If the agent assigns a $cr_j(P)$ value but no $cr_i(P \mid E)$ value, that's fine, as long as there is some way of completing her cr_i distribution to yield $cr_i(P \mid E) = cr_j(P)$ without violating any of our other Bayesian rules.

12.2.2 Logical confirmation and logical learning

Yet even once we move to partial credence distributions, Normality still gives us problems. For one thing, an agent will often view one logical (or mathematical)[13] truth as supported by another. Suppose I am certain of the proposition S that 692 is an even number, certain of the proposition D that every integer ending in an even digit is even, and certain of the former *because* I'm certain of the latter. It would be nice for our theory of confirmation to capture the sort of support that D lends to S. But a positive relevance account will indicate no confirmation here, because $cr(S \mid D) = 1 = cr(S)$. The only way to get relevance confirmation of S by D would be if S's prior were less than 1. But if I assign $cr(S) < 1$, I will violate Normality (even the version of Normality that allows for partial distributions).

Perhaps we've already indicated the correct response to this problem. In Section 12.1.4 I suggested that the Bayesian probabilistic relevance account of confirmation may capture confirmation only as a *process*, not as a *standing relation*. If my certainty in S is justified by my certainty in D, that's a standing relation among my doxastic attitudes. Bayesian confirmation theory may simply not be up to tracking such relations.

Fine then—let's think about the *process* of logical confirmation. Perhaps before you read this chapter, you had never considered whether the trillionth digit of π was a 2. When I asked you to consider the betting ticket on page 428, you then assigned a credence to that proposition. I'm guessing it was a nonextreme credence—perhaps 0.1. About a paragraph later, I relayed Talbott's report that the trillionth digit of π is indeed a 2. Presumably this testimony increased your confidence in the proposition (perhaps all the way up to 1).

That confidence increase seems a very rational response to the evidence with which you were presented. Yet a Bayesian model of rationality cannot recognize it as such. If our Bayesian model allows for partial distributions, it will be all right with your initially adopting no attitude towards the π proposition. But the moment you adopt any degree of belief in that proposition with a value less than 1—the moment you assign any fair price to the π betting ticket other than \$1—your credences violate the Normality axiom.

This is an issue for at least two reasons. First, it seems like you do a rationally good thing when you increase your confidence in the π proposition upon learning of Talbott's testimony. But a Bayesian epistemology based on probabilism will negatively evaluate your credal sequence for violating Normality. Second, let's compare you with someone who *decreased* her confidence in the π proposition upon learning of Talbott's testimony. Unless she possesses some evidence you don't (say, that Talbott is a pernicious liar about logical truths), her reponse to what she's learned is irrational while yours is not. Yet since the Bayesian system evaluates you each negatively the moment you assign anything other than certainty to that proposition, there's no way for it to discriminate between the two of you on rational grounds.

It seems that to get an adequate model of rational logical learning, we need a Bayesianism that allows for deductive non-omniscience—not just the failure to adopt any attitude toward a logical truth, but the ability to rationally assign less-than-certainty to such truths. We'll now investigate whether such a Bayesianism might be available.

12.2.3 Allowing logical uncertainty

To sum up the discussion thus far: We can modify Normality to allow for partial credence distributions. Instead of requiring an agent to assign certainty to all logical truths, it can simply ban assigning less-than-certainty to any logical truths. But that ban still stands in the way of our modeling logical learning (and logical confirmation, should we wish to do so). Moreover, the ban seems unreasonable on its face—intuitively, when a logical or mathematical truth is highly complex and/or distant from an agent's cognitive reach, we shouldn't fault her for failing to be certain of it. This seems to me the central problem of logical omniscience: We want a Bayesian theory that permits agents to rationally assign credences less than 1 to logical truths.

How might we do that? One approach begins by re-expressing Kolmogorov's axioms in a slightly different way:

Non-Negativity: For any proposition P in $\mathcal{L}$, $cr(P) \geqslant 0$.

Normality: For proposition T in $\mathcal{L}$, if $\vDash \mathsf{T}$ then $cr(\mathsf{T}) = 1$.

Finite Additivity: For any propositions P and Q in $\mathcal{L}$, if $\vDash \sim(P \mathbin{\&} Q)$ then $cr(P \vee Q) = cr(P) + cr(Q)$.

In Section 2.1.1, we defined "$\vDash X$" to mean that X is true in every possible world. We also said that P and Q are mutually exclusive just in case there's no possible world in which both of them are true. Well, if no possible world makes $P \mathbin{\&} Q$ true, then every possible world makes $\sim(P \mathbin{\&} Q)$ true. So given our definition of "$\vDash$", all I've done here is rewrite the standard Kolmogorov axioms from Section 2.2 using different notation.

But what if we gave "$\vDash$" a different interpretation? For example, we might stipulate that "$\vDash X$" means X is a truth-functional truth. (A **truth-functional truth** is a proposition true by virtue of its truth-functional form—by the arrangement of truth-functional connectives within it. $\sim(P \mathbin{\&} \sim P)$ is a truth-functional truth, while $\sim[(\forall x)Fx \mathbin{\&} \sim Fa]$ is not. The latter's status as a logical truth depends not just on the truth-functional connectives but also on the quantifier within it.) We could then define **Truth-Functional Bayesianism** as a theory that requires rational credence distributions to be extendable to distributions satisfying the axioms above, with "$\vDash$" interpreted to indicate only truth-functional truths.

Truth-Functional Bayesianism isn't very plausible as a normative theory, for reasons I'll get to in a moment. But it's a useful toy theory for thinking about versions of Bayesianism that don't require logical omniscience. Given how Truth-Functional Bayesianism re-interprets "$\vDash$", we now have $\vDash \sim(P \mathbin{\&} \sim P)$, but $\nvDash \sim[(\forall x)Fx \mathbin{\&} \sim Fa]$. So the Normality axiom will fault an agent who assigns less-than-certainty to $\sim(P \mathbin{\&} \sim P)$, but not an agent who is uncertain of $\sim[(\forall x)Fx \mathbin{\&} \sim Fa]$. (And it certainly won't fault an agent who isn't certain of the trillionth digit of π!) Moreover, Finite Additivity will still treat P and $\sim P$ as mutually exclusive, so it will require their credences to add up to 1. But since $\nvDash \sim[(\forall x)Fx \ \& \ \sim Fa]$, Finite Additivity will not require $cr((\forall x)Fx \lor \sim Fa) = cr((\forall x)Fx) + cr(\sim Fa)$.[14]

Truth-Functional Bayesianism is implausible because it draws the line between what an agent's rationally required to be certain of and what she's not required to be certain of in a fairly artificial place—at the boundary between truth-functional and other forms of logic. This is problematic for two reasons. First, we might find Truth-Functional Bayesianism too demanding. There are some lengthy, intricate truths of truth-functional logic that it might be unreasonable to demand agents recognize as such. Yet Truth-Functional Bayesianism will fault an agent who is less than certain of even such recherché truth-functional truths. Second, Truth-Functional Bayesianism may not be demanding *enough*. Some logical relations—such as the mutual exclusivity of $(\forall x)Fx$ and $\sim Fa$—while not truth-functional, nevertheless seem simple and obvious enough to rationally fault an agent who doesn't recognize them. Truth-Functional Bayesianism is capable of modeling an agent who recognizes non-truth-functional logical relations; the theory *allows* an agent to assign $cr(\sim[(\forall x)Fx \ \& \ \sim Fa]) = 1$. But it doesn't *require* this certainty, and therefore won't fault a distribution that assigns that proposition a credence less than 1.[15]

Borrowing some helpful terminology from Skipper and Bjerring (2022), we want a Bayesianism that requires agents to be *logically competent* without requiring that they be *logically omniscient*. Truth-Functional Bayesianism misses the mark by ham-handedly offering the boundary between truth-functional and other logics as the dividing line between what an agent is and isn't required to know. But we could reinterpret "$\vDash$" in another, subtler way, so that our Bayesianism faults agents who miss obvious logical truths without requiring them to catch *all* logical truths. A number of philosophers have tried various interpretations—for example, Skipper and Bjerring draw the line by conceiving obviousness in terms of how many proof steps it would take to derive a particular logical conclusion. (For examples of other approaches, see Cherniak 1986 and Gaifman 2004.) To my mind, none of these formal approaches quite succeeds in drawing the line in a plausible way. But there's also a deeper problem that inevitably affects any such approach.

It's easy to think of logical omniscience as simply a problem about agents' attitudes toward logical truths. But Normality wasn't included among Kolmogorov's axioms to make trouble; it's used to derive a number of crucial Bayesian results, from Bayes's Theorem to the theory's applications in confirmation and decision theory.[16] Moreover, logical omniscience isn't idiosyncratic to Bayesianism: requirements of logical omniscience can be derived in other formal epistemologies, such as AGM and ranking theory; they can also be derived from such simple intuitive principles as Belief Closure or Comparative Entailment (Chapter 1). Logical omniscience requirements arise in all these systems because logical omniscience is deeply tied to the *other* things such epistemologies try to achieve. The fruitful applications that make these approaches appealing have logical omniscience as a side effect.

To see what I mean, consider what happens when Truth-Functional Bayesianism reinterprets "⊨" so that $\nvDash \sim[(\forall x)Fx \& \sim Fa]$. Because the reinterpretation affects Normality, Truth-Functional Bayesianism permits a rational agent to be less than certain of $\sim[(\forall x)Fx \& \sim Fa]$. But this result—along with the reinterpretation's effects on Finite Additivity—has knock-on consequences. For instance, Truth-Functional Bayesianism will permit a rational agent to be more confident of $(\forall x)Fx$ than she is of Fa. Usually this is forbidden because the former logically entails the latter. But since that entailment isn't truth-functional, it isn't recognized by Truth-Functional Bayesianism. When it comes time to make decisions, Truth-Functional Bayesianism will allow an agent to Dutch Book herself by paying $0.70 for a ticket that pays $1 if $(\forall x)Fx$, while also selling for $0.60 a ticket that pays out $1 if Fa. And perhaps worst of all, when it comes to confirmation theory, Truth-Functional Bayesianism will not require evidence that $\sim Fa$ to disconfirm the hypothesis that $(\forall x)Fx$ (see Exercise 12.4).

These examples feel particularly egregious because the entailment of Fa by $(\forall x)Fx$ is so obvious. But the general point stands, even if we go with one of the theories mentioned above that strikes the balance between competence and omniscience more plausibly. Logical truths are explicit manifestations of logical relations; when you allow an agent to be ignorant of some logical truth, you also allow her reasoning to run roughshod over those relations. When we weaken the traditional Bayesian axioms to allow some degree of logical non-omniscience, we lose some of the applications to decision and induction that made Bayesianism appealing in the first place.

12.2.4 Logical omniscience reconsidered

In his (1986), Harman maintained that "there is no clearly significant way in which logic is specially relevant to reasoning" (p. 20). But when an agent draws a conclusion through a logically invalid inference, isn't that a mistake of reasoning—a kind of rational error? Opposing Harman, Christensen writes, "There is a clear intuitive basis in our ordinary conception of rationality for distinguishing logical lapses from ordinary cases of factual ignorance.... Much of the point, after all, of thinking about rationality is to understand the idea of reasoning well" (2004, p. 155).

One way to recognize this connection between logic and good reasoning—and the resulting asymmetry between logical and empirical truths—is to defend logical omniscience as a requirement on ideally rational agents. Ideally rational agents (whom we already mentioned in Section 11.1.1) never make any mistakes in their reasoning. Given the connections we've just considered, this also means they are never less than certain of logical truths. On this approach, the Kolmogorov axioms (including Normality!) hold for the credences of ideally rational agents. But no actual agents are ideally rational, so less demanding standards apply to us, especially when it comes to logical omniscience.

This notion that the probability axioms apply to ideal agents rather than actual agents has been very popular in the Bayesian literature. It's been discussed a great deal, and I've criticized it at length elsewhere.[17] Briefly, as a philosophical move it strikes me as a cop-out. Ultimately we are interested in rational requirements on ordinary agents' attitudes, and what it is for a real person to go right or go wrong with respect to rationality. For agents like us, failures to recognize logical truths go hand in hand with failures to make good inferences. And such inference failures are genuine rational mistakes. So we should recognize them as such.

For a point of comparison, think back to the Conjunction Fallacy (Section 2.2.4) and the Base Rate Fallacy (Section 4.1.2). When confronted with the story of Linda the bank teller, or some data about the frequency of a disease and the reliability of a test for it, ordinary agents often make poor probabilistic inferences. Now suppose it emerged that those agents had false beliefs about probability theory, or were ignorant of the relevant axioms and theorems. Would this change your mind about whether the inferences in question were rational? Would it somehow make those inferences correct? Ignorance of the rules of correct reasoning doesn't let poor reasoning off the hook, even for agents who are non-ideal. Similarly, ignorance of logic doesn't make the poor logical or probabilistic reasoning that results any less irrational.

Thus authors such as (Smithies 2015) and (Titelbaum 2015b) have recently suggested that rationality may forbid less-than-certainty in logical truths even for ordinary agents. How does this position overcome our intuitions that agents who fail to grasp complex logical truths aren't being irrational? One crucial response is that recognizing a rational flaw in an agent's doxastic state doesn't require us to cast aspersions on the agent—much less call *her* "irrational". de Finetti put the point as follows:

> To speak of coherent or incoherent (consistent or inconsistent) individuals has been interpreted as a criticism of people who do not accept a specific behavior rule.... It is better to speak of coherence (consistency) of probability evaluations rather than of individuals, not only to avoid this charge, but because the notion belongs strictly to the evaluations and only indirectly to the individuals. Of course, an individual may make mistakes sometimes, often without meriting contempt.   (1937/1964, p. 103)

When a credence distribution violates Normality, the flaw is in that distribution. Depending on how it arose, we need not criticize or blame the agent who possesses the distribution. I think this alleviates some of the intuitive pushback against logical omniscience requirements, while recognizing that there is a genuine problem embedded in the way this agent sees the world. It also squares with how we hope the agent will assess her own former views, once the difficulty is pointed out and corrected. As Jeffrey wrote, "The point is merely that whether he knows it or not, his beliefs suffer from a logical failing.... Presumably the fellow, being rational, will not want decision theory to be so permissive as to neglect to classify the situation we have been envisaging as a fault, on the ground that it was not his fault" (1970, p. 165).

So perhaps logical omniscience requirements are appropriate components of a theory of rationality. Even if that's so, we still might want a tool for modeling and assessing episodes of logical learning. Well, here's a brief suggestion for how that might be done: Suppose we grant that any credence distribution that assigns less-than-certainty to a logical truth is thereby rationally flawed. We might nevertheless think some rational errors are worse than others. In fact, various authors have developed formalisms for measuring just how far a given nonprobabilistic distribution is from rational ideality.[18] We might think, then, that logical learning involves moving from a less rational state to a more rational one. An agent who learns a logical truth isn't doing something rationally good in the sense that both her prior and posterior distributions satisfy all rational requirements. Instead, she's doing something rationally good by getting herself (measurably) closer to a rational ideal.

12.3 Exercises

Unless otherwise noted, you should assume when completing these exercises that the credence distributions under discussion satisfy the probability axioms and Ratio Formula. You may also assume that whenever a conditional credence expression occurs or a proposition is conditionalized upon, the needed proposition has nonzero unconditional credence so that conditional credences are well-defined.

Problem 12.1. 🥢 Prove that if $cr(E) = 1$, then for any $H \in \mathcal{L}$, $cr(H \mid E) = cr(H)$.

Problem 12.2. 🥢🥢 Consider the following probabilistic credence distribution:

B	E	H	cr_1
T	T	T	0.015
T	T	F	0.135
T	F	T	0.045
T	F	F	0.005
F	T	T	0.36
F	T	F	0.04
F	F	T	0.04
F	F	F	0.36

(a) Show that on this distribution, B is positively relevant to E, E is positively relevant to H, but B is negatively relevant to H.[19]

(b) Suppose that between t_1 and t_2, this agent has an experience whose credal effects originate in the $B/\sim B$ partition. In particular, it sets $cr_2(B) = 0.4$. Use Jeffrey Conditionalization to create a probability table for the agent's full cr_2 distribution. (Hint: You may want to re-read page 159 for instructions on how to do this.)

(c) Show that while the agent's confidence in E increased between t_1 and t_2, her confidence in H decreased.

(d) In this example, the condition for confirmation specified by the Confirmation and Probability principle is satisfied, while the condition for confirmation given by Confirmation and Increase is not. (E is positively relevant to H, yet not every Jeffrey Conditionalization that increases confidence in E also increases confidence in H.) Show that the converse is also possible: Create a distribution on which every Jeffrey

Conditionalization on the $B/\sim B$ partition that increases confidence in E also increases confidence in H, yet on which E is negatively relevant to H.

Problem 12.3. 𝄞𝄞
(a) Show that given a regular prior, any Jeffrey Conditionalization update that originates in the $E/\sim E$ partition will keep the value of $s(H, E)$ constant. (<u>Hint</u>: Recall that Jeffrey updates always maintain Rigidity.)
(b) Do any of the other confirmation measures (d, r, l, z) defined in Section 6.4.1 display this property? Explain/prove your answer.

Problem 12.4. 𝄞 In this exercise we will assume Truth-Functional Bayesianism, and show why on that theory $\sim Fa$ may fail to disconfirm $(\forall x)Fx$.
(a) Construct a probability table using the following partition:

$$\{(\forall x)Fx\ \&\ Fa, (\forall x)Fx\ \&\ \sim Fa, \sim(\forall x)Fx\ \&\ Fa, \sim(\forall x)Fx\ \&\ \sim Fa\}$$

For the time being don't fill in any numerical values in the right-most column.
(b) Now fill out the numerical values in the right-most column so that it's *not* the case that $cr((\forall x)Fx\,|\sim Fa) < cr((\forall x)Fx)$. Keep in mind that while your numerical values still must be non-negative and sum to 1, in Truth-Functional Bayesianism the proposition $(\forall x)Fx\ \&\ \sim Fa$ may receive a nonzero value.

Problem 12.5. ✐ Does it seem to you that when an agent assigns less than certainty to a logical truth, this attitude is rationally mistaken? Provide arguments for your position.

12.4 Further reading

INTRODUCTIONS AND OVERVIEWS

John Earman (1992). *Bayes or Bust? A Critical Examination of Bayesian Confirmation Theory.* Cambridge, MA: The MIT Press

Chapter 5 distinguishes versions of the Problem of Old Evidence, then critically assesses solutions available, including formal models of logical learning.

CLASSIC TEXTS

Clark Glymour (1980). *Theory and Evidence*. Princeton, NJ: Princeton University Press

Contains "Why I Am Not a Bayesian", the essay that brought the Problem of Old Evidence to Bayesians' attention, as well as the canonical Einstein GTR example.

Leonard J. Savage (1967). Difficulties in the Theory of Personal Probability. *Philosophy of Science* 34, pp. 305–10
Ian Hacking (1967). Slightly More Realistic Personal Probability. *Philosophy of Science* 34, pp. 311–25

Classic exchange in which Savage raises the Problem of Logical Omniscience (among others), and Hacking responds with a first attempt at a solution.

Daniel Garber (1983). Old Evidence and Logical Omniscience in Bayesian Confirmation Theory. In: *Testing Scientific Theories*. Ed. by John Earman. Vol. 10. Minnesota Studies in the Philosophy of Science. Minneapolis: University of Minnesota Press, pp. 99–132
Ellery Eells (1985). Problems of Old Evidence. *Pacific Philosophical Quarterly* 66, pp. 283–302

Two attempts to address the Problem of Old Evidence by offering formal models allowing for logical non-omniscience.

EXTENDED DISCUSSION

Julia Staffel (2019). *Unsettled Thoughts: Reasoning and Uncertainty in Epistemology*. Oxford: Oxford University Press

Book-length examination of assessing the degree to which agents depart from fulfilling Bayesian norms.

Notes

1. For the non-astronomers in the crowd: Roughly speaking, classical Newtonian gravitational theory predicts that each planet will orbit in an ellipse, with the sun at one focus of that ellipse. Yet Mercury's orbit is not a perfect ellipse: its perihelion (the point at which it's closest to the sun) gradually rotates around the sun at a rate of 574 seconds of arc per century (where a second of arc is 1/3,600 of a degree). The classical theory can explain much of this precession—531 arc-seconds of it—by taking into account the gravitational influences on Mercury of the other planets. But the remaining 43 seconds of arc in the advance in Mercury's perihelion is anomalous, and could not be plausibly explained by any mechanism available in Newton's framework.

2. Olav Vassend points out to me that frequentists and likelihoodists (Chapter 13) also suffer from a version of the Problem of Old Evidence. If $\Pr(E) = 1$ then $\Pr(E \mid H) = 1$ for any $H \in \mathcal{L}$, so E will have the same likelihood on every available hypothesis.

3. In the mid-nineteenth century, Urbain le Verrier proposed the existence of a small planet, which he called Vulcan, lying between Mercury and the Sun that would explain the former's orbital pattern. This wasn't such a stretch as it might now seem, since le Verrier had already helped astronomers discover Neptune by analyzing anomalies in the orbit of Uranus.

4. D.H. Mellor notes that this challenge to Conditionalization dates back at least as far as C.S. Peirce: "Peirce raises other problems [for conditionalization].... Our conceptual limitations ... limit the outputs as well as the inputs of conditionalization: without a concept of money, for example, we could neither see nor infer anything about the prices of goods. So however we get *new* concepts, such as new concepts of God induced by seeing an apparently miraculous event, it cannot be by conditionalization" (Mellor 2013, p. 549, emphasis in original).

5. As I mentioned in Chapter 4, note 18, authors such as Bartha and Hitchcock (1999, p. S349) sometimes refer to the distribution an agent would have assigned under counterfactual conditions as her "hypothetical prior". This is *not* the way I use "hypothetical prior" in this book. On my usage, a hypothetical prior is a probability distribution representing an agent's abstract evidential assessment tendencies, not a credence distribution she assigns in this world or any other.

6. See the philosophical literature on "finks" and "masks", and Shope (1978) on this general problem for counterfactual accounts.

7. Proposals in this vicinity are discussed by Skyrms (1983), Eells and Fitelson (2000), and Meacham (2016, §3.7).

8. See Section 12.2.2 for an example and further discussion.

9. Compare Christensen (1999, pp. 437–8).

10. I've often wondered, for instance, if agents' standing support relations could be usefully modeled by Bayes Nets (Section 3.2.4), which use probabilistic correlations in more sophisticated ways to establish asymmetric, irreflexive relations.

11. As usual, this issue with Bayesianism was anticipated by Ramsey: "There are mathematical propositions whose truth or falsity cannot as yet be decided. Yet it may humanly speaking be right to entertain a certain degree of belief in them on inductive or other grounds: a logic which proposes to justify such a degree of belief must be prepared

actually to go against formal logic; for to a formal truth formal logic can only assign a belief of degree 1" (1931, p. 93).

12. In the terminology of Broome (1999), we might say that this paragraph recommends thinking of Bayesian rules as wide-scope rather than narrow-scope norms. We need to be a bit careful, though, because it's difficult to see how one could think of either Normality or Non-Negativity—considered in isolation—as a wide-scope norm. I prefer to think of Kolmogorov's axioms as working *together* to create the notion of a probabilistic distribution, and then the relevant norm as saying that rational distributions must be probabilistic. (After all, the notion of a probabilistic distribution may be axiomatized in many different extensionally equivalent ways.) It's the norm of *probabilism* that should be interpreted as wide-scope—once suitably adjusted according to the third point I'll make below. (For more on all this, see Titelbaum (2013a, §4.2).)

13. Throughout this book I have treated mathematical truths as a subspecies of logical truths. Or, to put it another way, I have assumed that mathematical truths are true in all logically possible worlds. While it often goes unmentioned, this is a typical working assumption among Bayesians, and underlies many of their most routine calculations. For example, when calculating a rational agent's credences involving the outcome of a die roll, I assume that her credence in even conditional on six is 1. Strictly speaking, this works only if she is certain that six is an even number, a simple mathematical truth we treat as covered by Normality's requirement of certainty in logical truths.

   Whether all mathematical truths are indeed logical truths is a matter of philosophical controversy. (See Tennant 2017.) One might also wonder whether other types of non-empirical truths—such as conceptual truths—should be covered by Normality, depending on how one views their relationship to logical truths. While these are interesting questions, I will continue to set them aside for our purposes.

14. In Section 2.1 we associated each proposition with a set of possible worlds, and said that $P \vDash Q$ just in case there is no possible world in which P is true but Q is not. When we reinterpret "$\vDash$", how do we understand that relation in terms of possible worlds?

   In Section 2.2.3 we reinterpreted "$\vDash$" slightly by relativizing the probability axioms to an agent's set of *doxastically* possible worlds, which we thought of as a proper *subset* of the logically possible worlds. Hintikka (1975) proposed accommodating logical nonomniscience by relativizing "$\vDash$" to a *superset* of the logically possible worlds. An agent uncertain of $\sim[(\forall x)Fx \mathbin{\&} \sim Fa]$ could be modeled as entertaining a possible world in which both $(\forall x)Fx$ and $\sim Fa$ are true. This would be what Hintikka called an "impossible possible world". Logically nonomniscient agents entertain both logically possible and logically impossible worlds among their doxastic possibilities; $P \vDash Q$ for such an agent just in case that agent entertains no worlds in which P is true and Q is false.

   The impossible possible worlds approach remains controversial. Max Cresswell has been urging for decades that it's difficult to understand just what an impossible world could be. What is a world like in which every object bears property F but object a doesn't? As Gaifman (2004, p. 98) puts it, "It is possible to have false arithmetical beliefs, since we may fail to comprehend a sufficiently long sequence of deductive steps; but the false belief cannot be realized in some possible world, for such a possible world would be conceptually incoherent."

15. Truth-Functional Bayesianism is a simplification of a system proposed by Garber (1983). Eells (1985, p. 241) argues that Garber's system is too demanding in the sense described by this paragraph. Hacking (1967) suggested reinterpreting "⊨" to apply to only those logical truths that the agent recognizes as such. Eells (p. 217) criticizes that proposal for not being demanding enough.

16. If you've been completing this book's exercises as you go along, ask yourself how many times you've invoked Normality in proving some result, or have invoked a result that depends on Normality in proving some further result.

17. See Titelbaum (2013a, pp. 72–5).

18. See Staffel (2019) for a thoroughly worked-out proposal, as well as references to such precursors as Lyle Zynda and Schervish, Seidenfeld, and Kadane.

19. Another example in which positive probabilistic relevance isn't transitive!

13

The Problem of the Priors and Alternatives to Bayesianism

A typical high-school statistics course offers a package of familiar techniques: significance tests, regressions, etc. While the high-school student may think these have been around forever (along with the teacher offering the instruction), most of the mathematics presented is less than a century old. Moreover, a student who pursues more advanced statistics at the university level may encounter a recent vogue for "likelihoodist" and "Bayesian" statistical methods. Why, then, this fixity of "frequentist" methods at the lower levels of instruction, and (until very recently) in the publishing practices of almost all scientific journals?

As always with these questions, part of the answer has to do with historical contingencies. But it's also important that for a long time (and in many quarters still), frequentism was thought to be the only statistical regime by which the quantitative assessment of hypotheses by data could be put on a firm, objective footing. Bayesianism, in particular, was thought to depend on subjective prior commitments (such as an agent's hypothetical priors—Section 4.3) that could not be defended in a scientifically responsible way. This was the Problem of the Priors, and it seemed for a long time to justify the dominance of frequentist statistics.[1]

This chapter begins with the Problem of the Priors, usually cited as the most important objection to Bayesian epistemology and its theory of confirmation. We consider where the problem originates in the Bayesian formalism, how best to understand it, and whether it can be overcome by mathematical results about the "washing out" of priors. Special attention is paid to how exactly we're meant to understand the problem's demand for objectivity.

Then we consider frequentist and likelihoodist statistical paradigms that claim to offer accounts of evidential support free of subjective influences. After offering brief introductions to some mathematical tools these paradigms employ, I describe some of the problems that have been pointed out for each by partisans of rival camps. My central question is whether these approaches can yield the kinds of confirmation judgments Bayesians are after without sneaking in subjective influences themselves.

Fundamentals of Bayesian Epistemology 2: Arguments, Challenges, Alternatives. Michael G. Titelbaum, Oxford University Press. © Michael G. Titelbaum 2022. DOI: 10.1093/oso/9780192863140.003.0013

13.1 The Problem of the Priors

Suppose you're a scientist, and you've just run an experiment to test a hypothesis. Let H represent the hypothesis, and E represent the outcome of the experiment. According to Bayesian orthodoxy, your new degree of belief in H at time t_2 (just after you've observed E) should be generated from your old, t_1 degrees of belief by Conditionalization. So we have

$$\text{cr}_2(H) = \text{cr}_1(H \mid E) \tag{13.1}$$

Bayes's Theorem, which can be derived from the probability axioms and Ratio Formula, then gives us

$$\text{cr}_2(H) = \text{cr}_1(H \mid E) = \frac{\text{cr}_1(E \mid H) \cdot \text{cr}_1(H)}{\text{cr}_1(E)} \tag{13.2}$$

What determines the values on the far right of this equation? Perhaps H deductively entails that when an experiment like yours is run, the result will be E (or $\sim E$, as the case may be). In that case your likelihood $\text{cr}_1(E \mid H)$ should be set at 1 (or 0, respectively). Yet the implications of H need not be that strong; well-defined scientific hypotheses often assign intermediate chances to experimental outcomes. In that case, something like the Principal Principle (Section 5.2.1) will require your likelihood to equal the chance that H assigns to E.

So the likelihood $\text{cr}_1(E \mid H)$ may be set by the content of the hypothesis H. But what drives your values for $\text{cr}_1(H)$ and $\text{cr}_1(E)$? These unconditional initial credences are your priors in H and E. $\text{cr}_1(E)$ can actually be eliminated from the equation, by applying the Law of Total Probability to yield

$$\text{cr}_2(H) = \text{cr}_1(H \mid E) = \frac{\text{cr}_1(E \mid H) \cdot \text{cr}_1(H)}{\text{cr}_1(E \mid H) \cdot \text{cr}_1(H) + \text{cr}_1(E \mid \sim H) \cdot \text{cr}_1(\sim H)} \tag{13.3}$$

We now need only the prior $\text{cr}_1(H)$ and the likelihoods $\text{cr}_1(E \mid H)$ and $\text{cr}_1(E \mid \sim H)$. (Your prior for the catchall hypothesis $\sim H$ is just $1 - \text{cr}_1(H)$.) But eliminating $\text{cr}_1(E)$ from the expression and adding the likelihood of the catchall—$\text{cr}_1(E \mid \sim H)$—may not be much of an improvement. If, for instance, our hypothesis is the General Theory of Relativity (GTR), it's not clear what would follow about experimental results from the *negation* of GTR. So it's difficult to determine how confident we should be in a particular experimental outcome E on the supposition of $\sim H$.

In some special cases, we have an idea what to think if H is false because it's one member of a finite partition of hypotheses $\{H_1, H_2, \ldots, H_n\}$. Bayes's Theorem can then be rewritten once more for a given hypothesis H_i to yield

$$cr_2(H_i) = cr_1(H_i \mid E) =$$
$$\frac{cr_1(E \mid H_i) \cdot cr_1(H_i)}{cr_1(E \mid H_1) \cdot cr_1(H_1) + cr_1(E \mid H_2) \cdot cr_1(H_2) + \ldots + cr_1(E \mid H_n) \cdot cr_1(H_n)}$$
$$(13.4)$$

In this case, your posterior credence $cr_2(H_i)$ after observing the outcome of an experiment can be determined entirely from the likelihoods of the hypotheses—which are hopefully clear from their contents—and those hypotheses' priors.[2]

Whatever mathematical manipulations we make, an agent's priors will still play a role in determining her credences after conditionalizing. And that role may be significant. In Section 4.1.2, on the Base Rate Fallacy, we considered a highly reliable medical test for a particular disease. A positive result from this test multiplies your odds that the patient has the disease by 9. If you start off 50/50 whether the patient has the disease (odds of $1:1$), the test will leave you 90% confident of illness. But if your prior is only 1 in $1,000$ confident of disease, then your posterior will be under 1%.

This ineliminable influence generates the **Problem of the Priors** for Bayesian epistemology: Where are an agent's priors supposed to come from? Obviously an agent's priors are just the credences she has in various hypotheses before the experiment is run. But the challenge is to justify those credences, and the role they play in Bayesian inference.[3]

Of course, an agent's opinions about various hypotheses before she runs an experiment are influenced by other experiments she has run in the past. Presumably those opinions were formed by earlier conditionalizations on earlier experimental outcomes. But the problem recurs when we ask what provided the priors for those earlier conditionalizations. Ultimately the Bayesian recognizes *two* influences on an agent's credences at any given time: the agent's total evidence accumulated up to that time, and her epistemic standards. The agent's epistemic standards capture how she assigns attitudes on the basis of various bodies of total evidence; by definition, they are independent of the evidence itself.

As we saw in Chapter 4, the Bayesian formalism provides a convenient mathematical representation of each of these influences. An agent's total evidence at a given time is a set of propositions to which she assigns credence 1.

Her epistemic standards are represented by a hypothetical prior distribution, which can be conditionalized on her total evidence at a given time to yield her credences at that time. So the question "Whence the priors?" ultimately becomes a question about the origins of hypothetical priors, representing epistemic standards.

An Objective Bayesian (in the normative sense, as defined in Section 5.1.2) responds to the Problem of the Priors by maintaining that only one set of hypothetical priors is rational. All rational agents apply the same epistemic standards, in light of which their evidence uniquely dictates what to believe. The Objective Bayesian then has to work out the details of her view: she has to specify something like a Principle of Indifference (Section 5.3) or Carnapian logical probability approach (Section 6.2) that yields plausible results. If that can be accomplished, the Problem of the Priors is met.

For the Subjective Bayesian, however, the problem looms larger. On this view, more than one set of epistemic standards is rationally permissible. Two rational agents may observe the same outcome of an experiment, and assign wildly different credences to the same hypothesis as a result. They may even disagree about whether the outcome confirmed or disconfirmed that hypothesis.[4] In many cases, this difference of opinion will be traceable to relevant differences in the agents' total evidence before they arrived at the experiment. But according to the Subjective Bayesian, there will also be cases in which the difference results entirely from differences in the agents' epistemic standards. Each agent will be rational in her conclusions, and there will be no difference in either the evidence from this experiment or evidence from previous experiments that accounts for the difference between those conclusions. This is a problem for our epistemology and philosophy of science—isn't it?

13.1.1 Understanding the problem

Let's see if we can be a bit more clear about why the Problem of the Priors is supposed to be a *problem* for Subjective Bayesians. The flat-footed Subjective Bayesian position is just that an agent should base her updates on whatever prior opinions she has about the hypotheses in question (as long as those satisfy requirements of rationality such as the probability axioms), plus whatever new evidence she's updating upon. Why is this position considered insufficient, and what's the worry about letting prior opinions affect inductive inference?

Characterizing the Problem of the Priors, Howson and Urbach write:

The prior distribution from which a Bayesian analysis proceeds reflects a person's beliefs before the experimental results are known. Those beliefs are subjective, in the sense that they are shaped in part by elusive, idiosyncratic influences, so they are likely to vary from person to person. The subjectivity of the premises might suggest that the conclusion of a Bayesian induction is similarly idiosyncratic, subjective and variable, which would conflict with a striking feature of science, namely, its substantially objective character.

(2006, p. 237)

Here Howson and Urbach mention a couple of strands that are nearly ubiqitous in Problem of the Priors discussions. First, the concern is almost always about the role of priors in scientific reasoning. Second, that concern contrasts the subjectivity of priors with a desired objectivity of science. As Reiss and Sprenger (2017) put it, "Objectivity is often considered as an ideal for scientific inquiry, as a good reason for valuing scientific knowledge, and as the basis of the authority of science in society." Yet we've already seen that "subjective" and "objective" admit of diverse philosophical interpretations (Section 5.1). What does it mean in this context for priors to be subjective, and why would it be bad for science to be subjective in that sense?

A third strand in the Howson and Urbach quote is that (Subjective) Bayesianism allows "elusive, idiosyncratic influences" on scientific practice. Elliott Sober pursues this line in his critique of Bayesianism:

It is important to recognize how important it is for prior probabilities to be grounded in evidence. We often calculate probabilities to resolve our own uncertainty or to persuade others with whom we disagree. It is not good assigning prior probabilities simply by asking that they reflect how certain we feel that this or that proposition is true. Rather, we need to be able to cite reasons for our degrees of belief. Frequency data are not the only source of such reasons, but they are one very important source. The other source is an empirically well-grounded theory. When a geneticist says that Pr(offspring has genotype Aa | mom and dad both have the genotype Aa) = 1/2, this is not just an autobiographical comment. Rather, it is a consequence of Mendelism, and the probability assignment has whatever authority the Mendelian theory has. That authority comes from empirical data.   (2008, p. 26)

For Sober it's crucial that scientific conclusions be based on empirical data. But Bayesians allow priors to influence scientific conclusions, and the choice among priors is not entirely determined by empirical evidence. (Bayesian priors combine both evidence and epistemic standards, and the latter are independent of evidence by definition.)

It's important not to take this critique of Subjective Bayesianism too far. Critics of Bayesianism often depict subjective priors as *mere* opinions (Sober's "how certain we *feel* that this or that proposition is true"). But priors—even hypothetical priors—need not be groundless, expressions of nothing beyond the agent's whim. They also need not be a vehicle for political or moral values to interfere with science, threatening the value-neutrality some scientists have sought to secure for their discipline. An agent's prior in a hypothesis may reflect commonly accepted practices of theorizing in her culture, or her discipline; it may reflect a considered trade-off among such purely epistemic values as simplicity and strength. It's just that the priors aren't entirely driven by her *evidence*; instead, they shape how she responds to that evidence.

Why reject extra-evidential influences on scientific reasoning? One concern is that such influences will not be shared among practitioners, and so will lead to undesirable variation in opinion. Characterizing his opponents' views,[5] Savage writes:

It is often argued by holders of necessary and objectivistic views alike that that ill-defined activity known as science or scientific method consists largely, if not exclusively, in finding out what is probably true, by criteria *on which all reasonable men agree*. The theory of probability relevant to science, they therefore argue, ought to be a codification of *universally acceptable criteria*. Holders of necessary views say that, just as there is no room for dispute as to whether one propopsition is logically implied by others, there can be no dispute as to the extent to which one proposition is partially implied by others that are thought of as evidence bearing on it, for the exponents of necessary views regard probability as a generalization of implication.

(1954, p. 67, emphases added)

Similarly, E.T. Jaynes declares:

The most elementary requirement of consistency demands that two persons with the same relevant prior information should assign the same prior probability…. The theory of personalistic probability has come under severe criticism from orthodox statisticians who have seen in it an attempt to destroy the "objectivity" of statistical inference by injecting the user's personal opinions into it.   (1968, Sect. I)

There is certainly a strong current of opinion in modern science that evidence should be (at least in principle) shareable. But epistemic standards are presumably shareable as well—they needn't be inscribed in some mysterious, Wittgenstinian private language. Beyond being share*able*, why insist that they be *shared*?

Here we enter deep waters in the philosophy of science. I will simply comment that the absence of dispute is not obviously a hallmark of reasonable or even desirable scientific inquiry. Good scientists disagree about how to draw conclusions from their experiments just as much as they disagree about the conclusions themselves. So it's unclear whether a proper account of science must provide a path for the ultimate reconciliation of all rational differences. There may be pressure to explain what consensus *does* exist, but the Subjective Bayesian can chalk this up to intersubjective agreement in the priors of human agents who share a biological and cultural background. Such commonalities of perspective can be real without being rationally required.

Going in a different direction, one might worry that if an agent's opinions are not properly grounded in something outside herself, she will have insufficient reason to remain committed to those opinions. If nothing objective like empirical evidence ties the agent to a particular hypothetical prior, then when the evidence leads her to an opinion she doesn't like, why not switch to another prior on which that evidence generates a rosier conclusion? (See White 2005.)

Again, we should emphasize that a Subjective Bayesian's adherence to a particular prior need not be based on *nothing at all*. But we should also point out that attitudinal flip-flopping is banned by Conditionalization. When an agent updates by Conditionalization, her final credences are entirely dictated by the combination of her priors and her new evidence. And as we saw in Chapter 4, an agent who always updates by Conditionalization will be representable as having stuck with a single set of hypothetical priors across those updates. If Conditionalization is a rational requirement, then it's rationally impermissible for an agent to jump from one set of epistemic standards to another. And notice that the arguments for Conditionalization in Chapters 9 and 10 (based on Dutch Books and accuracy considerations, respectively) were perfectly consistent with Subjective Bayesianism. Even if multiple priors are rationally acceptable, it will still be the case that an agent with a particular prior who violates Conditionalization will face an expected accuracy cost or a diachronic Dutch Book. So Bayesians can argue against switching among hypothetical priors without arguing that some such priors are objectively better than others.[6]

Finally, there's the concern that if objective evidence doesn't fully constrain an agent's opinions, she'll be free to believe any old crazy thing she wants. Just as moralists fear the rational defector, Bayesians have long been haunted by the rational counterinductivist. Suppes writes:

> Given certain prior information is one a priori distribution as reasonable as any other? As far as I can see, there is nothing in my or Savage's axioms which prevents an affirmative answer to this question. Yet if a man bought grapes at [a certain] store on fifteen previous occasions and had always got green or ripe, but never rotten grapes, and if he had no other information prior to sampling the grapes I for one would regard as unreasonable an a priori distribution which assigned a probability 2/3 to the rotten state.
>
> (1955, p. 72)

In the same vein, D.H. Mellor suggests that we:

> Take Othello, whom Iago makes so jealously suspicious of Desdemona that conditionalizing on whatever she then says only strengthens his suspicion: reactions whose Bayesian rationality makes him no less mad. If this does not make Bayesianism false, it does at least make it seriously incomplete as an epistemology.   (2013, p. 549)

Yet Subjective Bayesianism need not be anything-goes. Just because the Subjective Bayesian thinks rational constraints are insufficient to single out a unique hypothetical prior, this needn't prevent her from developing strong requirements on rational credence. Suppes and Mellor are right that a Bayesianism constrained only by the probability axioms would deliver severely restricted epistemological verdicts. But we saw in Chapter 5 that all sorts of further constraints are available and defensible for Bayesians. Agents may be required to respect frequencies, chances, their own future opinions, or even the opinions of others. So the Subjective Bayesian is as free to deride counterinductive inferences as anyone else.

We have now worked our way through a number of possible concerns about Subjective Bayesianism—and responses to those concerns available to Bayesians—without settling on a unitary Problem of the Priors. I think that's representative of the literature; rather than a single, well-defined problem, Subjective Bayesians face a cluster of interrelated issues felt to be problematic.[7] Yet despite this vagueness, "the Problem of the Priors" is consistently cited as a crucial challenge to Bayesian epistemology, and a key motivation for endorsing other statistical approaches. Later in this chapter, we'll consider approaches to evidential support that seek to secure objectivity by eliminating the influence of priors. First, however, we'll investigate the claim that Subjective Bayesians may recover the desired objectivity through the application of formal convergence results.[8]

13.1.2 Washing out of priors

We'll begin our discussion of Bayesian convergence results with a simple example. Suppose two investigators, Fiona and Grace, have been given an urn containing ten balls. Each ball is either black or white, and the investigators know there is at least one of each color in the urn. Fiona and Grace then have nine hypotheses to consider: the urn contains one black ball; the urn contains two black balls; etc. on up to nine black balls. Fiona and Grace do not have any further evidence about the urn's composition, but each of them has a prior biasing her toward an opinion about the contents of the urn. The left-hand graph in Figure 13.1 depicts Fiona's and Grace's prior distributions over the urn hypotheses; Fiona's with solid lines and Grace's with dashed. Fiona expects more of the balls to be white than black, while Grace expects the opposite.

Now some evidence is gathered: one at a time, a ball is drawn at random, shown to the investigators, then returned to the urn. Let's say that out of the first four balls drawn, two are black and two are white. Fiona and Grace update by conditionalization, generating the new distributions shown on the right-hand side of Figure 13.1.

Notice that the posterior distributions have begun to converge; Fiona's and Grace's opinions are closer together than they were before the evidence came in. Why is that? If an agent updates by conditionalization on some evidence E between two times t_i and t_j, her credences in hypotheses H_1 and H_2 will evolve as follows:

$$\frac{cr_j(H_1)}{cr_j(H_2)} = \frac{cr_i(H_1)}{cr_i(H_2)} \cdot \frac{cr_i(E \mid H_1)}{cr_i(E \mid H_2)} \qquad (13.5)$$

The final fraction in this equation—the ratio of the likelihood of the evidence on H_1 to the likelihood of the evidence on H_2—is known as the **likelihood**

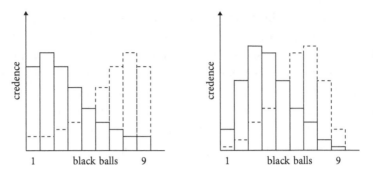

Figure 13.1 Convergence in response to evidence

ratio for H_1 and H_2.[9] When H_1 assigns E a higher likelihood than H_2 does, this ratio will be greater than 1, and the ratio of the agent's posteriors in H_1 and H_2 will be greater than the ratio of her priors in those hypotheses. So relatively speaking, the agent will become more confident in H_1 and less confident in H_2.

More generally, when a probabilistic agent updates by conditionalization, her credence shifts toward hypotheses that assign higher likelihoods to the evidence that was observed. In our urn example, the more equally a hypothesis takes the urn to be split between black and white balls, the higher a chance it will assign to randomly drawing two black balls and two white. So if Fiona and Grace satisfy the Principal Principle (or some other principle of direct inference), they will assign higher t_i likelihoods to a 2-2 observation on hypotheses that posit a fairly equal split. Once that observation is made, each investigator will shift some of her credence towards such hypotheses, which in turn will push their distributions closer together.

What happens if the observations continue? Let's suppose that, unbeknownst to our investigators, the urn is actually split 5-5 between black and white balls. The law of large numbers (Section 7.1) assures us that with probability 1, the ratio of black to white balls observed as the number of draws approaches the limit will approach 1:1. As this occurs, Fiona's and Grace's distributions will assign more and more credence to equal-split hypotheses and less and less credence to unequal splits. As a result, their distributions will move arbitrarily close to one another. In his (1954), Savage proved that with probability 1 (in a sense we'll clarify shortly), as the number of draws approaches infinity, the difference between the investigators' credences in any given urn hypothesis will approach 0.

It's important to understand why this convergence occurs. The underlying cause is that with probability 1, as the number of draws approaches infinity, each investigator's credence in the *true* urn hypothesis will approach 1. Since both investigators' opinions approach the truth (and the truth is the same for both of them), their opinions approach each other's as well.

This is the simplest of many formal convergence results that have been proven since Savage's work. The general thrust of all of them is that conditionalizing on more and more evidence will eventually **wash out** the influence of agents' priors. As the evidence piles up, its effect overwhelms any differences in the priors, and posterior opinions grow arbitrarily close together. If priors are a pernicious, subjective influence, the objectivity of evidence will cleanse scientific opinion of that influence if we just wait long enough.

Later results generalized Savage's work in a variety of ways. While we gave our investigators regular distributions over the available hypotheses, Regularity is not strictly required. Convergence results obtain as long as the investigators initially rule out the same possibilities by assigning them credence 0 (and of course don't assign credence 0 to the hypothesis that is true)[10]. Generalized results also tend to work with distributions over a continuous parameter, as opposed to the finite partition of hypotheses in our simple example. And the evidence-gathering process need not be so highly regimented as drawing balls with replacement from an urn. Gaifman and Snir (1982) proved a highly general convergence result which assumes only that the evidence and hypotheses are stated in a standard propositional language.

While Bayesian convergence results are technically impressive, their philosophical significance has come in for a lot of abuse over the years. I will present some of the criticisms in just a moment. But first I want to note that these results do establish something important. Observing the wide variety of priors permitted by the Subjective Bayesian approach, one might have worried that some priors could trap agents in a rut—by contingency and sheer bad luck, some agents might receive priors that forever doom them to high credences in wildly false hypotheses. (And then the skeptic asks how *you* know *you* aren't one of these doxastically tragic figures....) Yet the convergence theorems guarantee what Earman calls a "long-run match between opinion and reality, at least...for observational hypotheses" (1992, p. 148). And this guarantee makes minimal demands (the probability axioms, perhaps the Principal Principle) on the priors with which investigators begin.

Of course, we need to be careful what sort of guarantee we've been given. Both Savage's result and the law of large numbers provide for convergence "with probability 1". What sort of probability do we mean here; is it something like objective chance, or is it subjective credence? For the simple sampling situations Savage considered—like our urn example—either interpretation of "probability" will suffice. Applying the law of large numbers to chance values shows that in the limit there's a chance of 1 that the ratio of black to white balls in the sample will approach the true ratio in the urn (and therefore that conditionalizing agents will converge on the true hypothesis). A similar conclusion can also be drawn from within the scientists' point of view. Assuming she meets the minimal requirements for the theorem, each scientist will assign credence 1 that as the number of samples approaches the limit, her opinions will ever more closely match both the truth and the opinions of her peer.[11] Savage writes, "To summarize informally, it has now been shown that, with the observation of an abundance of relevant data, the person is almost certain to become highly convinced of the truth, and it has also been shown that he himself knows this to be the case" (1954, p. 50).

Since we're dealing with an infinitistic case here (number of observations approaching the limit), "probability 1" does not mean "necessarily must occur". As we saw in Section 5.4, when it comes to the infinite, probability-0 outcomes may be possible, and probability-1 outcomes may fail to occur. But many of us will be content to have probability 1 of reaching the truth even without an *absolute* guarantee. What's more problematic is that once we get past Savage and into the more advanced convergence results, the "probability 1" guarantee admits of only the subjective interpretation. It can no longer be proven to be a fact out in the world that the agent will converge on the truth with objective chance 1; we can only prove that the agent should have confidence 1 that she will approach the truth. It's unclear how reassuring we should find such predictive self-satisfaction, especially in the face of skeptical challenges.

One might also worry that many of the advanced results rely on Countable Additivity, to which we saw objections in Section 5.4. But the biggest problems for Bayesian convergence results center around their being results in the limit. The theorems demonstrate that in the long run, accumulating evidence will eventually overcome our priors and send us to the truth. But as Keynes (1923, p. 80) famously quipped, "In the long run we are all dead." (Woody Allen concurs that "eternity is very long, especially towards the end.")[12] Some of the simpler convergence results offer hopeful tidings for the shorter term: In straightforward sampling situations, we can assign a precise probability that opinions will converge to a certain degree within a certain number of samples. But for wider-ranging results, no such bound may be available. In that case we know only that objective influences will overtake our subjective priors if we hold out long enough.

While we wait, science may appear a subjective, conflicted mess. Given any body of total evidence, any hypothesis neither entailed nor refuted by that evidence, and any nonextreme posterior value, one can write down a probabilistic prior assigning that hypothesis that posterior in light of that total evidence. So no matter how much evidence we collect, it will be possible to find (or at least imagine) a probabilistic agent who disagrees with us by any arbitrary amount in light of that evidence. As Savage puts it, "It is typically true of any observational program, however extensive but prescribed in advance, that there exist pairs of opinions, neither of which can be called extreme in any precisely defined sense, but which cannot be expected, either by their holders or any other person, to be brought into close agreement after the observational program" (1954, p. 68).

In practice such disagreement may be cut down by Subjective Bayesians' imposing stronger constraints on hypothetical priors than just the probability axioms. But the basic issue is that no matter how much evidence we gain, until we pass to the limit our hypothetical priors' influence will never *entirely* disappear. Subjective Bayesians must concede that as long as our evidence is finite, the total objectivity and perfect scientific consensus sought by those who press the Problem of the Priors will never be achieved. de Finetti concludes:

> When the subjectivistic point of view is adopted, the problem of induction receives an answer which is naturally subjective but in itself perfectly logical, while on the other hand, when one pretends to *eliminate* the subjective factors one succeeds only in *hiding* them (that is, at least, in my opinion), more or less skillfully, but never avoiding a gap in logic. It is true that in many cases ... these subjective factors never have too pronounced an influence, provided that the experience be rich enough; this circumstance is very important, for it explains how in certain conditions more or less close agreement between the predictions of different individuals is produced, but it also shows that discordant opinions are always legitimate.   (1937/1964, p. 147)

13.2 Frequentism

We will now discuss two statistical approaches to data analysis distinct from Bayesianism: frequentism and likelihoodism. Frequentists and likelihoodists often contrast their approaches with Bayesianism by noting that their methods do not invoke priors. Instead, they assess the significance of evidence by making various calculations involving likelihoods. As we saw earlier, likelihoods can often be established by objective, non-controversial methods: likelihoods for diagnostic tests can be obtained by trying them on individuals whose condition is known; likelihoods associated with well-defined scientific hypotheses can be read off of their contents; etc. So these approaches claim to secure a kind of objectivity for scientific inference unavailable to Bayesians.

Assessing such claims is made difficult by the fact that each camp offers a variety of statistical tools, designed to do different things. Here it's helpful to follow Richard Royall (1997, p. 4) in distinguishing three questions one might ask about a particular observation:

1. What do I believe, now that I have this observation?
2. What should I do, now that I have this observation?
3. What does this observation tell me about *A* versus *B*? (How should I interpret this observation as evidence regarding *A* versus *B*?)

Distinguishing these questions helps us clarify the aims of various approaches. For instance, the frequentists Neyman and Pearson explicitly motivate their techniques with prudential considerations of what to *do* after receiving some data. On the other hand Royall (a likelihoodist) sets out to address the contrastive question of what the evidence says about *A* versus *B*.

We've seen over the course of this book that Bayesianism offers tools for answering all three questions. What an agent should believe after making an observation is her posterior credence upon conditionalization. What to do is addressed by decision theory, as described in Chapter 7. And whether the observation tells more strongly in favor of hypothesis *A* or *B* is revealed by the degree to which it confirms each (Chapter 6).

But addressing all three of Royall's questions doesn't automatically make Bayesianism the victor among statistical schools. Each approach has some tasks it accomplishes better than the others. For example, while frequentism and Bayesianism each offer tools for interpreting data after it's been collected, frequentism historically has had much more to say about the proper design of experiments to generate that data.

To describe every purpose one might try to achieve with statistical analysis, then assess the strengths and weaknesses of all of an approach's offerings with respect to each of those purposes, would be far too vast a project for this book.[13] Instead, for each broad approach I will focus on one or two of the tools it offers. For each tool, I will ask what it reveals about relations of evidential support. I do this for two reasons: First, this book is primarily about epistemology, and questions of evidential support (or justification) are central to epistemology. Second, frequentists and likelihoodists attack Bayesian confirmation theory on the grounds that evidential support is an important concept in science and so must be suitably objective. In light of this attack, it's fair to ask whether the methods promoted by frequentists and likelihoodists offer promising accounts of evidential support to take Bayesianism's place.

13.2.1 Significance testing

Frequentist methods are the methods taught in traditional statistics classes—significance tests, regression analyses, confidence intervals, and the like. If you've taken such a class, you'll know that frequentism is not a single, unified mathematical theory; instead, it's a grab-bag of tools for the analysis of data. Some of these tools address different questions from each other (say, finding a correlation coefficient versus significance testing), while other tools offer multiple ways of accomplishing roughly the same task (for instance, the variety of significance tests).

What do these tools have in common, such that they embody a common statistical approach? The term "frequentism" provides a clue. While many frequentists have historically adopted a frequency interpretation of "probability" (see Section 5.1.1), that isn't really what the "frequentist" moniker means. Also—contrary to what one sometimes hears—frequentism isn't really about basing one's opinions exclusively on observed frequency data. Instead, the hallmark of **frequentism** is to assess a given inference tool by asking, were it repeatedly applied, how frequently it would be expected to yield verdicts with a particular desirable (or undesirable) feature.

To illustrate, I'll focus on a particular frequentist statistical tool: the p-value significance test, most commonly associated with the work of R.A. Fisher. This type of test is applied when we have collected a body of data with a particular attribute. The significance test helps us determine whether the data's displaying that attribute was merely the result of chance, or whether it instead indicates something important about the underlying process that generated the data.

Let's take a specific example: We perform IQ tests on each of the sixteen members of Ms. B's second-grade class. We find that the average (mean) IQ score in the class is 110. This is interesting, because IQ tests have a mean in the general population of 100. We start to wonder: Is there some underlying explanation why Ms. B's class has a high average IQ score? Perhaps Ms. B is doing something that improves students' scores, or perhaps students were assigned to Ms. B's class because they possess certain traits that are linked to high IQs. On the other hand, the high average may have no explanation beyond the vagaries of chance. Even if a teacher's students are selected randomly from the general population, and nothing that teacher does affects their IQs, the luck of the draw will sometimes give that teacher a class with a higher mean IQ than average.

This "luck of the draw" hypothesis about the high IQ scores will be our **null hypothesis**. As a matter of definition, the null hypothesis is whatever hypothesis one sets out to assess with a significance test. In practice, though, the null is usually a hypothesis indicating "that the probabilistic nature of the process itself is sufficient to account for the results of any sample of the process. In other words, nothing other than the variation that occurs by chance need be invoked to explain the results of any given trial, or the variation in results from one trial to another" (Dickson and Baird 2011, p. 212).

How might we test the null hypothesis in this case? Here it helps that IQ scores are calculated so as to have a very specific statistical profile. In the general population, IQ scores have a normal distribution, with a mean of 100 and a standard deviation of 15. I won't explain what all that means—you can find an explanation in any traditional statistics text—but suffice it to say that this statistical profile allows us to calculate very precisely the probability of various outcomes. In this case, we can calculate that under the null hypothesis, the probability is less than 0.4% that the sixteen members of Ms. B's class would have an average IQ score at least as high as 110.[14]

This statistic is called the **p-value** of the null hypothesis on our data. The p-value records how probable it would be, were the null hypothesis true, that a sample would yield at least as extreme an outcome as the one that's observed. In our case, we add up the probability on the null hypothesis that the average IQ in Ms. B's class would be 110, the probability that it would be 111, that it would be 112, and so on for every IQ score at least as great as 110. The sum of these probabilities is the p-value of the null. The p-value in our example is quite low; if we drew a class of sixteen students at random from the general population, we should expect to get an average IQ score at least as high as that of Ms. B's class less than 0.4% of the time.[15]

Once we've calculated our p-value, we apply the significance test. Before examining the data, a statistician will usually have picked a numerical threshold α—typically 5% or 1%—to serve as the **significance level** for the test. If the p-value for the null hypothesis determined from the data is less than α, the result is deemed significant and we reject the null. In the case of Ms. B's class, a statistician working with a significance level of 1% will notice that the p-value associated with our sample is less than that. So the statistician will recommend "rejecting the null hypothesis at the 1% significance level."

Notice that the calculations leading to this recommendation worked exclusively with the data and with likelihoods relative to the null hypothesis. In our case, the null hypothesis says that the IQ scores in Ms. B's class are the chance result of a normal distribution with mean 100 and standard deviation 15. That hypothesis yields a particular likelihood that a class of sixteen will have an average score of 110, a likelihood that such a class will average 111, a likelihood of 112, etc. The *p*-value is the sum of these likelihoods. And all of them are determined directly from the content of the hypothesis being tested; no priors or catchalls are required.

Moreover, these calculations tell us something important about the frequency profile of the test. Suppose we make it our general policy to reject the null hypothesis if and only if our data exhibits a *p*-value for that hypothesis less than 1%. Focusing on cases in which the null hypothesis is true, in what percentage of those cases should we expect this policy to make the mistake of rejecting the null? Answer: less than 1% of the time. Call any body of data that leads us to reject the null hypothesis under this policy "rejection data". If the null hypothesis is true, the probability is less than 1% that a given sample will yield rejection data. So given the law of large numbers, we should expect in the long run to reject the null hypothesis in less than 1% of cases where the null hypothesis is true.

A frequentist statistical test is one for which we can use uncontroversial probabilities to generate an expectation for how frequently the test will display a particular desirable or undesirable feature. The previous paragraph performs that analysis for a Fisherian significance test. The probabilities in question are likelihoods derived from a well-defined null hypothesis. The (undesirable) feature in question is rejecting the null hypothesis when it's true.

13.2.2 Troubles with significance testing

Perhaps the biggest real-world problem with significance tests is how often their results are misinterpreted. Misinterpretation of *p*-values is so common that in February of 2016 the American Statistical Association (ASA) felt compelled to publish a "Statement on Statistical Significance and P-values" (Wasserstein and Lazar 2016), the first time in its 177-year history the association had taken an official position on a specific matter of statistical practice. The statement listed six "widely agreed upon principles underlying the proper use and interpretation of the p-value":

1. P-values can indicate how incompatible the data are with a specified statistical model.
2. P-values do not measure the probability that the studied hypothesis is true, or the probability that the data were produced by random chance alone.
3. Scientific conclusions and business or policy decisions should not be based only on whether a p-value passes a specific threshold.
4. Proper inference requires full reporting and transparency.
5. A p-value, or statistical significance, does not measure the size of an effect or the importance of a result.
6. By itself, a p-value does not provide a good measure of evidence regarding a model or hypothesis.

The misconceptions this list warns against are common not only among scientific practitioners in the field[16] but even in introductory statistics texts.[17] In elaborating the second principle, the ASA warned in particular:

> Researchers often wish to turn a p-value into a statement about the truth of a null hypothesis, or about the probability that random chance produced the observed data. The p-value is neither. It is a statement about data in relation to a specified hypothetical explanation, and is not a statement about the explanation itself.   (Wasserstein and Lazar 2016, p. 131)

The point here is that the p-value associated with the null hypothesis is not a statement about that hypothesis by itself—it doesn't, for example, establish the unconditional probability of the null hypothesis. Instead, the p-value establishes a specific relation between the data and the null hypothesis, which we can roughly characterize with the following conditional probability:

$$p\text{-value} = \Pr(\text{outcome at least as extreme as what was observed} \mid$$
$$\text{null hypothesis}) \tag{13.6}$$

Perhaps it's unfair to criticize frequentism based on the behavior of those who misuse it. So what's the proper way to understand a significance test's conclusions? A p-value doesn't say anything directly about the truth of the null hypothesis, but can we *use* it to adopt some attitude toward the null? When data yields a very low p-value for the null, a careful statistician will recommend "rejecting the null hypothesis at the 1% significance level". What does "reject" mean in this context? Is this a recommendation that we *disbelieve* the null hypothesis, or *believe* the null to be false? Probably not, for two reasons: First, it's unclear what it would mean to believe or disbelieve a proposition "at the 1% significance level".[18] Second, in many statistical paradigms "reject" contrasts with "accept" rather than "believe". Philosophers of science and statisticians have taken pains to clarify that "acceptance" is a technical term in this context, distinct from "belief". We should distinguish rejection from disbelief as well.

So if a low *p*-value doesn't justify *believing* the null hypothesis is false, what does it tell us? Fisher (1956, p. 39) suggested we conclude from a low *p*-value that *either the null hypothesis is false, or something very unlikely occurred*. For instance, in the case of Ms. B's third-grade class, if her students' IQs were random samples from the general population, the probability is less than 0.4% that an average IQ score at least as great as 110 would be observed. So if the null hypothesis is true, observing such a score is extremely unlikely. When the observation actually comes to pass, we conclude that either the null hypothesis is false, or something very unlikely occurred.[19]

Given this disjunction, we might apply Cournot's Principle (attributed to the nineteenth-century philosopher and mathematician Antoine Augustin Cournot), which recommends treating sufficiently unlikely outcomes as practical impossibilities. That would license a disjunctive syllogism allowing us for practical purposes to treat the null as false.[20] But this inference looks dubious on its face, as illustrated by an example from Dickson and Baird (2011, pp. 219–20). Consider the hypothesis that John plays soccer. Conditional on this hypothesis, it's unlikely that he plays goalie. (For the sake of the example we can make this as unlikely as we want.) But now suppose we observe John playing goalie. By the logic of significance testing, we know that either John doesn't play soccer or something unlikely has happened. Yet it would be a mistake to conclude that John doesn't play soccer.

A frequentist might reply that while the relevant inference pattern will sometimes yield this kind of mistaken rejection, for true hypotheses that will happen very infrequently. (After all, we started off by stipulating that goalies are rare soccer players.) But this reply misses the underlying issue: unlikely events occur, and occur all the time. We can even find ourselves in a situation in which we're *certain* some unlikely event has occurred. Consider a case in which we start with a partition of hypotheses, then observe some data that receives a low *p*-value on *every* hypothesis in the partition. Whichever hypothesis is true, something unlikely has occurred. In such a case it would be oddly biased to reject the null because of *its* low *p*-value.

Or consider a case in which we're comparing the null hypothesis with some alternative, our experimental result receives a low p-value on the null, but the alternative hypothesis had a low probability going into the test. Should we reject the null in such a case?[21] Admittedly, the framing of this example brings prior probabilities into the discussion, which the frequentist is keen to avoid. But ignoring such considerations encourages us to commit the Base Rate Fallacy (Section 4.1.2). Suppose a randomly selected member of the population receives a positive test result for a particular disease. This result may receive a very low p-value relative to the null hypothesis that the individual lacks the disease. But if the frequency of the disease in the general population is even lower, it would be a mistake to reject the null.

Perhaps when the frequency of a disease in the general population is known, the frequentist can factor that into her testing methodology. But reflecting on the role of priors can lead us to question significance testing in more subtle ways. Many professional scientific journals select a particular threshold (often 5%), then consider an experimental result reliable enough to publish just in case it's significant to that level. Of course reliability of a result is not the only criterion for publication—the hypothesis offered as an alternative to the null must be sufficiently novel, germane to scientific topics of interest, etc. But presumably scientists can determine whether a hypothesis has those sorts of qualities without empirical testing. So let's imagine that a group of scientists, without a strong sense of where the truth lies in their discipline, generates 1,000 hypotheses meeting the criteria of novelty, level of interest, etc. for publication. And let's imagine that, to be generous, roughly 5% of the hypotheses generated are true. There will be approximately 950 cases in which the suggested hypothesis is false and the null hypothesis is true. When the scientists run empirical experiments in these cases, a significance test at the 5% level will reject the null around forty-seven or forty-eight times. On the other hand, in the cases involving the fifty true hypotheses, not every empirical test will yield results significant enough to reject the null. Taking all these cases togther, the scientists will clear the bar for publication on something like ninety to 100 of their hypotheses, and almost half of those hypotheses will be false.

In this example I simply stipulated that about 5% of the scientists' hypotheses are true. In real life there'd be no way of knowing. But it's worth thinking about what percentage of hypotheses scientists put to the test actually are true, and the downstream effects this has on the percent of results published in significance-test journals that actually support the hypothesis claimed. Many commentators suggested that journals' reliance on significance tests was responsible for psychology's recent "replication crisis". A group of psychologists (Open Science Collaboration 2015) set out to replicate randomly selected results published in three prominent psychology journals. Of the ninety-seven results with p-values under 5% they attempted to replicate, they obtained only thirty-five outcomes significant at that α level.[22]

To be clear: The mathematics of p-values is entirely uncontroversial. I'm not challenging the relevant math, nor am I challenging the objectivity of the likelihoods underlying frequentist calculations. I'm trying to determine whether these mathematical calculations have any *epistemological* significance. (Recall that frequentists raise a similar question about Bayes's Theorem, which they're happy to concede is a theorem.) Perhaps it's a mistake to use significance tests for hypothesis rejection (in *any* sense). Could we find a more conservative—yet still epistemologically important—lesson to draw from low p-values? Suppose the null hypothesis makes it highly unlikely that something as extreme as a particular observation would occur. When that observation is actually recorded, doesn't this *disconfirm* the null, at least to some extent?

The prospects for this confirmational interpretation of significance testing dim when we think back on the case in which every available hypothesis receives a low p-value from the data. In such a case it may be that *none* of the hypotheses is disconfirmed by the data. But there's an even deeper problem with reading low p-values as disconfirmatory: p-values exhibit a worrying sort of language dependence.

Suppose we have a coin of unknown bias, and our null hypothesis says that the coin is fair. We flip the coin twenty times, and observe the following string of outcomes:

HHHTHHHHHHTHHHHTHHHHH

This string of twenty flip outcomes is extremely unlikely on the null. But then again, any particular string of twenty Hs and Ts will be extremely unlikely on the null. To run a meaningful significance test, we need to set aside some of our total evidence and describe the data in a coarse-grained fashion. We might select the number of heads as our test statistic, then calculate a p-value on the null hypothesis for the observation "seventeen heads out of twenty flips." The resulting p-value is under one-half of one percent.

Yet number-of-heads isn't the only statistic we might use to summarize flip data. Instead of asking how many heads were observed in a string of twenty tosses, we might ask for the length of the longest run of heads observed. For example, the string of outcomes above has a longest-heads-run of five.

There's no obvious reason to prefer number-of-heads as a test statistic over longest-heads-run. They each carve up the possibilities with the same fineness of grain (in each case the possible values run from zero to twenty), and each is positively correlated with the coin's bias toward heads. But these statistics yield different p-values for the null hypothesis. As I mentioned above, number-of-heads yields a p-value for this string of outcomes sufficient to reject the null at the 1% significance level. But the probability on the null of getting a longest-heads-run of at least five is almost exactly 25%. So should we count this observation as significant enough to reject the null hypothesis?

It's easy to cook up examples that work in the other direction, as well: If my flips came out ten heads followed by ten tails, the longest-heads-run p-value for this observation would be 0.59%.[23] But using number-of-heads, the null hypothesis would receive a p-value over 50%. Set aside questions of rejection for a moment—if p-values are our tool for assessing confirmation, should we take this evidence to disconfirm the null hypothesis or not?[24]

It's time to sum up. Despite the purportedly objective character of frequentist methods, we have identified a number of subjective influences on the use of significance tests. First, while a significance test is meant to assess a single null hypothesis in light of experimental data, the epistemological import of such a test can depend heavily on features of the alternative hypotheses with which the scientist contrasts the null.[25] Second, significance tests can operate only when the data is summarized by some test statistic. Yet as we saw above, the very same data can yield wildly divergent p-values for the null depending on which statistic the experimenter uses to summarize her data.[26] Finally, while we haven't delved into this issue here, the choice of a significance level α (5%? 1%?) sufficient for rejection seems to have little objective basis.[27]

Fisherian significance testing is hardly the only tool in the frequentist's toolbox. Other frequentist tools, such as Neyman-Pearson statistics, may fare better on the particular examples I've mentioned, but ultimately admit subjective influences for very similar reasons. (The details are spelled out at great length in this chapter's Further Readings.) Moreover, frequentists' abstention from priors leaves them open to the Base Rate problem I discussed above, as well as a bigger-picture problem I'll discuss at the end of Section 13.3.1. One is reminded of the de Finetti remark I quoted earlier:

> When one pretends to *eliminate* the subjective factors one succeeds only in *hiding* them (that is, at least, in my opinion), more or less skillfully, but never avoiding a gap in logic. (1937/1964, p. 147)

13.3 Likelihoodism

Earlier, in Equation (13.5), we saw that when an agent conditionalizes on some evidence E, her relative credences in hypotheses H_1 and H_2 change as follows:

$$\frac{cr_j(H_1)}{cr_j(H_2)} = \frac{cr_i(H_1)}{cr_i(H_2)} \cdot \frac{cr_i(E \mid H_1)}{cr_i(E \mid H_2)} \tag{13.7}$$

The left-most ratio in this equation captues how the agent compares H_1 to H_2 after the update. The middle ratio expresses how the agent made that comparison before the evidence came in. To see how the comparison has changed, we focus on the right-most ratio—the likelihood ratio. The likelihood ratio expresses the evidence's effect on the agent's relative credences in the hypotheses. If the likelihood ratio is greater than 1, the agent's relative confidence in H_1 compared to H_2 increases; less than 1, H_1 fares worse with respect to H_2; and a likelihood ratio of exactly 1 yields no change.

Likelihoodists take this equation *very* seriously: according to them, the likelihood ratio is our best tool for understanding what a piece of evidence says about one hypothesis versus another. They endorse Hacking's (1965)[28]

Law of Likelihood: Evidence E favors hypothesis H_1 over H_2 just in case $Pr(E \mid H_1) > Pr(E \mid H_2)$. In that case, the degree to which E favors H_1 over H_2 is measured by the likelihood ratio $\frac{Pr(E \mid H_1)}{Pr(E \mid H_2)}$.

The Law of Likelihood has two components: First, it makes the *qualitative* claim that E favors H_1 over H_2 just in case the likelihood ratio is greater than 1; second, it makes a *quantitative* claim that the likelihood ratio measures the degree of favoring.[29]

Notice that while Equation (13.7) worked with credence values (cr), in stating the Law of Likelihood I switched to generic probability notation (Pr). That's because likelihoodists think the likelihood values that drive personal changes in relative confidence should ultimately be grounded objectively. And it's objectively grounded likelihoods that are supposed to figure in the Law of Likelihood. Where do these likelihoods come from? As we've discussed, probabilities of experimental outcomes can be read directly off of well-defined scientific hypotheses; similarly, diagnostic test likelihoods may be based on frequency data from past trials. Likelihoodists stress that in such situations, likelihoods will be independent of priors; the reliability profile of a diagnostic test doesn't depend on how frequently the disease it tests for appears in a population.[30] So all the values in the Law of Likelihood can be established on a purely objective basis; likelihoodists think this gives their approach to evidential support a strong advantage over the subjectivity of Bayesianism.

Likelihoodists also think they succeed better than frequentists at achieving objectivity. Recall that in order to apply a significance test, the frequentist must summarize her observations with a test statistic. Different test statistics may yield different verdicts (about rejection of the null, etc.), so the choice of test statistic seems to introduce a subjective element into significance testing. The Law of Likelihood, on the other hand, works with the full description of an observation, honoring the Principle of Total Evidence and eschewing summary statistics.

This difference between likelihoodism and frequentism springs from a deeper, underlying difference: When we apply a significance test to a null hypothesis based on some particular observation, we ask how likely the null would make that observation *or any other observation more extreme*. The calculated *p*-value thus depends on what other observations might have occurred. This is the point at which different methods of partitioning the possible observations make a difference. The Law of Likelihood, however, operates only on the two hypotheses under comparison and the observation that was actually made. So likelihoodism has no troubles with language dependence. Moreover, interpreting the Law of Likelihood's favoring results does not require us to ask what further hypotheses beyond H_1 and H_2 might have been under consideration. Contrast this with significance tests on a hypothesis; we sometimes felt we had to know how other hypotheses might have fared before we could interpret the meaning of a significant *p*-value.

Richard Royall neatly summed up this case for likelihoodism in his (1997), when he wrote:

> Fortunately, we are not forced to choose either of these two evils, the sample-space dependence of the frequentists or the prior distributions of the Bayesians. Likelihood methods avoid both sources of subjectivity.
>
> (p. 171)

Bayesians and frequentists retort that whatever the Law of Likelihood's objectivist *bona fides*, it fails at a fundamental level: it gets basic cases of evidential support wrong. To borrow one of Royall's own examples (1997, Sect. 1.7.1), suppose you draw an ace of diamonds from a shuffled deck of cards. Consider the hypothesis that this is an ordinary fifty-two-card deck, versus the hypothesis that the deck is all aces of diamonds. The ordinary-deck hypothesis confers a probability of 1/52 on drawing an ace of diamonds, while the trick-deck hypothesis gives a likelihood of 1. So by the Law of Likelihood, drawing an ace of diamonds favors the trick-deck hypothesis over an ordinary deck. This has struck many commentators as problematic; it seems like no matter what card we draw from an ordinary deck, that card will count against the hypothesis that the deck is ordinary.

To understand the likelihoodist response to this case, we need to clarify a couple of features of likelihoodism. Like Bayesians and frequentists, likelihoodists offer an account of evidential support. But there's a key difference between their approaches. Bayesians and frequentists are happy to analyze two-place relations between a body of evidence and a single hypothesis—whether the evidence suffices to reject the hypothesis, whether the evidence confirms the hypothesis, etc. Likelihoodism restricts itself to *three*-place relations, between a body of evidence and *two* hypotheses.[31] Likelihoodists emphasize that their view is *contrastive*.

It may feel like drawing an ace of diamonds shouldn't be any evidence against the ordinary-deck hypothesis. But the likelihoodist never said it was, because the Law of Likelihood doesn't say whether a particular observation is evidence for or against a particular hypothesis. The likelihoodist asks whether the observation offers stronger support for the ordinary-deck hypothesis *than some other particular hypothesis*. If the two hypotheses you're entertaining are ordinary-deck versus trick-deck-of-diamond-aces, drawing an ace of diamonds does look like it speaks in favor of the latter over the former. Notice, by the way, that if we consider other trick-deck hypotheses (trick-deck-of-diamond-kings? trick-deck-of-spade-deuces?), the Law of Likelihood has an ace of diamonds observation favoring the ordinary-deck hypothesis over those. So the observation favors ordinary-deck over some hypotheses, but not over others.

One still might protest that having drawn the ace of diamonds, we would feel no inclination to *believe* the deck was all aces of diamonds or to abandon an ordinary-deck assumption. But this brings us to the second key feature of likelihoodism: likelihoodism isn't about what you should believe. Royall distinguished three questions we can ask about an observation (Section 13.2) specifically to point out that likelihoodism addresses question three (What does this observation tell me about *A* versus *B*?), *not* question one (What do I believe?). What to believe (or what unconditional credence to assign) is the kind of question a Bayesian answers, by calculating a posterior using Bayes's theorem. When first reading the ace of diamonds example, you probably assumed some kind of background setup that gave the ordinary deck a much higher prior than the trick-deck-of-diamond-aces. And this probably accounts for your high credence in the ordinary deck even after the observation. But that's separate from the question of which hypothesis the observation favors more strongly when considered alone.[32]

13.3.1 Troubles with likelihoodism

There are, however, more worrisome counterexamples to likelihoodism. Fitelson (2007) asks us to consider a card drawn at random from a deck known to be ordinary. Suppose you are told that the card is a spade. Which of two hypotheses is favored by this evidence: that the card is the ace of spades, or that it's black?

Most of us think the answer is obviously the latter. Yet the likelihood of drawing a spade on the ace-of-spades hypothesis is 1, while the likelihood of spade on black is 1/2. So the Law of Likelihood goes in the opposite direction. Here we've focused on a purely contrastive question, about what the evidence says (not about what you should believe), that likelihoodism gets dead wrong. Moreover, we can back up our intuition about this case with a general principle: the Logicality principle from Section 6.4.1. In that section we offered Logicality as an adequacy condition on measures of (two-place) confirmation, but it can be read as a purely comparative principle: Any observation favors a hypothesis it entails over a hypothesis that it doesn't (and treats equally any two hypotheses that are both entailed). The general idea is that deductive entailment is the strongest kind of evidential support we can get. But likelihoodism says that the card's being a spade favors its being the ace of spades (which it doesn't entail) *over* the card's being black (which it does).

Likelihoodists have some available responses to this counterexample. First, some likelihoodists (such as Chandler 2013 and Gandenberger ms) think hypotheses can be *competing* only if they are mutually exclusive, and so apply the Law of Likelihood only in that case. In the present example, the hypothesis that the card is the ace of spades and the hypothesis that it's black are not mutually exclusive, so the Law of Likelihood yields no verdict (much less a counterintuitive one). Second, a likelihoodist might say that examples invoking Logicality are necessarily special cases. Likelihoodism is driven by a quest for objectivity in scientific data analysis. Deductive entailment relations are clearly objective, so likelihoodists may be happy to concede an intuitive condition on favoring that can be applied entirely using entailment facts. Perhaps the likelihoodist will grant that, when the evidence deductively entails exactly one of the hypotheses, the favoring relations are as Logicality says. But in the vast majority of cases—where Logicality doesn't settle the matter—we should revert to the Law of Likelihood. (We'll presently discuss an interpretation of likelihoodism as this kind of "fallback" position.)

These responses are best addressed by a counterexample that avoids both of them. We want an example in which the hypotheses are mutually exclusive, neither is entailed by the evidence, yet the Law of Likelihood still gets the favoring wrong. We might create such an example by taking Fitelson's spade case and introducing a bit of statistical noise. (Perhaps you *thought* you heard that the card is a spade, but aren't *entirely* certain... etc.) Instead, let's work with a new, cleaner example:

> We're playing the card game Hearts (with an ordinary deck). The goal of Hearts is to score as few points as possible. At the beginning of the game, the player to my right passes me one card, face down. Some cards I might receive give me no risk of scoring points. The two of hearts exposes me to some risk, but that risk is fairly low. So if I'm passed the two of hearts, I'm mildly annoyed. However if I receive a different heart, or the queen of spades, those represent real trouble, and I get really pissed off.
>
> Now suppose you catch a glimpse of the card passed to me, and see only that it's a heart. Does this evidence favor the hypothesis that I'm mildly annoyed or that I'm pissed off? (Assume these hypotheses are mutually exclusive.)

In this example the evidence that the card is a heart entails neither hypothesis, and by stipulation the hypotheses are mutually exclusive. So neither likelihoodist response above applies. Yet I submit that the Law of Likelihood gets this case wrong. The probability of heart-passed given mildly annoyed is 1, while the probability of heart-passed given pissed-off is 12/13. So by likelihoodist lights the former is favored. But intuitively, catching a glimpse of a heart favors the hypothesis that I'm pissed off over the hypothesis that I'm mildly annoyed.[33]

One can respond to both this counterexample and Fitelson's spade case using a different likelihoodist tack. Notice that in both examples, Bayesian confirmation calculations track our intuitions. If we apply the log likelihood ratio to measure confirmation (Section 6.4.1), drawing a spade in Fitelson's example favors black over ace-of-spades, while glimpsing a heart favors pissed-off over mildly annoyed.[34] Elliott Sober suggests we accept the Bayesian's confirmation verdicts—and her methods of reaching them—for these particular examples.

Sober is a likelihoodist because he wants scientific analyses to proceed on purely objective grounds. In these card-playing examples the distribution of cards in the deck is available, and the draws are assumed to be random, so it's easy to calculate any prior, likelihood, or catchall values we might desire. When all these values can be established objectively, Sober is happy to let the Bayesian employ all of them in making a judgment of evidential support. But when objective priors are not available, likelihoodism comes into its own. Sober writes:

> These examples and others like them would be good objections to likeli-hoodism if likelihoodism were not a fallback position that applies only when Bayesianism does not. The likelihoodist is happy to assign probabilities to hypotheses when the assignment of values to priors and likelihoods can be justified by appeal to empirical information. Likelihoodism emerges as a statistical philosophy distinct from Bayesianism only when this is not pos-sible. The present examples therefore provide no objection to likelihoodism; we just need to recognize that the ordinary words "support" and "favoring" sometimes need to be understood within a Bayesian framework in which it is the probabilities of hypotheses that are under discussion; but sometimes this is not so. Eddington was not able to use his eclipse data to say how probable the [General Theory of Relativity] and Newtonian theory each are. Rather, he was able to ascertain how probable the data are, given each of these hypotheses. *That's* where likelihoodism finds its application.
>
> (2008, p. 37, emphasis in original)

There are three problems with Sober's "likelihoodism is a fallback position" response, one technical and two more methodological. First, it's easy to modify our examples so that objective priors are unavailable. Imagine that in either Fitelson's spade example or the Hearts example a few cards from the deck have been lost, but you have no idea which cards they are (except that in the spades example you know the ace of spades is still there, while in the Hearts example the two of hearts and the queen of spades remain). In that case we don't know the distribution of the deck, and so can't calculate any objective priors. According to Sober's fallback position, we should therefore abandon Bayesian confirmation measures and judge favoring relations using the Law of Likelihood. But I submit that our intuitions about favoring in these lost-card cases are the same as in the original examples, and continue to run counter to the Law of Likelihood.

Second, thinking about Sober's fallback methodology yields even bigger problems. Here's an analogy: There's a very popular international magazine I used to read to learn about goings-on in far-flung reaches of the world. Then one day I read an article in that magazine about a local issue I knew quite well. The article got both the facts and their significance fairly starkly wrong. And so I began to wonder: If the magazine is so unreliable about this particular case I understand, can I really trust what it says about matters I know little about?

Sober agrees with the Bayesian's take on favoring in examples where the priors are objective. In some of those examples (Fitelson's spade, the Hearts example), the Law of Likelihood disagrees with Bayesian confirmation theory on where the favoring falls. So Sober must grant that in those examples, the Law of Likelihood gets favoring wrong. Why, then, should we trust the Law of Likelihood to get favoring relations right in the cases where objective priors are unavailable?

Third, and finally, we should have *learned* something from the problems likelihoodism confronts when objective priors are available. If we're willing to grant that Bayesianism gets those cases right, we should see if there's any clear explanation of why likelihoodism sometimes gets them wrong. And here I think the answer is straightforward: likelihoodism relies on tools that are fundamentally backward.

Positive probabilistic relevance is symmetrical between E and H: conditioning on E increases the probability of H just in case conditioning on H increases the probability of E. So if all we want is a judgment on whether E supports H, it doesn't matter whether we work with $\Pr(E \mid H)$ or $\Pr(H \mid E)$. But when it comes to measuring degree of support, or comparing whether evidence supports one hypothesis over another, the order in which those propositions appear can make all the difference. Ask yourself: When it comes to evidential support, should the question of favoring be settled by asking whether one hypothesis entails the evidence (as occurs in both our counterexamples)? Or is it more relevant whether the evidence entails one of the hypotheses?

Royall justifies the Law of Likelihood by writing, "The hypothesis that assigned the greater probability to the observation did the better job of predicting what actually happened, so it is better supported by that observation" (1997, p. 5). This sounds intuitive at a first pass, but we should attend to the transition across that "so". Royall's third question asks what the *evidence* says about the *hypotheses*; likelihoodism examines what the *hypotheses* say about the *evidence*. It would be nice if you could discern the former by examining the latter. But that's exactly the bit that fails in the spade and Hearts examples.[35]

In the end, likelihoodists and frequentists have a common problem. They demand statistical tools built from quantities with a certain sort of objectivity. Sometimes they admit priors and catchalls, but often only likelihoods are sufficiently objective. So they build tools from the pieces they've allowed themselves. (The Law of Likelihood combines likelihoods in a simple ratio; frequentist methods make more baroque likelihood calculations.) Yet these pieces point in the opposite direction from the kinds of conclusions they want to draw. No matter how you combine them, likelihoods don't suffice to measure evidential favoring. You need priors or catchalls to truly capture evidential support.

13.4 Exercises

Unless otherwise noted, you should assume when completing these exercises that the distributions under discussion satisfy the probability axioms and Ratio Formula. You may also assume that whenever a conditional probability expression occurs or a proposition is conditionalized upon, the needed proposition has nonzero unconditional probability so that conditional probabilities are well defined.

Problem 13.1. ✒ What do you think it means when someone says that science is objective? Do you think it's important that science be objective in that sense?

Problem 13.2. ☽ Some of the exercises below demonstrate how Bayesianism, frequentism, and likelihoodism evaluate a simple coin-toss scenario. To set you up for those exercises, here are a few basic probability calculations you might want to make first:

   (a) Suppose I toss a fair coin six times. What is the chance it comes up heads exactly five times? More generally, for each n from zero through six, what is the chance that a fair coin tossed six times will come up heads exactly n times?

   (b) Now suppose instead that the coin is biased toward heads, such that its chance of coming up heads is 3/4 (though results of successive tosses are still independent of each other). For this biased coin, what is the chance that it will come up heads exactly n times in six tosses, for each n zero through six?

Problem 13.3. 🌶 Suppose you entertain exactly two hypotheses about a particular coin. H_1 says that the coin is fair. H_2 says the coin is biased toward heads: it has a 3/4 chance of coming up heads on any given flip, though outcomes of successive tosses are probabilistically independent. Let's say that at t_i, you assign $cr_i(H_1) = 0.2$ and $cr_i(H_2) = 0.8$, and you satisfy the Principal Principle.

(a) Between t_i and t_j, I flip the coin six times, and it comes up heads exactly five times. If you update on this evidence by conditionalizing, what will be your $cr_j(H_1)$?

(b) Imagine that—unbeknownst to you—this coin is actually fair. Out of all the outcomes I might have generated by flipping the coin six times between t_i and t_j, which outcomes would have increased your credence in the true hypothesis?

(c) Given that the coin is in fact fair, when I set out to flip it six times between t_i and t_j, what was the chance that the result of those flips would increase your credence in the true hypothesis?

(d) Explain what this has to do with the idea of priors' "washing out" from Section 13.1.2.

Problem 13.4. 🌶 Imagine you've just seen me flip a coin six times; it came up heads exactly five times. You suspect that I might be flipping a coin that's biased toward heads.

(a) Let's take it as our null hypothesis that the coin is fair. If the null hypothesis is true, what's the chance that flipping the coin six times would yield *at least* five heads?

(b) Does seeing five heads on six flips suffice to reject the null hypothesis at a 5% significance level? At a 1% significance level? Explain why or why not for each.

(c) Imagine all six of my flips had come up heads. Would that data recommend rejecting the null at a 5% significance level? How about 1%? Explain why or why not for each.

(d) In order to answer parts (b) and (c), did you need to consider the prior probability of the null hypothesis?

Problem 13.5. 🖉 Returning to page 462, go through each of the ASA's six widely agreed-upon principles one at a time, and explain why it's true.

Problem 13.6. 🌶 Consider the hypotheses H_1 and H_2 about a particular coin described in Exercise 13.3.

(a) Suppose that between t_i and t_j I flip the coin six times, and it comes up heads exactly five times. According to the Law of Likelihood, does this evidence favor H_1 over H_2?

(b) What possible outcomes of flipping the coin six times between t_i and t_j would have favored H_1 over H_2, according to the Law of Likelihood?

(c) In order to answer parts (a) and (b), did you need to consider $cr_i(H_1)$ or $cr_i(H_2)$?

Problem 13.7. 🪶🪶 Suppose we have a probability distribution Pr, two mutually exclusive hypotheses H_1 and H_2, and a body of evidence E. Here are two conditions that might obtain for these relata:

(i) Conditional on the disjunction of the hypotheses, E is positively relevant to H_1. That is,

$$\Pr(H_1 \mid E \,\&\, (H_1 \lor H_2)) > \Pr(H_1 \mid H_1 \lor H_2)$$

(ii) According to the Law of Likelihood, E favors H_1 over H_2 on Pr.

Prove that these two conditions are equivalent. That is, the relata satisfy condition (i) just in case they satisfy condition (ii).[36]

Problem 13.8. 🖋 Considering Bayesianism, frequentism, and likelihoodism, which do you think offers the best tools for assessing relationships of evidential support? Explain why you think so.

13.5 Further reading

INTRODUCTIONS AND OVERVIEWS

John Earman (1992). *Bayes or Bust? A Critical Examination of Bayesian Confirmation Theory*. Cambridge, MA: The MIT Press

Chapter 6 presents and evaluates a variety of Bayesian convergence results, admittedly at a high level of mathematical sophistication.

Elliott Sober (2008). *Evidence and Evolution*. Cambridge: Cambridge University Press

Chapter 1 provides a highly accessible introduction to Bayesianism, likelihoodism, frequentism, and challenges faced by each—albeit from a likelihoodist's perspective.

Classic Texts

Leonard J. Savage (1954). *The Foundations of Statistics*. New York: Wiley

Contains Savage's basic washing out of priors result, and a keen assessment of its significance.

Ronald A. Fisher (1956). *Statistical Methods and Scientific Inference*. Edinburgh: Oliver and Boyd

J. Neyman and Egon Pearson (1967). *Joint Statistical Papers*. Cambridge: Cambridge University Press

Canonical sources in the development of frequentism.

Ian Hacking (1965). *The Logic of Statistical Inference*. Cambridge: Cambridge University Press

A. W.F. Edwards (1972). *Likelihood: An Account of the Statistical Concept of Likelihood and its Application to Scientific Inference*. Cambridge: Cambridge University Press

Richard M. Royall (1997). *Statistical Evidence: A Likelihood Paradigm*. New York: Chapman & Hall/CRC

Canonical books making the case for likelihoodism.

Extended Discussion

James Hawthorne (2014). Inductive Logic. In: *The Stanford Encyclopedia of Philosophy*. Ed. by Edward N. Zalta. Winter 2014. URL: http://plato.stanford.edu/archives/win2014/entries/logic-inductive/

Presents state-of-the-art Bayesian convergence results, many of which have marked advantages over previous efforts in the area. For instance, some of these results eschew Countable Additivity, and some avoid concerns about the irrelevance of the long-run.

Colin Howson and Peter Urbach (2006). *Scientific Reasoning: The Bayesian Approach*. 3rd edition. Chicago: Open Court

Statistically sophisticated defense of the Bayesian perspective, with extensive, detailed critiques of frequentist methods.

Deborah Mayo (2018). *Statistical Inference as Severe Testing: How to Get beyond the Statistics Wars*. Cambridge: Cambridge University Press

An extended defence of frequentism, which (among many other things) offers responses to the objections raised in this chapter.

Prasanta S. Bandyopadhyay and Malcolm R. Forster (2011). *Philosophy of Statistics*. Vol. 7. Handbook of the Philosophy of Science. Amsterdam: Elsevier

A collection of excellent philosophical essays—many accessible to the beginner—exploring and assessing a wide variety of techniques in statistics.

Notes

1. Efron's (1986) article "Why Isn't Everyone a Bayesian?" begins by pointing out that "Everyone used to be a Bayesian. Laplace wholeheartedly endorsed Bayes's formulation of the inference problem, and most 19th-century scientists followed suit." When Efron reaches, at the end of his article, his "summary of the major reasons why [frequentist] ideas have shouldered Bayesian theory aside in statistical practice," he cites as crucially important that "The high ground of scientific objectivity has been seized by the frequentists."
2. Some experiments test the value of a continuous physical parameter. In that case we have an infinite partition of hypotheses—one for each possible value of the parameter. Assuming we can determine the likelihood of E on each possible setting of the parameter, and we have a prior distribution over the possible parameter settings, a calculation like that of Equation (13.4) can be accomplished with integrals.
3. Notice that switching from Conditionalization to Jeffrey Conditionalization (or the other Bayesian updating rules discussed in Chapter 11) would make no difference here. Priors play a role in determining the posterior credences generated by all of those rules.
4. Looking back at the list of historically significant Bayesian confirmation measures in Section 6.4.1, each one invokes either the agent's prior in the hypothesis or some credence involving the catchall.

5. Like Howson and Urbach, Savage was a Subjective Bayesian. It's a consistent theme throughout this part of the book that objections to Bayesianism are often stated more clearly by the targets of those objections than by the objectors themselves.

6. There's an interesting comparison here to the action theory literature on intentions. Most authors agree that there are situations in which either of two incompatible intentions is equally rationally permissible for an agent. (Consider Buridan's ass, for instance.) Once the agent forms one of those permitted intentions, is there rational pressure for the agent to stick with that choice? The lack of an objective reason for the initial selection needn't entail that that selection may be rationally abandoned later on. (See Bratman 1987 and the literature that followed.)

7. A similar point could be made about the general concern for "objectivity" in science. See Reiss and Sprenger (2017).

8. In Chapter 14 we will discuss another approach to recovering objectivity for Bayesianism: modeling agents' confidence using intervals of values instead of single real numbers.

9. The Bayes factor, which we discussed in Section 4.1.2, is a special case of the likelihood ratio in which the two hypotheses are H and $\sim H$. Equation (13.5) generalizes Equation (4.8) from our discussion there. (Recall that the likelihood ratio also furnished our measure of confirmation l in Section 6.4.1.)

10. Actually, some Bayesian convergence results (Lele 2004) work even when the true hypothesis isn't included among those the investigators consider. In that case, opinion converges on the hypothesis in the agents' partition that is *closest* to the truth, in a well-defined sense. (Thanks to Conor Mayo-Wilson for the reference.)

11. One slight difference between the two interpretations of "probability" here: On the chance interpretation, there is one particular composition of the urn (5 black, 5 white), such that the chance is 1 that both scientists' opinions will converge on the hypothesis describing that composition. On the other hand, there is no particular hypothesis such that either scientist assigns credence 1 that the two of them will converge on *that* hypotheses. The scientists are certain that a truth exists, assign credence 1 that they'll eventually converge on whatever it is, but are uncertain where exactly it lies.

12. Quoted at Bandyopadhyay and Forster (2011, p. xi), which in turn attributes the quotation to Rees (2000, p. 71).

13. Though it is arguably accomplished by another book, Howson and Urbach (2006).

14. For those who understand such things, the z-score of our sample is 8/3.

15. If you're new to significance testing, it might seem awfully odd to assess the probability on the null hypothesis that our sample would yield a result *at least as extreme as* what was observed. Why not assess the probability that the sample would yield *exactly* what was observed? Well, consider that in our IQ example, even though IQ scores are reported as whole numbers, Ms. B's class contains 16 students, so the average IQ observed could have been any fraction with a denominator of 16—not just the integers I mentioned in the text. That means there are a *lot* of exact averages that might have arisen from our sample, which in turn means that none of them has a particularly high probability. Even the probability of getting a 16-student sample with an average IQ of exactly 100—the most probable average on the null given the way IQ scores are calculated—is extremely low. Now generalize that thought to more common statistical

tests, which often have hundreds or thousands of individuals in their samples, and you'll see that the probabilities attached to getting a body of data with the *exact* attribute observed grow so low as to be near-useless. (Never mind situations in which your data estimates the value of a continuous parameter, in which case all the probabilities may be 0!)

You might also wonder why we calculated the *p*-value by summing the probabilities of only IQ averages greater than or equal to 110, instead of the probabilities of all averages at least as unlikely as 110. (Given the IQ distribution in the general population, an average of 90 is just as unlikely as 110; an average of 89 is even less likely than 110, etc.—so why didn't we add the probabilities of those sub-90 averages into our sum?) Because the alternatives to the null hypothesis we proposed (Ms. B increases students' IQ scores, the class was drawn from a high-IQ population, etc.) all would push the observed IQ in the same direction (that is, upwards), we performed what's known as a one-sided test. If we had calculated the *p*-value using not just the probabilities of averages above 110, but also the probabilities of averages below 90, that would be a two-sided test. Differences between one-sided and two-sided tests—and reasons for using one kind rather than the other—are discussed in standard statistics texts.

16. For instance, a study by Carver (1978) found education researchers espousing the following "fantasies" about the *p*-value: (1) *p* is the probability that the trial result is due to chance; (2) $1 - p$ indicates the replicability of the trial; and (3) *p* is the probability that the null hypothesis is true.

17. Picking a popular introductory statistics textbook (Moore, McCabe, and Craig 2009) at random, I found all of the following:

   - "The smaller the *P*-value, the stronger the evidence against [the null hypothesis] provided by the data." (p. 377, in the *definition* of *P*-value)
   - "If the *P*-value is less than or equal to [the significance level], you conclude that the alternative hypothesis is true." (p. 380)
   - "This *P*-value tells us that our outcome is extremely rare. We conclude that the null hypothesis must be false." (p. 381)
   - "The spirit of a test of significance is to give a clear statement of the degree of evidence provided by the sample against the null hypothesis. The *P*-value does this." (p. 395)

   This isn't an instance of cherry-picking—Moore, McCabe, and Craig (2009) was literally the first text I examined in search of examples. I certainly don't mean to beat up on them; similar passages appear in many other statistics texts.

18. "Rejecting the null hypothesis at the 1% significance level" does *not* mean assigning the null a credence of 0.01 or less. To conflate these is a version of the second mistake on the ASA's list.

19. For what it's worth, it's not entirely clear how to derive Fisher's disjunction with a probabilistic disjunct from the kind of conditional probability we equated with *p*-value in Equation 13.6. It looks like we're moving from "the conditional probability of this kind of observation given the null hypothesis is low" to "if the null hypothesis is true, then the probability of what was observed is low" en route to "either the null is false or what was observed is improbable." In light of our discussion of conditionals and conditional probabilities in Section 3.3, the first of these steps is highly suspect.

20. Instead of describing the relevant inference as a disjunctive syllogism, some authors call it "probabilistic *modus tollens*". We begin with the conditional "If the null hypothesis is true, then something very unlikely occurred", take as our minor premise that nothing very unlikely occurred, then conclude that the null is false.

21. A wonderful xkcd comic (that I wish I had the rights to reproduce) involves a neutrino detector designed to determine whether the sun has gone nova. When the button on the device is pressed, it rolls two fair six-sided dice. If the dice come up double-sixes, the device buzzes no matter what. On any other dice result, the device buzzes only if the sun has exploded.

   Someone presses the button, and the device buzzes. A frequentist statistician says, "The probability of this result happening by chance is $1/36 = 0.027$. Since $p < 0.05$, I conclude that the sun has exploded." The Bayesian statistician replies, "Bet you \$50 it hasn't" (xkcd.com/1132, by Randall Monroe).

22. Besides suggesting journals decrease their reliance on significance tests, some scholars of scientific methodology have advocated scientists' reporting *all* the hypotheses they test, not just the ones for which they obtain significant results. This is one motivation for item #4 on the ASA's list of principles above.

   For more dramatic reactions, see Woolston (2015), about a psychology journal that stopped publishing papers containing p-values, and Amrhein, Greenland, and McShane (2019), a commentary with over 800 signatories calling "for the entire concept of statistical significance to be abandoned."

23. The longest-heads-run p-values in this section were calculated using a web application by Max Griffin, which in turn developed out of a discussion dated July 24, 2010 on the online forum askamathematician.com. For further examples of statistical redescriptions that flip a significance test's verdicts about a sequence of coin flips, see Fitelson and Osherson (2015).

24. Frequentists will of course have responses to this type of example. For instance, they might complain that longest-heads-run is not what statisticians call a "sufficient" statistic, while number-of-heads is. In that case, we can shift the example to compare two sufficient statistics. (In any particular experimental setup, many sufficient statistics will be available—for one thing, combining a sufficient statistic with any other statistic will automatically yield another sufficient statistic.) Fisherian statisticians may now complain that only one of the statistics used is a "minimal" sufficient statistic, but it's not clear that there's a good justification for confining our attention to those. (Thanks to Conor Mayo-Wilson for extended discussion on this point.)

25. Also, when more advanced statistical methods are applied, it can sometimes make a difference *which* among the rival hypotheses the scientist chooses to designate as the null.

26. As I hinted earlier, significance testing a statistic that captures only some features of one's data violates the Principle of Total Evidence (Section 4.2.1). Beyond the language-dependence problem discussed in the main text, this generates another problem for significance tests having to do with the optional stopping of experiments. It's sometimes suggested that on a frequentist regime, one can interpret the results of an experiment only if one knows whether the experimenters would have kept collecting data had they

received a different result. This seems to inject the experimenters' subjective intentions into the interpretation of objective results. (See, e.g., Berger and Berry 1988).

27. Deborah Mayo (2018) has defended what she sees as a non-subjective approach to selecting significance levels. She also offers thoughtful, epistemologically sensitive responses to many of the other critiques of frequentism I describe in this section. Unfortunately, engaging with her proposals goes well beyond the scope of this book.

28. Hacking is oft-cited by likelihoodists for having articulated the Law of Likelihood. Yet Reiss and Sprenger (2017) trace the origins of likelihoodism back to Alan Turing and I.J. Good's work on cracking the Enigma code during World War II.

29. The Law of Likelihood should not be confused with another likelihoodist commitment, the **likelihood principle**. The Law of Likelihood considers one experimental observation, and explains how it bears on the comparison between two hypotheses. The likelihood principle considers two different observations, and explains when they should be taken to have the same evidential significance. Birnbaum (1962) gave the latter principle its name, and showed how it could be proven from two other, commonly accepted statistical principles. (See also Gandenberger 2015.)

30. The claimed independence between likelihoods and priors isn't supposed to be a *mathematical* independence; likelihoods and priors have well-defined mathematical relationships captured by equations such as Bayes's Theorem. Instead, the idea is that likelihoods can be *established* without consideration of priors, usually because their values are determined by different physical systems than the values of priors. The likelihood profile of a medical diagnostic test is determined by the biological mechanics of the test and the human body; the prior probability that a given subject has the tested-for condition is determined by the broader health situation in the population. The likelihood that a particular experiment will yield a particular outcome if the General Theory of Relativity is true is determined by the content of that theory; who knows what determined the prior probability that General Relativity would be true in the first place.

31. Perhaps all these relations should include yet another relatum—a background corpus—in which case Bayesians and frequentists would go in for three-place relations while likelihoodists would hold out for four. I will suppress any mention of background corpora in what follows.

32. Royall drives the point home by considering a case in which we start out with two decks, one ordinary and one full of aces of diamonds, then a fair coin flip determines which deck a card is drawn from. Now which hypothesis does the ace of diamonds observation incline you to believe? Is the evidential favoring in this case really any different than it was in the original?

33. A bit of history on this example: I originally proposed it to Greg Gandenberger, who posted it to his blog along with a poll (Gandenberger 2014). Seventy-two percent of respondents to the poll agreed with my intuition about favoring in the example, though admittedly the sample was small and perhaps not broadly representative. Jake Chandler responded to the example with a principle he proposed in Chandler (2013): that a piece of evidence favors one hypothesis over another just in case it favors that hypothesis when we first conditionalize on the disjunction of the hypotheses. If we first conditionalize on mildly-annoyed-or-pissed-off, then heart-passed intuitively favors

mildly annoyed over pissed-off, so Chandler's principle has the Law of Likelihood getting the case right. Steven van Enk, however, replied that Chandler's principle cannot be used to support the Law of Likelihood, because unless one has already accepted the Law, assuming the hypotheses' disjunction looks like it "change[s] our background knowledge in a way that is not neutral between the two hypotheses" (2015, p. 116). van Enk's paper also contains the only published discussion of the Hearts example of which I'm aware.

34. It's easy to get the log likelihood ratio measure of confirmation confused with likeli-hoodists' use of likelihood ratios, so let me distinguish the two approaches. Bayesians use the log likelihood ratio to calculate a numerical answer to a question about two relata: to what degree does this body of evidence support this hypothesis? They do this by comparing $\Pr(E \mid H)$, the probability of the evidence on the hypothesis, to $\Pr(E \mid \sim H)$, its probability on the negation of the hypothesis (the catchall). If a Bayesian wants to know whether E favors H_1 over H_2, she calculates the log likelihood ratio separately for each hypothesis, then checks which numerical result is greater.

   A likelihoodist, on the other hand, answers the comparative favoring question directly by comparing the likelihood of the evidence on one hypothesis to the likelihood of that evidence on the other (that is, by comparing $\Pr(E \mid H_1)$ to $\Pr(E \mid H_2)$). This can yield different results than the Bayesian approach—as we've already seen, comparisons of log likelihood ratios satisfy Logicality while the Law of Likelihood does not. Also, the likelihoodist approach yields no answer to two-place questions about, say, how strongly E supports H_1 straight out.

35. Frequentism is sometimes described as a probabilified Popperian falsificationism. One might say I'm suggesting that likelihoodism is a probabilified hypothetico-deductivism.

36. This equivalence is reported by Chandler (2013), who says it was pointed out to him by Branden Fitelson.

14

Comparative Confidence, Ranged Credences, and Dempster-Shafer Theory

Through most of this book we've considered one kind of doxastic attitude—degree of belief—and represented it with one kind of formal device—a numerical distribution over a propositional language. This is the descriptive basis of traditional Bayesian epistemology. On the normative side, we started off by articulating five core normative Bayesian rules. We then experimented with supplementing them, and perhaps tweaking them a bit (especially in Chapters 5 and 11). But our main business has been to consider the arguments for, applications of, and challenges to this basic formalism.

In this chapter I will describe three alternative formalisms for representing an agent's confidence in propositions. First, I will give a much more in-depth treatment of comparative confidence rankings than I provided in Chapter 1. Second, I will consider the ranged credence approach, which models an agent's doxastic state using a *set* of numerical distributions. And finally, I will consider Dempster-Shafer theory, which employs one distribution to model an agent's evidence and a second, associated distribution to model her doxastic attitudes.

For what it's worth, under the definition offered in Section 1.2.2 I would consider ranged credences a genuinely Bayesian epistemology, and the other two Bayesian-adjacent. But I don't think the labels are terribly important. Instead, we will focus on how these formalisms depart from—and hope to improve upon—the traditional approach of assigning a single real number to each proposition in a language. Comparative confidence rankings claim to better match the coarseness of actual agents' psychologies. Ranged credences claim to represent a wider variety of doxastic states, including some that respond more appropriately to certain types of evidence. Dempster-Shafer theory is more supple in its representations of evidence, and offers (among other things) an elegant solution to the Problem of the Priors (Chapter 13).

For each formalism I will describe: the formal elements it uses to represent doxastic states; the formalism's normative rules for rational doxastic states; claimed advantages of the formalism over the traditional Bayesian approach; and some challenges for and potential extensions to the formalism. I will also

Fundamentals of Bayesian Epistemology 2: Arguments, Challenges, Alternatives. Michael G. Titelbaum,
Oxford University Press. © Michael G. Titelbaum 2022. DOI: 10.1093/oso/9780192863140.003.0014

compare the three new formalisms to each other as we go along. My treatment will necessarily be abridged (especially in the case of Dempster-Shafer theory), but hopefully will explain the attractions of these approaches and make their affiliated literatures more accessible should you wish to explore further.

14.1 Comparative confidence

This weekend my colleague from the philosophy department is going to a conference. The conference is somewhere on the East Coast, but I'm not sure where. However, I'm more confident that she's going to New York than I am that she's going to Boston.

How might we represent this comparative confidence? Suppose we have a language $\mathcal{L}$ of propositions. For the time being, we'll assume that I entertain only a finite number of doxastic possibilities, so the language $\mathcal{L}$ will contain a finite number of atomic propositions. For any two propositions P and Q in $\mathcal{L}$, we write

$$P \succeq Q \tag{14.1}$$

to indicate that I am at least as confident of P as Q. When I am *equally* confident in P and Q, we write

$$P \sim Q \tag{14.2}$$

Notice that when I am equally confident in P and Q, I am at least as confident of P as Q, and at least as confident of Q as P. So we can define $\sim$ in terms of $\succeq$: $P \sim Q$ just in case $P \succeq Q$ and $Q \succeq P$.

What if I am strictly more confident in P than Q? Then we write

$$P \succ Q \tag{14.3}$$

This relation holds just in case $P \succeq Q$ but not $Q \succeq P$.[1]

When discussing comparative confidence, it's easy to slip into ways of talking that suggest an underlying numerical scale. Perhaps I say to you, "my confidence that my colleague is going to New York is greater than my confidence that she's going to Boston." That makes it sound like I have a confidence in New York, a confidence in Boston, each confidence has a number attached, and the New York number is greater than the Boston number. But when we write

$$N \succ B \tag{14.4}$$

the comparative doxastic attitude expressed need not be understood in terms of underlying numerical credences. It *might* be that I assign quantitative credences to each proposition, and $>$ simply expresses a greater-than relation between those numerical values. But the comparative relation might be more basic than the credal relation,[2] or it might obtain in cases in which I fail to assign precise credences to N and B at all. Patrick Suppes writes:

> The intuitive idea of using a comparative qualitative relation is that individuals can realistically be expected to make such judgments in a direct way, as they cannot when the comparison is required to be quantitative. On most occasions I can say unequivocally whether I think it is more likely to rain or not in the next few hours at Stanford, but I cannot in the same direct way make a judgment of how *much* more likely it is not to rain than rain.
>
> (2002, p. 226–7, emphasis in original)

This section examines norms for rational comparative confidence. We will make no assumptions to start about whether an agent's comparative attitudes are tied (descriptively or normatively) to numerical credences. Instead, we will treat confidence comparisons simply as a kind of doxastic attitude that an agent may assign, capable of being studied and normatively evaluated in its own right.

14.1.1 de Finetti's comparative conditions

As we pointed out a moment ago, the $>$ and $\sim$ relations may both be defined in terms of the $\geq$ relation. So we will focus on rational constraints for $\geq$ (while occasionally employing $>$ and $\sim$ for convenience).

Up until this chapter, I've talked loosely about comparisons (confidence comparisons, preference comparisons) introducing "rankings" of propositions, some of which are "complete". While I'll continue to use "ranking" in this intuitive fashion, I will now also employ more careful terminology. Strictly speaking, $\geq$ introduces a **preorder** on $\mathcal{L}$ just in case it satisfies

Comparative Equivalence: For any propositions P and Q in $\mathcal{L}$, if $P \equiv\models Q$, then $P \sim Q$.

Comparative Transitivity: For any propositions P, Q, and R in $\mathcal{L}$, if $P \geq Q$ and $Q \geq R$, then $P \geq R$.

Comparative Equivalence is intuitively appealing for similar reasons as our probabilistic Equivalence rule; if two propositions are true in exactly the same possible worlds, it seems irrational to be more confident in one than the other.[3]

Comparative Transitivity is a bit more interesting. Transitivity meshes with the intuitive idea that comparative confidence is about *ranking* propositions. It orders that ranking in a single direction, ruling out the kinds of loops that would occur if an agent could be more confident of P than Q, Q than R, and R than P. Yet by itself, Comparative Transitivity still allows for gaps (propositions that are part of the language but aren't placed in the ranking), and branches (cases in which two propositions are both ranked above a single proposition, but aren't ranked with respect to each other). To create what mathematicians call a **total preorder**, we need to add

Comparative Completeness: For any propositions P and Q in $\mathcal{L}$, either $P \geq Q$ or $P \leq Q$ (or both).

Comparative Completeness says that for any two propositions in $\mathcal{L}$, the agent makes some confidence comparison between the two. Most commentators have found this implausible as a rational requirement, especially for agents working with realistically large propositional languages. However, assuming Comparative Completeness dramatically simplifies formal theorizing about the $\geq$ relation. So we will take Completeness on board for now, then consider formalisms for non-total confidence preorders in Section 14.1.3.[4]

Next we have three conditions on $\geq$ that we can think of as comparative correlates to Kolmogorov's probability axioms:

Comparative Non-Negativity: For any proposition P and contradiction F in $\mathcal{L}$, $P \geq F$.

Comparative Non-Triviality: For any tautology T and contradiction F in $\mathcal{L}$, $T > F$.

Comparative Additivity: For any propositions P, Q, and R in $\mathcal{L}$, if P and Q are each mutually exclusive with R, then $P \geq Q$ just in case $P \vee R \geq Q \vee R$.

"Non-Negativity" may be a bit of a misnomer for the first condition, because comparisons aren't numerical values. But this condition has the same effect as Kolmogorov's Non-Negativity: forbidding an agent from being less confident of some proposition than she is of a contradiction. A contradiction represents something the agent has entirely ruled out of her doxastic possibility space, so this requirement establishes impossibility as a confidence floor.

If we wanted, we could add a Comparative Regularity requirement that for every non-contradictory P in $\mathcal{L}$, $P \succ F$. (Compare Section 4.2.) But we won't assume that here. We will, however, require an agent to be more confident in a proposition that is *required* by her doxastic space than a proposition that is *ruled out* by it. That's Comparative Non-Triviality. Comparative Non-Triviality is also important because it forbids trivial preorders in which every proposition in the language is ranked the same.

Comparative Additivity is the most interesting of the three. It says that if you're at least as confident in P as Q, then adding the same mutually exclusive disjunct to each shouldn't change that ranking.[5] To bring out the intuition, consider this example of a preorder that violates Comparative Additivity: Suppose I tell you I'm certain my colleague is traveling to exactly one East Coast city for her conference, and that I'm more confident that the city is New York than I am that it's Boston. But then I also tell you I'm more confident she's traveling to either Boston or Washington than I am that she's traveling to New York or Washington. This combination of attitudes seems irrational.[6]

These conditions on comparative confidence have a variety of plausible consequences, which improve their case for representing rational constraints. Together, Comparative Equivalence, Transitivity, Completeness, Non-Negativity, Non-Triviality, and Additivity entail:

- For any proposition P in $\mathcal{L}$, $P \succeq P$. (Comparative Reflexivity)
- For any proposition P and tautology T in $\mathcal{L}$, $T \succeq P$. (Comparative Normality)
- For any propositions P and Q in $\mathcal{L}$, if $P \vDash Q$ then $P \preceq Q$. (Comparative Entailment)
- For any propositions P, Q, R, and S in $\mathcal{L}$, if $P \Dashv\vDash R$, $Q \Dashv\vDash S$, and $P \succeq Q$, then $R \succeq S$. (Comparative Substitution)
- For any contradiction F and mutually exclusive propositions P and Q in $\mathcal{L}$, if $Q \succ F$ then $P \vee Q \succ P$.
- For any propositions P and Q in $\mathcal{L}$, if $P \succeq Q$ then $\sim P \preceq \sim Q$.
- For any propositions P and Q in $\mathcal{L}$, if $P \succeq \sim P$ and $Q \preceq \sim Q$, then $P \succeq Q$.
- For any propositions P, Q, R, and S in $\mathcal{L}$, if $P \succeq Q$, $R \succeq S$, and P and R are mutually exclusive, then $P \vee R \succeq Q \vee S$.

Perhaps the intuitive plausibility of these conditions suffices to convince you that they are all rational requirements on confidence comparisons. But there's another argumentative route we could take. Surely there are at least some cases in which it's appropriate for an agent to assign precise numerical credences—if you're not convinced that that's *all* cases, focus on simple examples involving dice or urns in which clear objective chance information is available. So let's take an agent (call her Raina) in a situation like that, and represent her credences with a distribution $cr(\cdot)$ over $\mathcal{L}$. We devoted a great deal of attention in Chapters 8 through 10 to arguments that rationality requires Raina's credences to satisfy the probability axioms. What consequences does this normative claim have for Raina's confidence comparisons?

Presumably for cases in which an agent assigns numerical credences over the entirety of $\mathcal{L}$, rationality requires her to satisfy

Comparative Matching: For any propositions P and Q in $\mathcal{L}$, $P \succeq Q$ just in case $cr(P) \geq cr(Q)$.

This principle allows us to convert rational constraints on Raina's credences into constraints on her comparative confidence attitudes. For instance, from the Kolmogorov axioms we proved the Entailment rule that if $P \vDash Q$, then $cr(Q) \geq cr(P)$. From this, Comparative Matching allows us to quickly prove the Comparative Entailment condition above. In general, all of the previously mentioned comparative confidence conditions can be proven by combining the probability calculus with Comparative Matching (plus the assumption that cr assigns a value to every proposition in $\mathcal{L}$).[7]

To put the same point in a slightly different way, the comparative confidence conditions are necessary for probabilistic representability. An agent's confidence ranking is **probabilistically representable** just in case there exists a probabilistic distribution over her language that aligns with that ranking through Comparative Matching. For a preorder to be probabilistically representable, it must satisfy all the conditions on $\succeq$ we've described so far.

de Finetti (1949/1951) famously conjectured that these conditions are also jointly *sufficient* for probabilistic representability. More specifically, he conjectured that for any agent whose confidence ranking satisfies Comparative Equivalence, Transitivity, Completeness, Non-Negativity, Non-Triviality, and Additivity, we can write down a probabilistic distribution over that agent's language that matches the agent's confidence comparisons.[8] This was important to authors such as de Finetti and Savage because it suggested to them that everything they needed for a full theory of rational confidence could be accomplished in a purely comparative setting, without having to make the psychologically implausible assumption that agents actually assign numerical credence values. If those six conditions suffice for probabilistic representability, they capture all the consequences of the probability calculus (and the various arguments for the probability calculus) for comparative attitudes.

Think about Raina again: From the assumption that rationality requires Raina's credences to be probabilities, and Comparative Matching, we can deduce a number of rational constraints (Comparative Entailment, etc.) on her confidence ranking. If de Finetti's six conditions suffice for probabilistic representability, then *every* rational constraint on Raina's comparative confidences that can be deduced from the probability calculus follows from those six conditions. While it might sometimes be mathematically convenient to consider Raina's underlying numerical credences, strictly speaking everything the probability calculus could teach us about rational constraints on Raina's confidence comparisons could be recovered equally well by just working with the comparative conditions, and not thinking about numerical credence at all.

Unfortunately, de Finetti's sufficiency conjecture was incorrect—his six comparative conditions turn out *not* to be sufficient for probabilistic representability. This follows from a counterexample constructed by (Kraft, Pratt, and Seidenberg 1959). In order to work, the counterexample requires at least a five-member partition of the agent's doxastic space. So let's suppose our agent has a partition $\{A, B, C, D, E\}$. Using the members of this partition, we can express thirty-two nonequivalent propositions, all of which (except the contradiction) are disjunctions of partition members. For the sake of efficiency, and for this example only, we will abbreviate the disjunction $A \vee B \vee C$ as "ABC", and so forth. Using that notation, suppose the agent comparatively ranks the thirty-two propositions as follows:

$$F \prec A \prec B \prec C \prec AB \prec AC \prec D \prec AD \prec BC \prec E \prec ABC$$
$$\prec BD \prec CD \prec AE \prec ABD \prec BE \prec ACD \prec CE \prec BCD$$
$$\prec ABE \prec ACE \prec DE \prec ABCD \prec ADE \prec BCE \prec ABCE \tag{14.5}$$
$$\prec BDE \prec CDE \prec ABDE \prec ACDE \prec BCDE \prec T$$

Next, fill in the further comparisons required by Comparative Equivalence, Transitivity, and Completeness. (For example, the agent assigns $F \prec A$ and $A \prec B$, so add in $F \prec B$.) These additions give us a total preorder. Since F is the lowest element in the ranking, the preorder satisfies Comparative Non-Negativity, and it obviously satisfies Non-Triviality as well. The preorder also satisfies Comparative Additivity, though I will leave checking that condition as an exercise for the reader. (Just to give one example how the check would go: The preorder assigns $A \prec C$, DE is mutually exclusive with A and with C, so to satisfy Comparative Additivity the preorder must assign $ADE \prec CDE$, which it does.)

Now let's try to construct a probabilistic credence distribution cr that matches these confidence comparisons. Applying Comparative Matching to particular comparisons in Equation (14.5), we get:[9]

$$\mathrm{cr}(AC) < \mathrm{cr}(D) \qquad (14.6)$$

$$\mathrm{cr}(AD) < \mathrm{cr}(BC) \qquad (14.7)$$

$$\mathrm{cr}(CD) < \mathrm{cr}(AE) \qquad (14.8)$$

Because cr satisfies the probability axioms, and because the individual capital letters represent mutually exclusive propositions, we can apply Finite Additivity to obtain

$$\mathrm{cr}(A) + \mathrm{cr}(C) < \mathrm{cr}(D) \qquad (14.9)$$

$$\mathrm{cr}(A) + \mathrm{cr}(D) < \mathrm{cr}(B) + \mathrm{cr}(C) \qquad (14.10)$$

$$\mathrm{cr}(C) + \mathrm{cr}(D) < \mathrm{cr}(A) + \mathrm{cr}(E) \qquad (14.11)$$

Adding up the left- and right-hand sides of these three inequalities, then canceling terms that appear on both sides, we obtain

$$\mathrm{cr}(A) + \mathrm{cr}(C) + \mathrm{cr}(D) < \mathrm{cr}(B) + \mathrm{cr}(E) \qquad (14.12)$$

Finally, applying Finite Additivity (Extended) yields

$$\mathrm{cr}(ACD) < \mathrm{cr}(BE) \qquad (14.13)$$

So if cr is probabilistic, Comparative Matching will require $ACD \prec BE$. But checking Equation (14.5) above, we find the opposite. So while the ranking described by Equation (14.5) satisfies all of de Finetti's comparative conditions, it is impossible to construct a probabilistic distribution matching that ranking. Satisfying de Finetti's six conditions is not sufficient for being representable by a probabilistic distribution.

14.1.2 The Scott Axiom

Why do de Finetti's conditions fail to suffice for probabilistic representability? It turns out there's a further necessary condition for a preorder to match a probability distribution. I will first state the condition, then explain what the various parts of it mean:

Scott Axiom: For any two equinumerous, finite sequences of propositions drawn from $\mathcal{L}$, $\{A_1, A_2, \ldots, A_n\}$ and $\{B_1, B_2, \ldots, B_n\}$:

- *if* in each of the agent's doxastically possible worlds, the A-sequence contains the same number of truths as the B-sequence
- *then* if there exists some i such that $A_i \succ B_i$, there also exists some j such that $A_j \prec B_j$.

Let's take this one step at a time. Start with a total preorder $\succeq$ over language $\mathcal{L}$. Now create two finite sequences, each containing the same number of propositions drawn from $\mathcal{L}$. We'll call the members of the first sequence $\{A_1, A_2, \ldots, A_n\}$, and the members of the second sequence $\{B_1, B_2, \ldots, B_n\}$. The numerical subscripts allow us to pair off members of the sequences (creating what mathematicians call a "one-to-one mapping"); A_1 is paired with B_1, A_2 is paired with B_2, etc.

The *if* part of Scott's Axiom requires that for any possible world the agent entertains, the A-sequence and the B-sequence have the same number of truths in that world. For example, the following two sequences have the relevant property:

$$\{P, P \supset Q\} \qquad \{Q, Q \supset P\}$$

In any possible world where P and Q are both true, each of these sequences contains two true propositions. In every other possible world, each sequence contains exactly one true proposition. (This part of Scott's Axiom is sometimes expressed by saying that the two sequences are "logically guaranteed to contain the same number of truths".)

Finally, the *then* part of the axiom: Suppose we've selected our two equinumerous, finite sequences from $\mathcal{L}$, with their order establishing a one-to-one mapping between them. Suppose also that the sequences we've selected contain the same number of truths as each other in each doxastically possible world. Then no matter which sequences we picked, if the agent is strictly more confident of some member of the A-sequence than its paired B-proposition, there must be some compensating member of the B-sequence in which the agent is strictly more confident than its corresponding A-proposition. In the example above, P is paired with Q and $P \supset Q$ is paired with $Q \supset P$. So if the agent assigns $P > Q$, satisfying the Scott Axiom requires her to assign $P \supset Q < Q \supset P$.

Now let's try to get a grip on what the Scott Axiom says, and why one might think that it represents a rational requirement. The easiest way to get two sets of propositions that are guaranteed to have the same number of truths in each possible world is to take two partitions. (In each possible world, exactly one member of each partition will be true.) So suppose I am certain that my philosophy colleague traveling to the East Coast is going to exactly one of the following destinations: Boston, New York, or Washington D.C. At the same time, the chair of my department is traveling to the West Coast, and will be going to exactly one of these destinations: Los Angeles, Portland, or Seattle. Now we start comparing my confidences concerning my colleague's and my chair's destinations. I tell you I'm just as confident that my colleague is going to Boston as I am that my chair is going to Los Angeles, and I'm just as confident in New York as I am in Portland. But I am *more* confident that my colleague is going to Washington than I am that my chair is going to Seattle.

This combination of comparative attitudes violates the Scott Axiom. Suppose language $\mathcal{L}$ contains a bunch of propositions about who's traveling where. Draw from $\mathcal{L}$ an A-sequence containing the three propositions about my colleague's possible East Coast destinations, ordered alpabetically as I have above; then draw a B-sequence listing alphabetically my chair's three possible destinations. Since the members of each sequence form a partition, each sequence contains exactly one truth in each doxastically possible world. So these sequences satisfy the *if* part of the Axiom. But now consider the pairing generated by these sequences: Boston–Los Angeles, New York–Portland, Washington–Seattle. I'm strictly more confident in one of the A-propositions (Washington) than its paired B-proposition (Seattle), but there is no B-proposition in which I'm more confident than its corresponding A-proposition. So these sequences fail to satisfy the *then* part of the Axiom, and I've violated Scott's Axiom as a whole.

If you think about it for a bit, my combination of comparisons not only violates the Scott Axiom—it's also intuitively irrational. I'm just as confident in Boston as Los Angeles, and in New York as Portland. I know that neither Boston nor New York will happen just in case my colleague goes to Washington, and neither Los Angeles nor Portland will happen just in case my chair goes to Seattle. But I'm strictly more confident in Washington than Seattle? It seems like I should be equally confident in Seattle as Washington. Or if I insist on being more confident in Washington than Seattle, I should compensate by being more confident of one of the other West Coast destinations than its paired East Coast destination.

Beyond these intuitive considerations, pragmatic arguments (akin to money pumps, Dutch Books, and representation theorems) have been offered for the Scott Axiom. Fishburn (1986, pp. 337–8) considers an agent who violates the Scott Axiom by assigning $A_i > B_i$ for some paired propositions in the A- and B-sequences, but then assigns $A_j \geq B_j$ for every other pair. This agent views each A-proposition at least as favorably as she views its corresponding B-proposition, and views at least one of the A-propositions *more* favorably than its corresponding B-proposition. So presumably she would look favorably on a game in which she wins \$1 for each member of the A-sequence that's true, but loses \$1 for each true member of the B-sequence. In fact, the agent would rather be on this side of the game than the side that wins on B-truths but loses on A-truths. No matter how weak that preference might be, the agent should be willing to pay at least *some* small amount of money to be on her preferred side of the game. But then the agent is guaranteed to come out a financial loser, because once she's made her initial payment, the wins that follow will equal her losses no matter which possible world she's in. (The A- and B-sequences in the Scott Axiom are specifically selected to contain the same number of truths in each possible world!)[10]

So if we accept the Scott Axiom, what does that get us? In Exercise 14.3, you'll prove that satisfying the Scott Axiom is a necessary condition for a comparative confidence ranking to be probabilistically representable. As I mentioned at the beginning of this section, this necessary condition goes beyond what's required by de Finetti's conditions on comparative confidence. In fact, the Kraft et al. example in Equation (14.5) fails to be probabilistically representable because it violates the Scott Axiom. Consider the sequences $\{ACD, D, BC, AE\}$ and $\{BE, AC, AD, CD\}$. The reader may verify that on each member of the partition $\{A, B, C, D, E\}$, these sequences contain the same number of truths. Now pair up the members of the sequences in the order in which I've listed them. (So ACD pairs with BE, D pairs with AC, etc.) In the confidence ranking presented in Equation (14.5), the agent is strictly more confident in each member of the first sequence than in the corresponding member of the second sequence ($ACD > BE$, $D > AC$, etc.). The confidence ranking in Equation (14.5) satifies de Finetti's conditions, but does not satisfy the Scott Axiom. Thus de Finetti's conditions do not entail the Scott Axiom.

On the other hand, the combination of Comparative Completeness, Comparative Non-Negativity, Comparative Non-Triviality, and the Scott Axiom entails all of de Finetti's conditions. So, among other things, that combination entails Comparative Equivalence, Transitivity, Additivity, and all of the comparative confidence conditions from our list on page 488.[11] Moreover, in the same (1959) paper in which Kraft et al. provided their counterexample, they also proved a result that gives necessary and sufficient conditions for a ranking to be probabilistically representable. A few years later, Dana Scott (1964) situated their result in a more general mathematical framework, reformulating the crucial condition in a way that was easier to understand.[12] It's that reformulated condition we call the Scott Axiom; combining it with Comparative Completeness, Non-Negativity, and Non-Triviality yields a set of necessary and sufficient conditions for probabilistic representability. The Scott Axiom thus provides a crucial link between probabilistic theories of rationality and conditions for rational comparative confidence. If we want to list *all* the constraints imposed on an agent like Raina by the probability calculus and Comparative Matching, we have to include the Scott Axiom, not just the weaker Comparative Transitivity and Additivity.

Any confidence ranking that satisfies the Scott Axiom along with Comparative Completeness, Non-Negativity, and Non-Triviality will be representable by a probabilistic numerical distribution. But in most cases we won't get a *unique* probabilistic representation. For example, suppose I become certain my colleague is headed to exactly one of Boston and New York; I am more confident in New York than Boston; and I satisfy all of the conditions we have listed so far. Even then, there are many possible probabilistic credence distributions that match this ranking. I might assign

$$\text{cr}(N) = 2/3 \qquad \text{cr}(B) = 1/3 \qquad (14.14)$$

or I might assign

$$\text{cr}(N) = 0.51 \qquad \text{cr}(B) = 0.49 \qquad (14.15)$$

and so forth.

It's possible to supplement the comparative confidence constraints we've seen so far with additional constraints that ensure a unique probabilistic representation. For example, Patrick Suppes suggests the following condition, which (combined with Comparative Equivalence, Transitivity, Completeness, and Additivity) is sufficient but not necessary for probabilistic representability:

Suppes Continuity: For any propositions P and Q in $\mathcal{L}$, if $P \succeq Q$ then there exists an R in $\mathcal{L}$ such that $P \sim Q \vee R$.

Any confidence ranking that satisfies Suppes Continuity (and the conditions in the previous parenthetical) has exactly one probabilistic numerical representation. While this might seem an attractive feature, it comes about because Suppes Continuity forces an agent to assign equal confidence to each doxastically possible world she entertains (thereby leaving only one numerical representation available).[13] So Suppes Continuity should be worrying for anyone who gets nervous around the Principle of Indifference (Section 5.3).

One can also achieve unique probabilistic representability by moving to a language with infinitely many atomic propositions, and adopting various continuity and Archimedean conditions. For more on such approaches, see (DeGroot 1970) and (Krantz et al. 1971, §5.2.3).[14]

In closing, it's worth a quick reminder that just because an agent's confidence comparisons are represent*able* by a numerical, probabilistic distribution—even a unique such distribution—that may not suffice to establish that the agent has probabilistic credences. As we've discussed in Section 8.3.1, Section 10.4, and elsewhere, different levels of realism about psychological states generate different standards for attitude attribution; some philosophers hesitate to attribute quantitative doxastic attitudes to actual people at all. Also, whenever a probabilistic numerical representation of a confidence ranking is possible, multiple non-probabilistic distributions matching that ranking will be available as well. (We'll look at some details of this in the next section's discussion of comparative decision theory.) Nevertheless, having comparative conditions that are necessary and sufficient for probabilistic representability is *useful* in the following sense: Anything we can prove about an agent's confidence ranking by invoking those conditions can also be proven using the probability axioms and Comparative Matching—and vice versa.

14.1.3 Extensions and challenges

In this section we'll consider a number of distinct issues that either challenge the constraints on comparative confidence we've seen so far, or call on us to extend them.

Incommensurability. In ordinary English, "indifference" may connote nonchalance toward an event or possibility. But the indifference relation $\sim$ we've been considering is a very specific, committed attitude. $A \sim B$ means the agent is just as confident in A as she is in B.

What if the agent makes *no* comparison between propositions A and B, either because it's never occurred to her to do so, or because she views them as incomparable in some way? I might be more confident that it will rain tomorrow than shine, and I might be more confident that there is intelligent extraterrestrial life in the universe than not, but I might balk at saying whether I'm more confident in rain or alien intelligence.

To represent this kind of incommensurability, and the potential for lack of comparison between propositions in general, we need to drop Comparative Completeness as a constraint on comparative attitudes. In other words, we need to allow preorders that fail to be total. When an agent doesn't assign any kind of comparative attitude to a pair of propositions A and B, her confidence ranking will not include the relation $A \succeq B$, but it also won't include $A \preceq B$. (Again, this is distinct from the case of indifference, in which we have *both* $A \succeq B$ and $A \preceq B$.)

We saw in the previous section that a total preorder satisfying even very strong rational conditions may still be matched by many different numerical credence distributions. If we drop the assumption of Comparative Completeness—allowing our preorder to be "partial" rather than "total"—we allow for an even wider variety of numerical correlates. This makes preorders very flexible in representing incommensurability. We saw in Section 12.2 that it's possible to construct partial credence distributions—distributions that assign numerical values to some but not all of the propositions in a language. Such partial distributions can do some work in representing agents who don't make particular comparisons. Again, suppose I fail to compare the pair of propositions A (that there is intelligent alien life) and R (that it will rain tomorrow). We can represent this lack of comparison with a partial distribution that either doesn't assign a credence value to A or doesn't assign a value to R (or both). But suppose that while I fail to compare A to R, I nevertheless am more confident in A than $\sim A$ and more confident in R than $\sim R$. The only way to capture these comparisons in a credence distribution is to assign numbers to all four propositions. Yet the moment I do that, a comparison between A and R is implied.

A partial preorder, on the other hand, can assign $A \succ \sim A$ and $R \succ \sim R$ without assigning any comparison to the pair of A and R. Thus preorders are able to represent a wider variety of doxastic states involving incommensurability than can be represented by partial credence distributions.[15]

So suppose we've decided to work with partial preorders. How do we evaluate such structures for rationality? The standard approach holds that a preorder is rational just in case it is *extendable* to a total preorder satisfying some preferred set of comparative constraints. In other words, given a partial preorder, if there is some way to add $\geq$ relations to it so as to yield a total preorder that satisfies Comparative Completeness and other constraints (Comparative Non-Negativity, Additivity, etc.), then the partial preorder is rational despite being incomplete.[16]

To illustrate, suppose I tell you I'm more confident in rain tomorrow than shine, and I'm more confident in alien intelligence than not. Then I tell you I'm more confident in shine tomorrow than I am in alien intelligence. In other words, I make the following three comparisons:

$$R \succ {\sim}R \qquad\qquad A \succ {\sim}A \qquad\qquad {\sim}R \succ A \qquad\qquad (14.16)$$

Intuitively, there's something wrong with this set of comparisons. But to verify that the ranking is irrational, let's see what happens when we try to extend it to a rational total preorder. Any preorder containing the three comparisons above would be defined over a language containing R, ${\sim}R$, A, and ${\sim}A$ (among other propositions). To be total, the preorder would have to satisfy Comparative Completeness, and so would have to assign $R \geq {\sim}A$ or $R \leq {\sim}A$ (or both). If we think de Finetti's conditions (Comparative Non-Negativity, Non-Triviality, and Additivity) are at least minimal requirements for rationality, then the preorder would also have to satisfy the bulleted conditions in the list on page 488. The sixth of those conditions, combined with $R \geq {\sim}A$, gives us ${\sim}R \leq {\sim}{\sim}A$, which by Comparative Substitution yields ${\sim}R \leq A$. That contradicts the third comparison above. On the other hand, $R \leq {\sim}A$ combines with the first comparison above and Comparative Transitivity to yield ${\sim}A \geq {\sim}R$ (keeping in mind that $\succ$ entails $\geq$). With the third comparison and another Transitivity step, we get ${\sim}A \geq A$, which contradicts the second comparison.

If we accept at least de Finetti's rational requirements, there's no comparison between R and ${\sim}A$ rationally consistent with the three comparisons above. That set of comparisons cannot be extended to a rational total preorder. So assigning those three comparisons is irrational.

Perhaps some readers will resist this final inference. You might grant me that particular conditions are rational requirements for the cases in which an agent truly makes comparisons among every proposition in her language (cases likely to be few and far between), but resist the claim that a confidence ranking is rational only if it can be extended to a rational total preorder. Well, imagine that in the weather and aliens case above, I've assigned the three given comparisons, but have never really thought about how to compare R and ${\sim}A$. Or perhaps I have no idea of any basis on which to make that comparison.

Then you come to me and ask about my relative opinions concerning those propositions. If I am clever—and have read my de Finetti—I may soon realize that there is no way to fill out my confidence ranking without violating his conditions. So what do I do then? Simply refuse to compare R and $\sim A$, so as to avoid a rational inconsistency? (Keep in mind that we've granted *arguendo* that de Finetti's conditions do indeed represent rational requirements on total preorders.)

We formalize incomplete rankings because we recognize legitimate reasons why an agent might fail to compare particular propositions. But the cogency of an attitude set should not crucially depend on its incompleteness.[17] Among other things, such a set stands in a precarious position with respect to possible future courses of evidence. We might find out that in a short time we will receive evidence that strongly supports a particular comparison between R and $\sim A$. Even before receiving that evidence, I can tell that whatever comparison it supports will be rationally incompatible with my present attitudes. And notice that this isn't a case in which I currently have an attitude that I expect to alter as I become more informed. Instead, it's a case in which my other attitudes leave me no rationally viable option to fill in a missing attitude when I gain evidence that merits such completion.

Conditional comparisons. As we saw in Equations (14.14) and (14.15), it's often possible to find many different numerical credence distributions matching the same confidence ranking. Among other things, this means that confidence rankings usually underdetermine confidence *ratios*. If Equation (14.14) depicts the credence distribution underlying my confidence ranking, then I am twice as confident that my colleague is headed to New York as Boston. But the credence distribution in Equation (14.15), while yielding the same ranking of propositions, gives those propositions a confidence ratio of almost 1:1.

Why do confidence ratios matter, and why might we want to recover such information from confidence comparisons? For one thing, agents sometimes say things like, "I'm twice as confident in New York as Boston," and we might want to meaningully represent such declarations in our doxastic formalism.[18] But perhaps more importantly, confidence ratios are centrally implicated in the standard Bayesian approach to *conditional* confidence.

Just as agents can make unconditional confidence comparisons (more confident in New York than Boston), it seems natural for them to make conditional confidence comparisons. I might tell you that, while I'm still not certain whether my colleague is headed to Boston or New York, on the supposition that she's going to Boston I'm more confident that she'll try some seafood than I am that she'll avoid it entirely. We could represent this confidence comparison as

$$S \mid B \succ {\sim}S \mid B$$

This comparison indicates that I'm more confident in seafood given Boston than I am in no seafood given Boston.

How might such conditional comparisons be rationally linked to unconditional comparisons? One path runs through probabilistic representations and the Ratio Formula. Suppose we take an agent's unconditional confidence comparisons over language $\mathcal{L}$, and find a probabilistic distribution over $\mathcal{L}$ matching those comparisons. Given that unconditional probabilistic distribution, we can calculate conditional credences using the Ratio Formula. Presumably if the agent is rational, those conditional credences should match her conditional confidence comparisons according to

Conditional Comparative Matching: For any propositions P, Q, R, and S in $\mathcal{L}$, $P \mid R \succeq Q \mid S$ just in case $cr(P \mid R) \geq cr(Q \mid S)$.

In Chapter 3 we introduced the Ratio Formula as a principle relating a rational agent's unconditional credences to her conditional. Now we have a similar strategy for relating an agent's unconditional confidence comparisons to her conditional. But there's an important difference between the two cases. Given the Ratio Formula, a rational agent's conditional credences supervene on her unconditional credences: fully specifying the agent's unconditional credence distribution over $\mathcal{L}$ also suffices to specify her conditional credence distribution over all ordered pairs of propositions in $\mathcal{L}$. (Or at least, those ordered pairs in which the condition receives nonzero unconditional credence.) But in the comparative case, an agent's unconditional confidence ranking will often underdetermine her conditional confidence rankings.

We can see how this underdetermination plays out when we try to use probabilistic representations and the Ratio Formula to link conditional and unconditional comparisons in the manner described above. I imagined us taking the agent's unconditional confidence ranking and finding a probabilistic distribution matching those comparisons. But as we saw in Section 14.1.2, even given plausible constraints on rational comparative confidence, a total unconditional preorder on a propositional language may not generate a *unique* matching probabilistic distribution. (Not to mention the non-probabilistic matching distributions that will also be available.) So even if an agent makes unconditional confidence comparisons between every pair of propositions in her language, multiple probabilistic representations of her confidence ranking will often be available. These multiple representations will yield differing conditional credence distributions, which may in turn match up with different conditional confidence rankings. (For an example of this phenomenon, see Exercise 14.4.) So for a typical unconditional confidence ranking, there will be many different conditional rankings that can be matched up to it via probabilistic representation and the Ratio Formula.

This point isn't an artifact of our particular Ratio-Formula strategy for linking unconditional comparisons with conditional. In general, unconditional confidence rankings (even total preorders) underdetermine conditional confidence comparisons. Many authors have responded to this fact by starting their account of comparative confidence with conditional comparisons rather than unconditional.[19] Domotor (1969), for instance, carried out the analog of the Kraft/Pratt/Seidenberg program for conditional comparisons: he offered rational constraints on *conditional* confidence rankings necessary and sufficient to make them probabilistically representable.

Since then further work has been done on axiomatizing conditional confidence rankings (e.g., Krantz et al. 1971, §§5.6ff., Fishburn 1986, §7), but many questions remain unanswered and much remains unknown. Rather than delve into the details here, I want to highlight a couple of applications for which conditional comparisons might be crucial. First, on a traditional Bayesian approach, the concepts of evidential independence, relevance, and confirmation are all defined using conditional credences. So if someone working in a comparative framework wanted to employ these concepts, conditional comparisons would seem to be needed.[20] Second, conditional doxastic attitudes are central to the traditional Bayesian approach to updating: According to Conditionalization, an agent's unconditional attitude toward a proposition after an update should equal her earlier attitude toward that proposition conditional on whatever she learned in-between. If this idea is on the right track, then a rule for rationally updating confidence comparisons might have to work with conditional confidences as well.

Interpersonal comparisons. Interpersonal comparisons of *utility* have been a notorious problem for decision theorists since the birth of the field. My wife seems to enjoy chocolate more than I do—is there some meaningful sense in which she assigns more utils to a bite of chocolate than I? Following the representation theorem approach from Chapter 8, one might closely observe my preferences and derive a utility scale from them, then do the same for my wife. The trouble is that each of those scales will be unique only up to positive affine transformation. My preference ranking over acts determines a utility ranking over propositions, and requires my utility gap sizes to stand in particular ratios. But it doesn't pin me to any particular number of utils for any particular proposition, such that my utility assignment could be compared meaningfully to my wife's assignment to the same proposition. I might, for instance, be indifferent between an act that guarantees me a bite of chocolate and an otherwise-similar act yielding no chocolate, while my wife prefers the former act to the latter. Yet that information doesn't discriminate between these two ways that she and I might assign utilities to chocolate outcomes:

	me	wife		me	wife
bite of choc	100	2	bite of choc	0	200
no chocolate	100	1	no chocolate	0	100

On the other hand, when agents assign rational numerical *credences*, those might be comparable across persons. As Stefánsson (2017) points out, the crucial foundation for interpersonal credence comparison is that all rational agents share the same "top" and "bottom" propositions. An agent who satisfies the Kolmogorov axioms will assign identical, maximal credences to all tautologies, and identical, minimal credences to all contradictions. While the choice of numerical measurement scale is somewhat arbitrary past that point, it seems reasonable to represent every rational agent with the *same* scale—whether it's 0 for contradictions and 1 for tautologies, 0 for contradictions and 100 for tautologies, or whatever else. My certainty should be considered just as certain as your certainty, and vice versa.[21] Assuming Finite Additivity, representation theorem techniques will then allow us to fill in each agent's non-extreme credence values from her preference rankings, in a way that yields substantive interpersonal information when one agent assigns (say) credence 0.9 to a particular proposition while the other assigns 0.6.

Similar structural features offer hope for limited interpersonal comparisons of confidence rankings, depending on which axiom system we select. Most systems establish tautologies and contradictions as "top" and "bottom" propositions in rational confidence rankings. With a strong enough additivity axiom, we might be able to say of any two agents who are both as confident of rain tomorrow as not that they are equally as confident in rain as each other. Beyond that, the specifics of the axioms will dictate how many meaningful confidence comparisons can be made between arbitrary rational agents. Notice that the availability of such interpersonal confidence comparisons may be important to some applications, such as aggregating comparative attitudes across individuals to generate a group ranking of propositions.

Risk aversion. Daniel Ellsberg proposed a famous counterexample to the conditions we've surveyed, now known as the **Ellsberg Paradox**.[22] He imagines "an urn known to contain 30 red balls and 60 black and yellow balls, the latter in unknown proportion" (Ellsberg 1961, p. 653). A single ball is to be drawn at random from the urn. First, you are asked whether you would prefer a bet that yields a large prize on red or a bet that yields the same prize on black. Ellsberg reports that in systematic studies (and nonsystematic polls of his famous economist friends), many subjects report that they would prefer the bet on red. This seems to be driven by a kind of risk aversion: You're certain that at least thirty of the balls are red, while for all you know there may be no black balls in the urn. Now second, you are asked whether you would prefer a bet that pays out on red or yellow, or a bet that pays out on black or yellow. Ellsberg finds that many subjects apply the same risk-averse reasoning to prefer black-or-yellow (guaranteed sixty balls in the urn) over red-or-yellow (only guaranteed thirty balls).

Ellsberg suggests that this betting pattern indicates the following confidence comparisons:

$$R \succ B \tag{14.17}$$

$$R \vee Y \prec B \vee Y \tag{14.18}$$

Yet this combination of comparative attitudes violates Comparative Additivity. If assigning these comparisons is rationally permissible, then neither Comparative Additivity nor the Scott Axiom (which entails it) are rational requirements. Fishburn (1986, §3) lists additional intuitive counterexamples to the confidence conditions we've seen in this chapter. If any of these counterexamples holds up, we may need to explore alternative sets of comparative confidence conditions.[23]

Decision theory. Finally, the question of how agents might make practical decisions on the basis of confidence comparisons—rather than full-blown numerical credences—remains wide open. As we've noted since Chapter 1, an agent's decisions may depend not only on her confidence ranking of propositions but on how much more confident she is in one than another. Suppose I get an opportunity to travel to either New York or Boston, and I consider it desirable to go to the same place my colleague went. (Perhaps she can give me good restaurant recommendations.) But suppose a ticket to New York is more expensive than a ticket to Boston. If I have the credence distribution about my colleague's possible destinations described in Equation (14.14), it might be worth the extra expense. But if my credences are those of Equation (14.15), my gap in confidence between New York and Boston is so small it probably isn't worth paying more. The two credence distributions match the same confidence ranking, but rationalize different decisions. Rational decisions may depend on the relative sizes of confidence gaps.

Suppes Continuity can give us some leverage on this problem. Any total preorder that satisfies Suppes Continuity and Comparative Transitivity, Completeness, and Additivity is numerically representable by a unique probabilistic distribution, every positive affine transformation of that distribution, and nothing else. So we can choose whether to represent such a confidence ranking using a probabilistic distribution assigning values from 0 to 1, a percentage-style distribution from 0 to 100, a Zynda-style distribution from 1 to 10 (Zynda 2000), or whatever we want.[24] But, because they are positive affine transformations of each other, all of these distributions will agree on the relative sizes of confidence *gaps*. Thus we may conclude that, at least for decision-making purposes, there is no significant difference between them. Perhaps in the end these representations differ only in the numerical conventions they stipulate for measuring a single, underlying reality—much the way we view the Kelvin, Celsius, and Fahrenheit scales for measuring temperature. Given an arbitrary choice among such measurement scales, we might pick the probabilistic one for mathematical convenience. But there's no real threat that we'd be *misrepresenting* the agent's attitudes by working with any of the matching distributions. So once we generate our preferred numerical representation of the agent's confidence ranking, we can plug it into a standard decision theory to determine how the agent should behave on the basis of her confidence comparisons.[25]

Yet many Bayesians would reject Suppes Continuity for its ties to Indifference. Unfortunately, if we have a preorder (over finitely many possibilities) that satisfies de Finetti's conditions and the Scott Axiom but not Suppes Continuity, a vast variety of numerical distributions will align with that preorder via Comparative Matching—even if the preorder is total. Typically, those distributions will not all be affine transformations of each other, and will not all agree on the relative sizes of confidence gaps. Return, for example, to the distributions in Equations (14.14) and (14.15). They both match the same preorder that satisfies de Finetti's conditions and the Scott Axiom. Yet in Equation (14.14), my confidence in N is as far from my confidence in B as the latter is from a contradiction; while in Equation (14.15) my confidences in the two propositions are much closer together.[26] Suppose all you know about my doxastic attitudes is that I assign the confidence ranking matching those two distributions. If you choose to represent me using one of the distributions rather than the other, you are making substantive assumptions about my confidences not in evidence from the preorder provided. If you then, say, try to predict my actions (which plane ticket will I purchase?) on the basis of the representation you've chosen, you might make an important error. (A similar point can be made about inferring confirmation relations from different numerical representations of the same confidence ranking; both gap sizes and confidence ratios can make a difference to confirmation judgments.)

So if an agent's confidence comparisons fail to satisfy a strong requirement like Suppes Continuity, can *anything* substantive be said about what decisions she should make on the basis of those comparisons? Fine (1973, §IIG) made some progress on a comparatives-based decision theory by relying heavily on dominance reasoning. This allowed him to rule out a number of possible decision rules involving comparative confidence, but not to choose an optimal rule among those that remained. In the end he conceded:

The approach to decision making we tentatively propose is intended more as an illustration of our assertion [that interesting decision problems can be resolved using comparative probability] than as a thoroughly considered "best" formulation of decision-making in [a comparative probability framework]. Much more research is needed before we can claim a satisfactory complete analysis of decision-making in [comparative probability].   (p. 37)

14.2 Ranged credences

In the previous section we considered an example in which I am certain that my philosophy colleague is traveling to either Boston or New York, and I am more confident in New York than Boston. If that's all you know about my doxastic state, then there are many possible probabilistic numerical distributions that would match what you know. If you assume that at all times, any agent's doxastic state is best represented by a particular numerical credence distribution, then the credence distribution representing my state must be one of the distributions compatible with those comparative constraints. You just don't know which distribution it is.

But there may be cases in which an agent's doxastic state is *not* best represented by a point-valued, numerical credence distribution. It may be that I view certain propositions as incommensurable, in a manner not representable by a numerical distribution. (See Section 14.1.3.) Or perhaps my state is simply not committal enough to merit specific numerical representation—I may be more confident of New York than Boston, but there may simply be no fact of the matter how *much* more confident I am.

If a numerical credence distribution is not the best formal tool for representing an agent's attitudes in such cases, what other options are available? In the previous section, we took an agent's confidence comparisons and used them to establish a ranking over a language of propositions. We then worked directly with this ranking, and various rational constraints to which the ranking might be subject.

But other formal options are available. For instance, instead of associating a single real number with each proposition in a language, we might assign a *range* of reals to each proposition. Take the proposition A that there is intelligent alien life in the universe. We might represent an agent's attitude toward this proposition with the numerical range $[0.6, 0.75]$. (0.75 is sometimes called the **upper probability** of A in this range, and 0.6 the **lower probability**.) We could then represent the agent's attitude towards $\sim A$ with the range $[0.25, 0.4]$. Because the lower probability of A's range is higher than the upper probability of the range for $\sim A$, we would be representing that agent as more confident in A than $\sim A$. But if the two ranges had some amount of overlap—say, $[0.3, 0.6]$ and $[0.4, 0.7]$—that would allow us to represent an agent who has not settled on a definite confidence comparison between those propositions.

Notice that this formalism allows us to represent incommensurability without any gaps in our representation. We don't go silent on some of the propositions, failing to assign anything numerical to them. We assign each of the propositions a numerical range; it's just that some pairs of ranges don't give rise to confidence comparisons. The formalism also allows us to represent point-valued credences, of the type discussed in this book prior to this chapter. All we have to do is assign some proposition a range containing only one value, like $[2/3, 2/3]$. When we assign a wider range, containing multiple values, this represents a different type of doxastic attitude, which I'll call a **ranged credence**.

Exactly what kind of attitude is a ranged credence? What kinds of views about a proposition are better represented by a range of reals than a single point value? We'll explore these questions in the sections to come, and talk about how this type of attitude connects to decisions, evidence, other attitudes, etc. First, though, I'll devote the rest of this section to the formal machinery behind ranged credences.

Many early authors who worked on ranged credences (such as Smith (1961), Kyburg (1961), and Good (1962)) represented them by simply assigning a range of reals to each proposition in a language. But in his (1974), Isaac Levi

A	R	Pr_x	Pr_y
T	T	1/2	2/5
T	F	1/4	1/5
F	T	1/6	4/15
F	F	1/12	2/15

Figure 14.1 A simple representor

argued for a richer formal representation, of which these numerical ranges were simply one aspect. The formalism Levi favored subsequently became standard in the field. I'll start by explaining this formalism, then show why it's an improvement over working exclusively with numerical ranges.

Start again with an agent's confidence comparisons. Consider all of the point-valued, probabilistic distributions that match those comparisons (in the sense of the Comparative Matching principle I articulated earlier). Now collect all of those distributions into a set. This *set* of probability distributions—what van Fraassen (1980) famously called a **representor**—might serve as our formal representation of the agent's doxastic state.[27]

How can a set of probability distributions represent an agent's doxastic state? The trick is to say that the agent's overall state displays certain properties only if those properties are shared by *every* distribution in the set. For example, take the case from Section 14.1.3 in which I am more confident in the proposition A (that there is intelligent alien life) than $\sim A$, and more confident in R (that it will rain tomorrow) than $\sim R$. Now suppose my representor—the set of probability distributions that represents my overall doxastic state—turns out to contain just the two distributions described in the probability table of Figure 14.1.[28] There, each of the right-hand columns contains enough information to specify a full probabilistic distribution over a language $\mathcal{L}$ whose atomic propositions are A and R. Those two distributions, Pr_x and Pr_y, are the members of the representor. With a bit of math, you can determine that in each of those distributions, $\mathrm{Pr}(A) > \mathrm{Pr}(\sim A)$. So this representor represents me as being more confident of A than $\sim A$. Similarly, both distributions assign $\mathrm{Pr}(R) > \mathrm{Pr}(\sim R)$, so the representor has me more confident in R than $\sim R$.

But notice that while the first distribution has $3/4 = \mathrm{Pr}_x(A) > \mathrm{Pr}_x(R) = 2/3$, the second assigns $3/5 = \mathrm{Pr}_y(A) < \mathrm{Pr}_y(R) = 2/3$. Since the two distributions in the representor do not agree on the relative rankings of A and R, this representor represents me as being neither more confident of A than R, nor more confident of R than A, nor equally confident in the two. According to this representor, A and R are incommensurable for me. So a set of probability distributions can represent a doxastic state that is committal on some comparisons (e.g., A vs. $\sim A$, R vs. $\sim R$) while remaining noncommittal on others (A vs. R).

A representor is a formal representation of an agent's *entire* doxastic state; it encapsulates how she views *all* the propositions in the language. If we're interested in how she views just one proposition, we can extract a numerical range from the representor. Notice that the lowest value assigned to A by any of the distributions in my representor is 0.6, while the highest value is 0.75. So we might summarize my attitude toward A on this representor with the range $[0.6, 0.75]$. Similarly, we can summarize my attitude toward $\sim A$ with $[0.25, 0.4]$. This is another way of seeing that this representor has me more confident in A than $\sim A$. On the other hand, we can calculate out that both the distributions in Figure 14.1 assign $\Pr(R) = 2/3$. So the range for R associated with this representor is $[2/3, 2/3]$, representing a point-valued credence. In this way, we can read numerical ranges for specific propositions off the rich representation in a full representor.

Warning

The idea of representing a state with a set of assignments that agree on particular properties but not on others resembles certain supervaluationist approaches to vagueness. Yet a ranged credence need not be a *vague* or *ambiguous* attitude towards a proposition.[29] Moreover, while we find it formally convenient to *represent* a ranged credence with a collection of point-valued numerical probability distributions, this does not mean that the ranged credence attitude itself is somehow a collection of point-valued credence attitudes. A ranged credence is a distinct type of doxastic attitude from a point-valued credence, regardless of whether we represent it formally with a numerical range, a set of point-valued distributions, or something else.

If we can convey an agent's attitudes toward individual propositions by associating each proposition with a numerical range, why go the extra step of building a full representor? Levi argued that a representor has much more expressive capacity than just a set of numerical ranges assigned to propositions. For example, suppose an agent assigns the range $[0.2, 0.5]$ to proposition J and $[0.6, 0.8]$ to proposition L. This agent is clearly more confident in L

than J. But now suppose the agent also assigns $[0.4, 0.7]$ to K. Is the agent more confident in K than J? It turns out we can build different representors assigning the propositions those very same ranges which answer that question differently. We can construct one representor associated with those ranges that contains only probability distributions with $\Pr(J) < \Pr(K)$. According to that representor, the agent is more confident in K than J. But there's another representor associated with those ranges in which some distributions have $\Pr(J) < \Pr(K)$, while at least one distribution has $\Pr(J) > \Pr(K)$. (See Exercise 14.5.) On that representor, J and K are incommensurable for the agent. So we have two representors associated with the same numerical ranges, which disagree on the answer to an important comparative question. Representors may contain more information than their affiliated numerical ranges.

Similarly, if I took the representor from Figure 14.1 and reported to you only the upper and lower probabilities it generates for various propositions, you would miss an important feature of that representor. If we use the Ratio Formula to calculate conditional probabilities for each of the two distributions in that representor, we'll find that each assigns $\Pr(A \mid R) = \Pr(A)$. While they disagree on the precise values assigned to A, the distributions are unanimous that A is probabilistically independent of R. So according to this representor, I take the prospects for rain tomorrow to be uncorrelated with the prospects for intelligent alien life. A representor may encode judgments of probabilistic independence, judgments of probabilistic correlation, and therefore also con-firmational judgments.[30] Such judgments are not conveyed by mere numerical ranges, and as we saw in Section 14.1.3, cannot typically be recovered from unconditional confidence comparisons.

The information encoded in a representor also enables a very natural updating scheme. Suppose I start with the representor of Figure 14.1, then learn the proposition $A \supset R$. We can update the representor by conditionalizing each of its individual distributions on the evidence gained. After the update, my new representor is the one described in Figure 14.2. In this representor, my range for A is $[1/2, 2/3]$, and my range for R is now the non-singleton $[5/6, 8/9]$. Once more, numerical ranges or unconditional comparisons alone wouldn't provide enough information to uniquely specify such an update.

One does, however, need to be careful when working with representors. Knowing *which* unanimous features of the distributions in the set to read as genuine features of the agent is a bit of an art form.[31] For example, every distribution in the set assigns a point value to each proposition; yet we use representors to represent attitudes that aren't best understood as point-valued.[32] Perhaps more interestingly, we apply probabilism to ranged credences by

A	R	Pr_x	Pr_y
T	T	2/3	1/2
T	F	0	0
F	T	2/9	1/3
F	F	1/9	1/6

Figure 14.2 Representor after an update

requiring each distribution in a representor to satisfy the Kolmogorov axioms and Ratio Formula. What consequences does this have for the numerical ranges we read off the representor for individual propositions?

The answer is simplest for Normality: Since each distribution in the set will assign 1 to tautologies, a rational agent will assign $[1, 1]$ to every tautology. Non-Negativity entails that no rational range assigned to a proposition will ever contain a negative value. The implications of Finite Additivity are more subtle. For example, in the post-update representor of Figure 14.2, the mutually exclusive propositions A and $\sim\!A$ have ranges of $[1/2, 2/3]$ and $[1/3, 1/2]$ respectively. Adding the upper probabilities assigned to A and $\sim\!A$ yields 7/6, while adding their lower probabilities yields 5/6. Yet the tautological disjunction $A \vee \sim\!A$ receives a range of $[1, 1]$. In general, whenever you have two mutually exclusive propositions, the sum of their upper probabilities will be *at least as great* as the upper probability of their disjunction. The sum of their lower probabilities will never be *greater than* the lower probability of the disjunction. Using the terminology we introduced in Chapter 2, lower probabilities are sometimes superadditive, while upper probabilities may be subadditive.[33]

14.2.1 Ranged credences, representation, and evidence

If, in Chapter 1, you found it unrealistic to characterize an agent as assigning a particular real number to each proposition she entertains, this ranged credence business might make you *really* nervous—now we need the agent to assign *two* numbers to each claim? Or sign on to a whole *set* of numerical distributions? You might also wonder about the phenomenology of ranged credences. We think we have some grasp of what it is to believe a proposition, and maybe even to be highly confident of one. But what is it *like* to assign a ranged credence?

Levi (1980, §9.1) thought an agent adopted a ranged credence towards a proposition when she saw multiple point-valued credences towards that proposition as permissible given her evidence, and wanted to suspend judgment among them. Other authors, though, think of ranged credences as more fundamental than point-valued credences; they characterize the latter in terms of the former, rather than vice versa. Moreover, there may be situations (we'll consider some candidates below) in which a ranged credence is rationally required while no point-valued credence is rationally permitted.

When I introduced credences in Chapter 1, I eschewed the idea that they required agents to have numbers in their heads. And I tried not to rely on any phenomenological characterization of credence. Instead, I asked whether agents' doxastic attitudes have a level of structure that is usefully represented by a numerical distribution over a propositional language. There seemed to be features of an agent's doxastic state at a time—or changes in her doxastic state over time—that couldn't be adequately captured by a purely classificatory or comparative representation. Now we've seen that point-valued credence distributions may be similarly inadequate. It seems very plausible that an agent could simultaneously satisfy all of the following conditions for a particular pair of propositions A and R:

(1) be more confident of A than $\sim A$;
(2) be more confident of R than $\sim R$;
(3) take A and R to be independent; while at the same time
(4) view A and R as incommensurable (that is, make no unconditional confidence comparison between the two)

There is no single probability distribution (complete or partial) that satisfies all four of these conditions. To capture the first two conditions, the distribution would have to assign precise numerical values to A, $\sim A$, R, and $\sim R$. But this would yield a comparison of A to R, which would ruin the fourth condition. On the other hand, we've seen a representor (in Figure 14.1) that depicts a doxastic state satisfying all of these conditions. So representors allow us to represent realistic doxastic states not properly representable by point-valued credences.[34]

As a formal tool, representors allow us to represent a broader range of doxastic states than point-valued distributions do. Every doxastic state that can be represented by a point-valued distribution can also be captured by a representor, but the converse does not hold. So working with representors may enhance the *descriptive* project of Bayesian epistemology. Interestingly, though, the case for using representors has most often been made by appealing to *normative* considerations. Levi once wrote that, "even if men have, at least to a good degree of approximation, the abilities Bayesians attribute to them, there are many situations where, in my opinion, rational men *ought not* to have ... precise probability judgments" (1974, pp. 394–5). In such situations, representors are offered as the best representation of a type of attitude rationally mandated *in place of* point-valued credences.

What kind of situation are we talking about? Adam Elga suggests the following example:

> A stranger approaches you on the street and starts pulling out objects from a bag. The first three objects he pulls out are a regular-sized tube of toothpaste, a live jellyfish, and a travel-sized tube of toothpaste. To what degree should you believe that the next object he pulls out will be another tube of toothpaste?
>
> The answer is not clear. The contents of the bag are clearly bizarre. You have no theory of "what insane people on the street are likely to carry in their bags," nor have you encountered any particularly relevant statistics about this. The situation doesn't have any obvious symmetries, so principles of indifference seem to be of no help ....
>
> It is very natural in such cases to say: You shouldn't have *any* very precise degree of confidence in the claim that the next object will be toothpaste. It is very natural to say: Your degree of belief should be *indeterminate* or *vague* or *interval-valued.*   (2010, p. 1, emphases in original)

Elga is clearly on to *something* here; a point-valued credence feels like an uncomfortable response to this body of evidence. But why? The *precision* of point-valued credences—the seeming arbitrariness of selecting 0.7 as opposed to 0.71, or 0.69, etc.—can't be the real problem. After all, when we move to ranged credences there will still be an arbitrariness concern about why the endpoints of a given range weren't 0.01 higher or lower.[35]

If there are cases (such as, perhaps, Elga's) in which an agent's evidence rationally requires her to assign something other than a point-valued credence to a particular proposition, we need to get clear on what kinds of cases those might be, and what exactly about them makes the point value inappropriate. Here Joyce (2005) helpfully distinguishes three features of a body of evidence that may call for different types of attitudinal responses. He calls them "balance", "weight", and "specificity". We will discuss each feature of evidence in turn, and ask whether it can be adequately reflected in point-valued credences or requires a ranged attitude.

The **balance** of a body of evidence relative to a particular proposition indicates which way that evidence leans on that proposition, and how far it leans. For example, suppose I show you an urn containing 100 balls, and tell you that each ball is either black or white. In a while your friend Raj will come into the room and draw one ball from the urn, but you have a chance to get a sneak peek first. So you pull out ten balls, one at a time, replacing each ball in the urn before drawing the next one. Let's suppose your draw is eight black balls and two white. I now ask how confident you are that Raj's ball will be black. The balance of this evidence is toward black rather than white. If your sample had been 5-5, the evidence would have been perfectly balanced between black and its negation.

The balance of an agent's total evidence with respect to a proposition is easily reflected in a rational point-valued credence adopted towards that proposition—perfectly balanced evidence merits a credence of 0.5, while imperfect balance is indicated by a higher or lower value (depending on which way the evidence leans).

The **weight** of a body of evidence relative to a particular proposition captures how *much* information about that proposition the evidence provides. This terminology originated in Keynes, who wrote:

> As the relevant evidence at our disposal increases, the magnitude of the probability of the argument may either decrease or increase, according as the new knowledge strengthens the unfavourable or the favourable evidence; but *something* seems to have increased in either case—we have a more substantial basis upon which to rest our conclusion. I express this by saying that an accession of new evidence increases the *weight* of an argument. New evidence will sometimes decrease the probability of an argument, but it will always increase its "weight."   (1921, p. 78, emphases in original)

For example, suppose that instead of drawing ten balls from the urn with replacement and observing a 5-5 split, you had drawn one thousand balls with replacement and found a 500-500 split. Each of these evidential samples is perfectly balanced between the propositions that Raj's draw will be black or white, but the larger sample bears more weight than the former.

While the weight of an agent's total evidence with respect to a particular proposition may not be reflected in the point-valued credence she assigns to that proposition, it might be captured by other point-valued credences she assigns. After both the 5-5 draw and the 500-500 draw, the most rational point-valued credence in the proposition that Raj's ball will be black seems

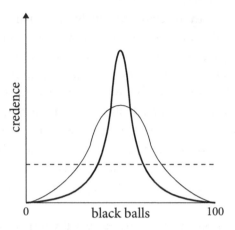

Figure 14.3 Increasing reslience

to be 0.5. So that particular value doesn't vary with the weight of the evidence. But consider your credences in various *chance* propositions about Raj's draw. Let's suppose that before you drew any balls from the urn, you were equally confident in each of the 101 available hypotheses about the urn's contents (running from 100 black/0 white to 0 black/100 white). This "flat" distribution over hypotheses is depicted by the dashed line in Figure 14.3. Drawing a sample of ten balls with a 5-5 split will rule out the most extreme hypotheses about the contents, make you less confident of hypotheses close to those, and make you more confident of compositions close to 50/50. The resulting credence distribution is represented by the thin solid curve in Figure 14.3. Finally, a perfectly split sample of 1,000 balls will make you even more confident of the urn compositions in the middle and even less confident of the options near the edges. This distribution is represented by the bold curve in the figure. While your credence that Raj's draw will be black is 0.5 in all three cases, the weight of the evidence behind that credence is reflected by the narrowness of your credence distribution over affiliated chance hypotheses.

In this case, your credence in the proposition that Raj's ball will be black develops out of your credences in a set of associated chance propositions. So changes in the weight of your evidence for the former proposition can be reflected by changes in your credences assigned to the latter. But there may be propositions for which an agent's credence is *not* constructed from affiliated chance propositions. How can we represent the weight of agents' evidence concerning such propositions? Brian Skyrms (1980a) pointed out that the three distributions depicted in Figure 14.3 vary in their **resilience**. The resilience of a credence in a particular proposition measures how dramatically that credence would change in the face of specific pieces of possible future evidence—with *more* resilient credences changing *less*. For example, suppose it is suddenly revealed that before you sampled any balls from the urn, another friend of Raj's came along and drew a sample of nine black and one white. How much will this new evidence *change* your credence that Raj's ball will be black? If your distribution over chance hypotheses is the flat one depicted with the dashes in Figure 14.3, your credence may move a great deal, landing near 0.9. But if you've already drawn your own sample that came out 5-5 (and so assign the thin-lined curve in Figure 14.3), adding evidence about the other friend's sample will move your credence less. Your new credence that Raj's draw will be black may land near 0.7. And if you've already drawn a sample with a 500-500 split (and therefore assign the bold curve), your credence in black will be highly resilient, and the new evidence will move it almost not at all.[36]

In other words, each of the distributions in Figure 14.3 yields the same value for cr(Raj's ball black), but assigns a different value to

$$cr(\text{Raj's ball black} \mid 9\text{-}1 \text{ sample}) \qquad (14.19)$$

The closer this conditional value is to the unconditional prior, the more resilient your credence. And it seems rational to have more resilient credences the weightier one's evidence is. (For another example of this effect, see Exercise 14.6.)

Finally, here's the kicker: Even when an agent's unconditional credence in a proposition is not driven by her credences in associated chance hypotheses, she may still assign conditional credences that govern how her attitude toward that proposition will change in light of possible future courses of evidence. These conditional credences dictate the resilience of her unconditional stance. So while an agent's current unconditional credence in a proposition reflects the balance of her total evidence with respect to that proposition, the agent's conditional credences that drive potential updates reflect not only the balance of her current evidence but also its weight.[37]

The responses I've offered to evidential balance and weight can both take place within a point-valued credence regime.[38] Of Joyce's three features of evidence, the only one that seems to *require* a ranged response is the evidence's degree of **specificity**. Joyce characterizes this feature as follows:

In the terminology to be used here, data is less than fully specific with respect to X when it is either *incomplete* in the sense that it fails to discriminate X from incompatible alternatives, or when it is *ambiguous* in the sense of being subject to different readings that alter its evidential significance for X. Both incompleteness and ambiguity are defined relative to a given hypothesis, and both are matters of degree.   (2005, p. 167, emphases in original)

Return to the urn of 100 either black or white balls, and consider the initial moment when you don't yet have *any* sampling information about its contents. How confident should you be that Raj's ball will be black? A Subjective Bayesian (in the normative sense) will say that this credence should be driven by your priors, and multiple priors are rationally acceptable in this case. Objective Bayesians, on the other hand, demand a unique rational response even to such scant data. As we saw in Section 5.3, the Principle of Indifference could be applied to set your credence at 0.5. But indifference principles face a number of objections. And in any case, Joyce and others argue that to assign a credence of 0.5 (or any other point-valued credence) is to act as if you have specific evidence about the urn's contents when you don't. As Knight (1921, Ch. VIII) put it, this is to confuse "risk" with "uncertainty".[39] You face risk when you sit down at a roulette wheel—the outcome is unknown, but you have specific evidence about frequencies and chances on which to base your attitudes. Elga's jellyfish bag, on the other hand, generates uncertainty—the evidence doesn't offer any similar basis for probability assignments.[40]

Point-valued credences are appropriate when your evidential situation involves only risk. But many authors would argue that to assign a point-valued credence in the face of uncertainty is to pretend like you're in a kind of evidential situation that you're not. Better in such cases to adopt a ranged credence. When you don't have any evidence about the contents of the urn, your representor should assign the proposition that Raj's ball will be black a range of values from 0 to 1.

This approach might be used to build an Objective Bayesianism without indifference principles. After all, it gives us a unique rational response to the initial urn situation. Even if one thinks that cases involving unspecific evidence rationally require ranged credences, one could maintain that in each such case exactly one ranged credence is permitted.[41] Or one might hold that some unspecific bodies of evidence allow for multiple permissible ranged responses (generated by alternative rational ranged priors), in which case one would be a Subjective Bayesian about ranged credences.[42]

14.2.2 Extensions and challenges

Now that we have some idea of the representor formalism and what sorts of work it's meant to accomplish, I will describe some challenges for the approach and places in which it might be further fleshed out.

Convexity. Consider again the representor of Figure 14.1. I suggested we associate that representor with the interval $[0.6, 0.75]$ for the proposition A. I chose that interval because it's the tightest interval that captures every real value assigned to A by at least one distribution in the representor.

As I mentioned in note 28, the representor described in Figure 14.1 is fairly unrealistic; it's hard to imagine a real-life situation that would rationalize the doxastic attitude represented by just those two probability distributions. But despite being unrealistic, the example raises an important question: Is it rationally acceptable for the probability distributions in a representor to assign just a smattering of values to a particular proposition? Or if the features of an agent's evidence (weight, specificity, etc.) merit including distributions that assign, say, both 0.6 and 0.75 to a particular proposition, shouldn't they also merit including some distributions that assign values in-between?

If we take this last suggestion to its logical extreme, we get what Weisberg (2009) calls the

Interval Requirement: Suppose Pr_x and Pr_y are two distributions in the representor, and P is some proposition in $\mathcal{L}$. If $Pr_x(P) = x$ and $Pr_y(P) = y$, then for any real number z between x and y, there is a distribution Pr_z in the representor such that $Pr_z(P) = z$.

The representor of Figure 14.1 fails to satisfy the Interval Requirement. That representor contains distributions assigning 0.6 and 0.75 to A, but no distribution assigning, for instance, a value of 0.7 (or any of the other values strictly between 0.6 and 0.75).

Why might we adopt the Interval Requirement as a constraint on representors? Here's one reason: Suppose we think of a representor as a convenient formal means of representing confidence comparisons. In the Figure 14.1 representor, both distributions assign $Pr(A) > Pr(\sim A)$. But if that inequality is satisfied by a distribution that assigns $Pr(A) = 0.6$, and by a distribution that assigns $Pr(A) = 0.75$, then it's also going to be satisfied by distributions that assign A values between 0.6 and 0.75 (not to mention a variety of others!). The set of values that satisfies an inequality forms a continuous range, not just a discontinuous collection of scattered reals.

One important way in which a representor can satisfy the Interval Requirement is for it to display

Convexity: If a representor contains two distributions $\Pr_x$ and $\Pr_y$, it also contains every linear combination of those two distributions.

Here's what we mean by a linear combination of two distributions: Given the distributions $\Pr_x$ and $\Pr_y$, pick any real number α between 0 and 1 (inclusive). Now generate a new distribution $\Pr_z$ as follows: For each proposition $P \in \mathcal{L}$, $\Pr_z(P) = \alpha \cdot \Pr_x(P) + (1-\alpha) \cdot \Pr_y(P)$. (Notice that once you've chosen a value for α, it's held fixed as you construct all of $\Pr_z$—you don't choose a new α for each proposition.) Convexity says that if a representor contains two distributions, any new distribution you can generate by mixing them in this fashion will also be in the representor. (As you'll prove in Exercise 14.7, the linear combination of two probability distributions is always a probability distribution, so we don't have to worry that Convexity will insert distributions into a representor that violate the Kolmogorov axioms.)

While a convex representor will always satisfy the Interval Requirement, Convexity is actually logically stronger than the Interval Requirement. (That is, it's possible for a representor to satisfy the Interval Requirement without satisfying Convexity.)[43] Should we hold that rationality demands something as strong as Convexity?

That depends on how we understand ranged credences. Levi thought that an agent adopted a ranged credence when she viewed multiple point-valued credence distributions as permissible, and wanted to suspend judgment among them. He wrote:

> The suspension of judgment between the two [distributions] is manifested as a sort of conlict between two systems of values.... Weighted averages of the two [distributions] have all the earmarks of potential resolutions of the conflict; and, given the assumption that one should not preclude potential resolutions when suspending judgment between rival systems of valuations, all weighted averages of the two [distributions] are thus to be taken into account.   (1980, p. 192)

When an agent is conflicted between distributions $\Pr_x$ and $\Pr_y$, a linear combination is a kind of compromise between the two. To be tolerant of all such possible compromises, a reasonable agent's representor will include all such linear combinations.

But as I noted earlier, not everyone understands ranged credences this way. Moreover, convex representors have an important formal problem. When you make a linear combination of two probability distributions, some features shared by both of the original distributions are maintained. For example, if both original distributions assign a higher value to one proposition than another, the linear combination will maintain that comparison. But other shared features—especially features involving probabilistic correlation and independence—may be lost. Jeffrey (1987), who pointed out this problem, provided an example in which two distributions each treat a particular pair of propositions as independent, yet a linear combination of those distributions does not. (You'll analyze his example in Exercise 14.8.) Generally, if we have an agent who views two propositions as probabilistically irrelevant to each other, and we want to represent that agent with a set of distributions unanimously agreed on that irrelevance, that set may need to violate Convexity.[44]

Closed intervals and sticky ranges. Now I'd like to turn our attention to another question about the range of values a representor should assign to a proposition. Return to our example of the urn containing 100 balls. Suppose you haven't yet learned of any samples from the urn, and so know nothing more about its composition than that each of the balls is either black or white. Due to the maximal unspecificity of your evidence in this case, a ranged credence enthusiast might say that you should assign the proposition that Raj's draw from the urn will yield a black ball a credence ranging from 0 to 1. For instance, Mark Kaplan writes:

> Giving evidence its due requires that you rule out only those [probability distributions] the evidence gives you *reason* to rule out. And when you have no evidence whatsoever pertaining to the truth or falsehood of a hypothesis P, then, for every real number n, $1 \geq n \geq 0$, your set of [probability distributions] should contain at least one assignment on which [$\Pr(P) = n$].
> (1996, p. 28, emphasis in original)

Notice that Kaplan says your representor should contain a probability distribution assigning P each real number between 0 and 1—including 0 and 1 themselves. In our urn example, that means the representor should contain at least one distribution that assigns 0 to black, and another that assigns 1 to black. We associate this representor with the **closed interval** $[0, 1]$ for black.

On the other hand, if your representor contained distributions assigning black every real number strictly *between* 0 and 1, but no distributions assigning the extreme values, we would associate your representor with the **open interval** (0, 1) for black.[45]

What's the difference between these two ranges? Here it may help to think about the values each distribution in the representor assigns to each of the chance hypotheses consistent with your evidence. In order for the representor to assign the closed interval to black, there must be a distribution in the representor that assigns a value of 1 to black. If each distribution satisfies the Principal Principle (Section 5.2.1), then each distribution's value for black will be a weighted average of all the chance hypotheses to which it assigns a positive probability. So the only way a distribution can assign a value of 1 to black is if it assigns probability 1 to the hypothesis that all the balls in the urn are black, and 0 to every hypothesis with at least one white ball. At the other end, the hard 0 at the bottom of the closed [0, 1] interval requires a distribution that assigns probability 0 to every hypothesis with at least one black ball.

These extreme distributions might be rationally ruled out for two reasons. First, recall that on the ranged credence approach, we apply a rational require- ment to an agent's attitudes by applying it separately to each of the distributions in that agent's representor. If we think that the Regularity Principle (Section 4.2) is a requirement of rationality, we might want to forbid any distribution in a representor from assigning probability 0 to any doxastically possible proposition.[46]

Second, a closed [0, 1] interval would wreak havoc on the updating scheme for ranges we outlined earlier. The idea there was to conditionalize each distri- bution in the representor on whatever new evidence is gained. But if our first sample from the urn contains at least one white ball, this outcome will initially have been assigned a probability of 0 by the 100%-black distribution. Given the Ratio Formula and the definition of Conditionalization, it's impossible to conditionalize a distribution on evidence with a prior of 0. So while an agent who assigns a range of [0, 1] to black certainly thinks it's possible that her first sample will contain at least one white ball, the standard updating scheme for ranged credences will leave her unable to update on that kind of evidence should it be received.[47]

So maybe the rational initial attitude toward black is best represented by an open (0, 1) interval. Here's one way to build a representor assigning that range: Take the 101 initially available hypotheses about the contents of the urn, and have one distribution in the representor for each possible assignment of values to those hypotheses—provided those values honor the Kolmogorov axioms and don't assign 0 to any of the hypotheses. The resulting representor seems particularly apt to the unspecificity of your evidence in this case, since you entirely lack data about the black/white breakdown in the urn.

If you adopt that representor, how will your attitudes change as the evidence comes in? As you sample from the urn—drawing one ball at a time, observing it, then replacing it—something interesting will happen to your ranged attitude. Once you've seen at least one black ball and one white ball come out of the urn, your ranged credence that Raj's ball will be black will narrow to the open interval $(1/100, 99/100)$. That's because your evidence will have eliminated hypotheses on which the chance of black is less than $1/100$ (namely, hypotheses in which the urn contains no black balls) and hypotheses on which the chance of black is greater than $99/100$ (hypotheses with no white balls). But as long as all you do is sample balls one at a time with replacement—you don't sample multiple balls at a time, or dump out the entire urn and count—contionalizing will then leave your range for black at $(1/100, 99/100)$ going forward. For any sequence of draws you might get (no matter how long), and for any x such that $1/100 < x < 99/100$, there will have been some distribution in your initial representor that assigned black a probability of exactly x conditional on observing that particular sequence of evidence.[48]

Intuitively this may seem wrong. Suppose you make 1,000 draws with a 500-500 split. This is weighty evidence, perfectly balanced between black and white. It highly confirms that the urn contains somewhere around fifty black balls. So shouldn't this sample narrow your ranged credence, drawing your upper and lower probabilities that Raj's ball will be black closer to 0.5?

Here we have to remember that on the ranged credence approach, rational requirements apply to each distribution in the representor considered one at a time. Requirements applied singly to distributions in a representor may not produce intuitively expected effects on the overall range.

Start with the representor described above assigning $(1/100, 99/100)$ to the proposition that Raj's ball will black. Each distribution in that representor will assign a particular (nonextreme) probability that there is one black ball in the urn, a probability that there are two black balls, etc. Now imagine taking a particular single one of those distributions, and conditionalizing it on a sample of 500 black balls and 500 white. That evidence will make that distribution more peaked around a 50-50 white/black split: the evidence will decrease the distribution's probability that there's only one black ball in the urn; increase its probability that there are fifty; etc. Since the sample increases the distribution's probability that there are fifty black balls in the urn, that sample is positively relevant to the fifty-black hypothesis, relative to that distribution. Moreover, each distribution in the representor will respond to the sample in a similar fashion—the sample will be positively relevant to the fifty-black hypothesis on every distribution. Since this property is shared by every distribution in the representor, we can attribute it to your overall doxastic state: a 500-500 sample confirms for you that the urn contains fifty black balls.[49]

But shouldn't the sample *also* narrow the range of values the distributions in your representor assign the proposition that Raj's ball will be black? Well, each individual distribution in your representor will respond to the sample by moving its overall probability that Raj's ball will be black closer to 0.5 (unless the distribution already assigned that proposition exactly 0.5). So the intuition that your attitude to black should move "toward" 0.5 is respected in some sense. But because your initial representor contained distributions that were arbitrarily far from assigning that proposition 0.5, moving each distribution's probability that Raj will draw black closer to 0.5 still leaves distributions that are as far from 0.5 as you like (within the confines of 1/100 to 99/100). Each individual distribution in the representor does the intuitive thing in response to your evidence, but your overall range for the proposition that Raj's ball will be black remains unchanged.[50]

Dilation. Earlier I described the traditional updating procedure for ranged credences: take each distribution in your representor, conditionalize it on the evidence learned, then compose an updated representor from the set of distributions that result. As an example, I took the two distributions described in Figure 14.1, updated them on the evidence $A \supset R$, and obtained the representor described in Figure 14.2.

There's something odd about that update, which you may not have noticed at the time. While the distributions in the representor prior to the update both assign proposition R a probability of 2/3, the distributions in the posterior representor assign distinct values to R. The update takes the ranged credence in R from [2/3, 2/3] to [5/6, 8/9]. This is an example of **dilation**, a phenomenon in which conditionalizing on new evidence widens the range of values an agent assigns to a particular proposition.[51]

A number of Bayesians have found dilation intuitively troubling. In fact, some Bayesians have found it so troubling that they reject either the conditionalization updating scheme for ranged credences or the ranged credence approach entirely! These Bayesians are especially troubled by cases in which an agent knows in advance exactly what kind of evidence she's going to get—she knows that she will soon learn whether some particular proposition P is true or false. At the same time, she knows that whatever evidence she gets about

P, it will dilate the ranged credence she assigns some other proposition *Q*. We can even build cases in which the agent knows that learning the truth about *P* will dilate her currently point-valued credence in *Q* all the way to a 0-to-1 range! (See Exercise 14.10 for an example.)

Seidenfeld and Wasserman illustrate the concern with a real-world example:

> To emphasize the counterintuitive nature of dilation, imagine that a physician tells you that you have probability 1/2 that you have a fatal disease. He then informs you that he will carry out a blood test tomorrow. Regardless of the outcome of the test, if he conditions on the new evidence, he will then have lower probability 0 and upper probability 1 that you have the disease. Should you allow the test to be performed? Is it rational to pay a fee not to perform the test? (1993, p. 1140)

Considering a similar case, van Fraassen (2005, p. 28) complains that if dilation is possible, then "reception of certain factual information can be damaging to one's epistemic health."

But we should question the assumption that widening your ranged credence in a proposition always represents an epistemic degradation. Seidenfeld and Wasserman report (though don't endorse) a "seeming intuition that when we condition on new evidence, upper and lower probabilities should shrink toward each other" (1993, p. 1140). This intuition conflates evidential weight and specificity[52]—and it's the latter that should drive the width of ranged credences. Sometimes evidence that provides us with additional information about a proposition (thereby increasing evidential weight) is nevertheless less specific than the evidence we had before. Brian Weatherson offers this example:

> Imagine we are playing a rather simple form of poker, where each player is dealt five cards and then bets on who has the best hand. Before the bets start, I can work out the chance that some other player, say Monica, has a straight. So my credence in the proposition *Monica has a straight* will be precise. But as soon as the betting starts, my credence in this will vary, and will probably become imprecise. Do those facial ticks mean that she is happy with the cards or disappointed? Is she betting high because she has a strong hand or because she is bluffing? Before the betting starts we have risk, but no uncertainty, because the relevant probabilities are all known. After betting starts, uncertainty is rife. (2002, p. 52)

The moment you learn the rules of Weatherson's poker game, you can work out the precise chance that Monica is dealt a straight, and assign a point-valued credence equal to that chance. When Monica looks at her cards, adopts a facial expression, then pushes some chips into the pot, all of this provides you with important information about her hand. The weight of your evidence increases, and in an important sense your epistemic situation improves. Yet this new information is unspecific, and may lead you to a ranged posterior credence. Nothing about this dilation seems to me the least bit irrational.

What about cases in which new evidence dilates your previously point-valued credence in a proposition to the full $(0, 1)$ range? For evidence E to change your credence in hypothesis H at all, at least some of the distributions in your prior representor must view E as probabilistically correlated with H. Dilation occurs when the distributions disagree on that correlation—on its direction (positive or negative?) and/or on its strength. Earlier we saw that when your evidence is unspecific about the chance of a particular event, ranged credence advocates require you to assign a ranged credence to that event's occurrence instead of acting as if your evidence had delivered a specific point value. Similarly, when your evidence is unspecific about the correlation between E and H, you shouldn't respond to learning E as if you know precisely how it's correlated with H. You should dilate to a posterior attitude toward H that reflects the unspecificity of what you know about its correlation with E. And if what you know about that correlation is maximally unspecific, the resulting posterior in H should be a maximal range.[53]

Decision theory. Since a representor often contains more information than a comparative confidence ranking—in particular, information about relative gap sizes between confidences—it provides a more promising basis for a decision theory. Still, the topic of how to make decisions on the basis of ranged credences is very unsettled, generating a great deal of controversy in the literature. Rather than trying here to capture all the back and forth, I'll simply indicate some of the basic issues that arise.

Suppose an agent confronts a decision between two acts. The expected utility she assigns to each act depends in part on her credence distribution over possible states of the world. So if the agent's representor contains multiple distributions, the acts may receive different expected utilities relative to different distributions. Following the idea of respecting any property shared by *all* the distributions in a representor, everyone in the literature[54] agrees that if every distribution in an agent's representor assigns the first act a higher expected utility than the second act, then the agent should prefer the first act.

For example, the representor of Figure 14.1 assigns proposition A a ranged credence of [0.6, 0.75]. Suppose you have the corresponding doxastic state, and I offer you a ticket that pays \$1 if A is true, and nothing otherwise. If my ticket costs you only \$0.50, then you clearly ought to buy it. The distributions in your representor generate different fair prices for that ticket: one yields \$0.60, the other \$0.75. But on both distributions, \$0.50 is below the fair ticket price. In general, you should be willing to buy a ticket that pays \$1 on A for any amount up to \$0.60. And you should be willing to *sell* such a ticket for any price of at least \$0.75.

It's fairly easy to prove[55] that if you follow this policy—buy a betting ticket on a proposition for up to your lower probability for that proposition, sell a betting ticket for at least your upper probability—and the distributions in your representor are probabilistic, then the policy will never endorse a set of bets at a given time that guarantees a sure loss come what may. In the terminology of Chapter 9, you will not be susceptible to a Dutch Book.

But the policy provides incomplete advice. What should you do if I offer to sell you a \$1 ticket on A for \$0.70? That's a good price for this ticket relative to *one* distribution in your representor, but relative to the other distribution it's too high. The ranged credence decision theory controversy is all about what you should choose when the distributions in your representor disagree about which act maximizes expected utility.[56]

One response is to leave the agent's options open. Mark Kaplan, for instance, holds that in this case you should remain "undecided" between purchasing the ticket and declining. He clarifies that:

> To say that you should be undecided in this case is not, of course, to advise paralysis. It is rather to say that there is no option you have reason to prefer to the others and that, among the options for which no superior option has been found, there is nothing left but to pick one.   (1996, p. 25, n. 20)

The idea (dating back at least to (Good 1952)) is that if an act maximizes expected utility relative to at least one of the distributions in your representor, then that act is rationally permissible, and in cases where multiple acts are permitted by this standard, rationality has nothing more to say about which act you should choose.

One question about this decision policy is whether an agent's choices that go beyond what rationality requires must remain consistent over time. Suppose you make all your betting decisions today based on a fair price of \$0.65 for A, then your decisions tomorrow based on a fair A price of \$0.70. If your ranged credence for A remains at $[0.6, 0.75]$ throughout, both prices are acceptable on the proposed policy. But if you sell an A-ticket for $0.65 today then buy that ticket back for $0.70 tomorrow, you will have been diachronically Dutch-Booked.

Sarah Moss (2015) wants to leave open the possibility of such rational mind-changes, in which the price you're willing to pay for a bet changes over time—even if these changes cost you some cash. Moreover, many authors point out that the problem of diachronic choice consistency is not unique to ranged-credence decision theory. Even with point-valued credences, two acts may be tied for optimal expected utility, in which case a rationally-arbitrary decision is required. (The canonical example is Buridan's Ass.) Such decisions may be repeated over time, and there's a question whether rationality requires you to make the same decision at every opportunity.

Nevertheless, we may want less rational slack in our decision theory—we may want specific advice about how to act when the distributions in your representor are at odds. Here two broad types of strategy are available. Strategies of the first type (what we might call "global" strategies) decide based on some feature of your *entire* representor. For example, when faced with a choice between two acts, you might calculate the lowest expected utility of the first act on any distribution in your representor, calculate the lowest expected utility of the second act, then choose whichever act has the best worst-case scenario. (This is sometimes called the "minimax" strategy.) Strategies of the second type ("local" strategies) specify a procedure for selecting *one* of the distributions in your representor, then direct you to make all of your decisions in line with its evaluations.

Global strategies have various strengths and weakenesses, depending on their particulars. But Elga (2010) points out a problem common to most global strategies: they instruct agents to pass up guaranteed gains.[57] Suppose I offer to sell you a ticket for $0.65 that pays $1 if A is true. A short time later, I offer to sell you a ticket for $0.30 that pays $1 if A is false. If you have the doxastic attitudes described by the representor of Figure 14.1, the minimax strategy will lead you to decline both purchases. (The details are in Exercise 14.11.) By declining both, you are guaranteed to neither gain nor lose any funds. But there's another course of action—purchasing both tickets—that is guaranteed to net you $0.05 in every possible world. Elga claims it's irrational to decline both bets when another option is available that guarantees a better result. And many global strategies direct agents to pass up pairs of bets like this.[58]

Local strategies don't have this problem: no matter which single distribution in your representor drives your betting behavior, it will direct you to purchase at least one of Elga's two bets. Yet local strategies seem to have another problem. After arguing that some epistemic situations demand a ranged credence rather than a point-valued attitude, the ranged credence advocate with a local decision theory now advises you to behave in such situations exactly as you would if you possessed a single point-valued distribution. It's not clear why this theorist went to the trouble to differentiate—let alone argue for—ranged credences to begin with. Yet Levi (1980), Walley (1991), and Joyce (2010) all emphasize that letting one distribution drive the bus for practical purposes is not the same as possessing a representor containing only that distribution. Doxastic states have many roles to play in our mental economy beyond their role in decision-making.[59]

14.3 Dempster-Shafer theory

Most of this book has concerned the traditional Bayesian formalism for representing an agent's doxastic attitudes: point-valued credence distributions over propositional languages. In this chapter, we have introduced two other formalisms: comparative confidence rankings and sets of probability distributions. Still, we have barely scratched the surface of the many formalisms available for representing agents' doxastic states. A more complete survey would describe AGM theory, dynamic epistemic logic, default logic, formal learning theory, and ranking theory (just to name a few). While I lack both the space and the expertise to do those other approaches justice, I want to end with a brief overview of a formal approach that has much in common with traditional Bayesianism, yet diverges from it in a number of interesting ways.

The theory of belief functions originated in the work of Arthur P. Dempster, starting with his (1966). It was then generalized and expanded by Glenn Shafer's pivotal (1976). In his (1981), Jeffrey A. Barnett dubbed the resulting theory "Dempster-Shafer theory"; belief functions are therefore sometimes known as **Dempster-Shafer functions**.

To understand how Dempster-Shafer theory differs from traditional Bayesianism, consider an example adapted from (Halpern 2003, ch. 2): In the next room is an urn containing balls, each of which is either red, blue, or yellow. A ball has just been drawn from the urn. An absolutely reliable witness walks in from the next room and tells you something in Urdu. Your Urdu is rusty: you think there's an 80% chance he just said, "The ball drawn was either blue or yellow"; but there's a 20% chance he said, "That's a very attractive urn."

To model the import of this piece of evidence, Dempster-Shafer theory starts with a propositional language, just like the traditional Bayesian approach. However, a Dempster-Shafer theorist's propositional language will typically not include *all* of the propositions that can be composed from its atomics using propositional connectives. For technical reasons the Dempster-Shafer theorist will work with a **Boolean algebra**—a proper subset of the full propositional language that contains as many propositions as possible built from the atomics without any two members' being logically equivalent. As we noted in Exercise 2.3, a Boolean algebra with n atomic propositions will contain 2^{2^n} propositions overall, instead of the infinitely many propositions of the languages we've been working with.

With this Boolean algebra in place, the Dempster-Shafer theorist will represent the evidential significance of the Urdu testimony using a **mass function**. Formally, a mass function is a non-negative distribution over a Boolean algebra that: (1) assigns 0 to the contradiction; and (2) has values summing to 1. Letting "R" stand for the proposition that the ball drawn was red, "B" stand for blue, and "Y" stand for yellow, the mass function representing the Urdu testimony would assign a value of 0.8 to $B \lor Y$ and a value of 0.2 to the tautology T.[60] The mass value of 0.8 represents the evidential support supplied by the Urdu testimony to the proposition that the ball drawn was either blue or yellow. The other 0.2 of evidential mass supplied by this testimony does not support any particular proposition about the ball's color, so it gets assigned to the informationally vacuous tautology.

This mass function represents the significance of a particular piece of evidence, the Urdu testimony. We can generate a **belief function** from the mass function, representing a doxastic state justified by that evidence. A belief function is another distribution over the Boolean algebra. If we represent the mass function as m, we can generate a belief function Bel from it as follows: for each proposition X in the Boolean algebra, Bel(X) sums the m-values of all of the propositions that entail X (including X itself). For example, the mass function in the previous paragraph yields a belief function assigning Bel($B \lor Y$) = 0.8, Bel(T) = 1 (because both $B \lor Y$ and T entail T), and a Bel-value of 0 to every other proposition in the algebra.

How do belief functions differ from traditional Bayesian credence functions? Like a probabilistic cr, Bel is a non-negative real-valued distribution over a set of propositions. It assigns 0 to contradictions and 1 to tautologies. So it satisfies Normality and Non-Negativity. But belief functions are superadditive: two mutually exclusive disjuncts may have Bel-values summing to less than the Bel-value of their disjunction. For instance, the Urdu example has

$\text{Bel}(B) = \text{Bel}(Y) = 0$, yet $\text{Bel}(B \vee Y) = 0.8$.[61] This is an important divergence from the Bayesian approach. For a Bayesian, credence in a disjunction arises from credence in its disjuncts. But on the Dempster-Shafer approach, belief in a disjunction need not distribute to its disjuncts.

This difference at the level of doxastic states reflects a deeper difference in the ways the two theories approach evidence. Under Finite Additivity, an agent may become more confident in a disjunction of mutually exclusive disjuncts only by growing more confident in at least one of those disjuncts. So on a Bayesian confirmation theory, evidence for such a disjunction is always also evidence for at least one of its disjuncts. Dempster-Shafer theorists, on the other hand, think that a piece of evidence may support a disjunction without supporting either of its disjuncts. (In our Urdu example, $B \vee Y$ receives a positive m-value while neither B nor Y does.) Logically stronger propositions require stronger evidence; you can get evidence for something general (the ball was either blue or yellow) without thereby getting evidence for something more specific (that the ball was blue/that the ball was yellow).

Now there is a special case in which belief functions become probabilistic. If all of the positive value in a mass function is assigned exclusively to state-descriptions, the corresponding belief function will be additive, and therefore be indistinguishable from a Bayesian credence function. (See Exercise 14.13.) A mass function like that represents evidence that is maximally precise; instead of making various sets of possible worlds likely, the evidence speaks to each doxastic possibility one at a time. The Dempster-Shafer theorist will admit that evidence like this is *possible*, and therefore that probabilistic belief functions are sometimes supported by an agent's evidence. But this is a rather special case, uncommon in the real world.

To get a better sense for how belief functions work, let's consider one more example. Suppose another witness to the ball draw comes in from the next room, and says something in Swahili. Your Swahili is even worse than your Urdu. You think there's a 25% chance he said, "The ball was blue"; a 25% chance he said "The ball was blue or yellow"; a 25% chance he said "The ball was red"; and a 25% chance he said, "Lunch was delicious today!" Given our earlier Boolean algebra, the mass function for this evidence would assign:

$$m(B) = 0.25$$
$$m(B \vee Y) = 0.25$$
$$m(R) = 0.25 \tag{14.20}$$
$$m(T) = 0.25$$

Franz Huber (2016, §3.1) helpfully suggests we think of belief functions as follows. Start with a particular proposition, say $B \vee Y$. We can divide the mass of a particular piece of evidence into three parts: the part that speaks in favor of our proposition, the part that speaks against, and the part that's neutral. In Equation (14.20), the masses of both B and $B \vee Y$ speak in favor of $B \vee Y$; the mass assigned to R speaks against (because it supports $\sim(B \vee Y)$); and the mass assigned to T is neutral. When we construct a belief function from this evidence, $\mathrm{Bel}(B \vee Y)$ simply sums up all the mass supporting $B \vee Y$; so $\mathrm{Bel}(B \vee Y) = \mathrm{m}(B) + \mathrm{m}(B \vee Y) = 0.5$.[62] We can define another function, which Dempster-Shafer theorists call a **plausibility function**, that tallies up all the evidential mass that either supports or is neutral on the proposition in question. For this evidence, $\mathrm{Pl}(B \vee Y) = \mathrm{m}(B) + \mathrm{m}(B \vee Y) + \mathrm{m}(\mathsf{T}) = 0.75$. The less one's evidence speaks directly against a proposition, the more plausible it leaves that proposition. In general, for any proposition X in the Boolean algebra, $\mathrm{Pl}(X) = 1 - \mathrm{Bel}(\sim X)$.

I hope it's becoming clear that a mass function is an exceedingly flexible tool for representing many kinds of evidence. When Bayesians update by Conditionalization, evidence is always represented as a single (perhaps conjunctive) proposition in the agent's language. Mass functions are capable of representing evidence like that—the proposition in question receives an m-value of 1, while everything else receives a 0. In Jeffrey Conditionalization (Section 5.5), the evidence spreads its support across a partition—0.70 that the cloth glimpsed by candlelight is green, 0.25 that it's blue, 0.05 that it's violet. We can define a mass function like that as well.[63] But mass functions don't have to spread their positive value over a partition; they can represent evidence that supports multiple, not mutually exclusive propositions at once.

Mass functions may also be used to represent the significance of information-bearing states that aren't pieces of evidence. For example, a mass function may represent the background knowledge an agent brings into an evidence-gathering situation. When that background knowledge contains *no* information relevant to the propositions in our Boolean algebra, we represent it with a *vacuous* mass function: a function that assigns $\mathrm{m}(\mathsf{T}) = 1$ and a mass of zero to everything else. This is a tidy solution to Bayesians' vexing Problem of the Priors (Section 13.1); we get a simple, elegant representation of total ignorance without invoking any subjective elements or indifference principles.

Once we have a mass function representing an agent's background knowledge, and a mass function for each piece of evidence she gains, we may want to combine these into an overall belief function. When the information represented by mass functions comes from unrelated sources, those functions can be assembled using Dempster's rule of combination. The rule of combination takes two mass functions m_1 and m_2, and combines them into a new distribution over the same algebra, $m_1 \oplus m_2$. Exercise 14.14 explains the details of how that combination works. For our purposes, it's important to note that $m_1 \oplus m_2$ will itself be a mass function. So it can be used to generate a belief function, or may be combined with another mass function to iterate the process. In general, an agent's overall belief function for a given situation may be determined by combining the mass function representing her background knowledge with the mass functions representing her pieces of evidence in that situation, then generating a belief function from the mass function that results.

Dempster's combination rule has further nice formal features. Like Conditionalization, it is commutative and associative. (Its results don't depend on the order in which you combine mass functions, or on how you group them together.) As a technique for combining disparate evidential sources, this gives it an advantage over successive updates by Jeffrey Conditionalization. Also, the vacuous mass function described above is the identity element for Dempster combination—combining the vacuous function with any mass function m_1 will simply output m_1 again. So when an agent comes into a situation with no relevant background knowledge, her belief function will be determined entirely by the contributions of her evidence in that situation.

I've already mentioned how belief functions relate to probabilistic credence distributions. We can also relate them to representors.[64] Given any belief function, we can construct a representor (containing only distributions satisfying the probability axioms) such that for every proposition X in the Boolean algebra, $Bel(X)$ equals X's lower probability on that representor, and $Pl(X)$ equals X's upper probability. Interestingly, the converse does not hold: there exist probabilistic representors for which it's impossible to find a belief function such that $Bel(X)$ equals the lower probability of X for every X in the Boolean algebra.[65]

This inability of belief functions to match some representors may create a problem for Dempster-Shafer theory. Walley (1991, §5.13.4) offers an example in which you have two coin flips that are each fair, but whose outcomes may be correlated, anti-correlated, or independent (you have no evidence speaking to that question).[66] On the ranged credence theory, it seems reasonable to represent this situation with a representor in which each of the distributions assigns 1/2 to heads on each flip, but one distribution takes the flip outcomes to be independent, another takes them to be perfectly correlated, another takes them to be perfectly anti-correlated, another takes them to be slightly correlated, and so on. Walley shows (as will you in Exercise 14.15) that the resulting representor cannot have its lower probabilities matched by any belief function. To the extent that this representor captures the appropriate rational response to this evidential situation, the lack of a corresponding belief function seems a strike against Dempster-Shafer theory.

Yet although traditional credence distributions can be formally matched to a subset of belief functions, and belief functions in turn can be formally matched to a subset of representors, Dempster-Shafer theorists warn against reading their belief functions as just another formalism representing the same doxastic attitudes Bayesians have been modeling all along. Dempster-Shafer theorists insist that belief functions are meant to model attitudes grounded in evidence, while Bayesian credences may have other types of influences. (Most obviously, subjective priors.) Shafer himself (1981, §1.2) repeatedly warns against demanding that Bel values equal the chances that various propositions are true. So the Walley example just described may not be immediately fatal to the theory.

A similar point should be made about ongoing difficulties matching Dempster-Shafer theory to a decision theory. Here the options resemble those for the ranged credence approach: (1) give a rule for accepting bets below the lower probability (Bel-value) and above the upper probability (Pl-value), then go silent on all the rest; (2) associate a specific probability distribution with the belief function, then let it drive the betting decisions; etc. None of these options works great, but Dempster-Shafer theorists are keen to disentangle their belief functions from Bayesians' behavioristic past. Shafer emphasizes theoretical goals (such as understanding) as primary to the formation of belief functions, while Liu and Yager (2008, p. 27) write, "The theory of belief functions is not meant to be a normative or descriptive theory for decision making." They emphasize that belief functions have many other important applications, in fields such as engineering, business, and medicine.

There are, however, other problems internal to the Dempster-Shafer approach, even relative to the goals its practitioners have set for themselves. For example, Dempster's rule of combination is supposed to be applied when the sources of information modeled by two mass functions are "unrelated". In practice, this condition is often read like a requirement of conditional probabilistic independence: the reports of each information source should be screened off from each other by the truth-value of the proposition reported upon. But strictly speaking, unrelatedness is supposed to be a primitive notion established antecedently to the application of theory. This leaves us to wonder what exactly unrelatedness means, and how officially to adjudicate it. We also need combination techniques for informational sources that are not unrelated.

While there has been progress on both of these fronts over the years, neither problem is entirely solved. Like all of the formal systems described in this book, Dempster-Shafer theory remains a work in progress.

14.4 Exercises

Throughout these exercises, you should assume that any distribution appearing in a representor satisfies the probability axioms and Ratio Formula—except when explicitly noted otherwise. You may also assume that whenever a conditional probability expression occurs or a proposition is conditionalized upon, the needed proposition has a nonzero unconditional value so that conditional probabilities are well defined.

Problem 14.1. 🌶️🌶️
- (a) Take the bulleted conditions appearing in the list on page 488, and prove each of them from the combination of Comparative Equivalence, Transitivity, Completeness, Non-Negativity, Non-Triviality, and Additivity. You need not prove them in the order in which they appear in the list. <u>Important</u>: For now, do not prove the very last bulleted condition.
- (b) 🌶️🌶️🌶️ Now prove the very last bulleted condition in the list, using Comparative Equivalence, Transitivity, Completeness, Non-Negativity, Non-Triviality, and Additivity, plus anything you proved in part (a). Beware: this proof is *very* difficult!

Problem 14.2. 🌶️ Suppose that given a five-proposition partition $\{A, B, C, D, E\}$, an agent assigns the following comparisons:

$$A \sim B$$

$$C \sim D$$

$$(A \vee B) \sim (C \vee D \vee E)$$

$$(A \vee E) \sim (C \vee D)$$

- (a) Find a probabilistic credence distribution that matches these comparisons as specified in the Comparative Matching principle.

(b) Is your distribution from part (a) unique, in the sense of being the only probabilistic distribution that matches these comparisons? Explain how you know.

(c) Create a numerical distribution that matches the comparisons, is an affine transformation of your answer to part (a), but does not satisfy *any* of the probability axioms.[67]

Problem 14.3. 🎵 In this exercise we will prove that satisfying the Scott Axiom is necessary for a total preorder to be probabilistically representable. To start, assume we have two equinumerous, finite sequences drawn from $\mathcal{L}$—$\{A_1, A_2, \ldots, A_n\}$ and $\{B_1, B_2, \ldots, B_n\}$—that contain the same number of truths as each other on each doxastic possibility.

(a) We can think of the agent's doxastic possibilities as being represented by the state-descriptions of $\mathcal{L}$. Suppose you took each proposition in the A-sequence and replaced it with its equivalent in disjunctive normal form, then did the same for the B-sequence. Now imagine choosing a particular state-description, counting how many A-sequence propositions contained it as a disjunct, then counting how many B-sequence propositions contained it. Explain why, no matter which state-description you chose, these counts would come out the same.

(b) Suppose you had a probability distribution cr over $\mathcal{L}$. Recall that on a probability distribution, the probability of a proposition equals the sum of the probabilities assigned to the state-descriptions in its disjunctive normal form. Use this fact to explain why if you added up the cr-values assigned to all of the propositions in the A-sequence, then added up the cr-values assigned to all the propositions in the B-sequence, you'd get the same result.

(c) Suppose some total preorder over $\mathcal{L}$ matches probabilistic distribution cr according to Comparative Matching, and $\mathcal{L}$ contains two equinumerous, finite sequences—$\{A_1, A_2, \ldots, A_n\}$ and $\{B_1, B_2, \ldots, B_n\}$—that contain the same number of truths as each other on each doxastic possibility. Explain why, if there exists some A_i and B_i such that $A_i \succ B_i$, then there must exist some A_j and B_j such that $A_j \prec B_j$.

Problem 14.4. 🎵 Suppose we have a language $\mathcal{L}$ containing two atomic propositions, P and Q. The probability table below specifies two distributions over $\mathcal{L}$, cr and cr′:

P	Q	cr	cr′
T	T	2/3	6/11
T	F	1/4	3/11
F	T	0	0
F	F	1/12	2/11

(a) Create a comparative confidence ranking over *all sixteen* nonequivalent propositions in $\mathcal{L}$ that matches cr according to the Comparative Matching principle. Then show that cr′ generates the *same* matching ranking over the members of $\mathcal{L}$.

(b) Now suppose you used the Ratio Formula to generate conditional credences from cr and cr′, then used those conditional credences to generate matching conditional confidence comparisons. Show that cr and cr′ differ over whether the agent should be more confident in Q given P than she is in P given ~Q.

Problem 14.5. 🌙 Imagine our language contains the propositions J, K, and L, and we want a representor that gives these three propositions the ranges described below:

$$J : [0.2, 0.5] \qquad K : [0.4, 0.7] \qquad L : [0.6, 0.8]$$

(a) Specify a representor associated with those ranges in which every probability distribution assigns $Pr(J) < Pr(K)$.

(b) Specify a representor associated with those ranges in which at least one probability distribution assigns $Pr(J) < Pr(K)$ and at least one probability distribution assigns $Pr(J) > Pr(K)$.

(c) What is the minimum number of distributions you could have in a representor that would allow it to fulfill the conditions described in part (b)? Explain why you need that many.

Problem 14.6. 🌙 You have a coin. You're certain that it's either fair, biased with a 1/4 chance of coming up heads, or biased with a 3/4 chance of heads. At t_0 you divide your credence equally among those three chance hypotheses, and you satisfy the Principal Principle. Then you start flipping the coin. Between t_0 and t_1 you flip it once and it comes up heads, between t_1 and t_2 it comes up tails, and then between t_2 and t_3 it comes up heads again.

(a) For each of the times t_1, t_2, and t_3, calculate your updated credence in each of the three chance hypotheses, and your overall credence that the *next* flip will come up heads.

(b) The first heads result you witness (between t_0 and t_1) changes your overall credence in heads by a certain amount. The second heads result you witness (between t_2 and t_3) also changes your overall credence in heads by a certain amount. How do those two amounts compare? (If you've been working in fractions, it might help to convert to decimals here.)

(c) How does your answer to (b) illustrate the concept of resilience, and the link between resilience and weight of total evidence?

Problem 14.7. 𝄂𝄂 Prove that if two distributions over the same language each satisfy the Kolmogorov axioms, any linear combination of those distributions will satisfy the axioms as well.

Problem 14.8. 𝄂𝄂 Jeffrey (1987) describes a representor containing two distributions, $\Pr_q$ and $\Pr_r$, over a language with atomic propositions A and B. These distributions assign the following values:

$$\Pr_q(A) = \Pr_q(B) = 1/3$$
$$\Pr_q(A \, \& \, B) = 1/9$$
$$\Pr_r(A) = \Pr_r(B) = 2/3$$
$$\Pr_r(A \, \& \, B) = 4/9$$

(a) Create a probability table for distributions $\Pr_q$ and $\Pr_r$.

(b) Show that $\Pr_q$ and $\Pr_r$ each treat propositions A and B as probabilistically independent.

(c) Use a probability table to specify a distribution $\Pr_s$ that: (i) makes $\Pr_s(A \, \& \, B)$ the average of $\Pr_q(A \, \& \, B)$ and $\Pr_r(A \, \& \, B)$; and (ii) is a linear combination of $\Pr_q$ and $\Pr_r$.

(d) Show that your distribution $\Pr_s$ does not treat propositions A and B as probabilistically independent.

(e) Use a probability table to specify another distribution $\Pr_t$ that: (i) makes $\Pr_t(A \, \& \, B)$ the average of $\Pr_q(A \, \& \, B)$ and $\Pr_r(A \, \& \, B)$; and (ii) makes A and B probabilistically independent.

(f) Is $\Pr_t$ a linear combination of $\Pr_q$ and $\Pr_r$? Explain why or why not.

Problem 14.9. 𝄂𝄂𝄂

(a) The distributions $\Pr_x$ and $\Pr_y$ specified in Figure 14.1 both make proposition A probabilistically independent of proposition R. Prove that they also have a very curious feature: Any linear combination of those two distributions *also* makes A independent of R.

(b) Suppose we have two probabilistic distributions $\Pr_q$ and $\Pr_r$ which agree that propositions X and Y are independent. Prove that if a linear combination $\Pr_s$ of those two distributions also makes X and Y probabilistically independent, then at least one of the following three conditions must hold:

   (i) $\Pr_s$ is identical to either $\Pr_q$ or $\Pr_r$.

  (ii) $\Pr_q(X) = \Pr_r(X)$.

 (iii) $\Pr_q(Y) = \Pr_r(Y)$.[68]

Problem 14.10. 🌶️🌶️ Suppose P is a claim about which you have no specific information—perhaps something obscure about the life of a long-dead author. Yet you're certain I know whether P is true (and am scrupulously honest). So you take a fair coin, write "P" on its heads side, and "$\sim P$" on its tails side. You then close your eyes, flip the coin, and ask me whether the side that landed up has a true proposition written on it. I report that it does.

(a) Let H be the proposition that the coin lands heads. Explain why my answer to your question is equivalent to telling you that $P \equiv H$.

(b) Suppose that before flipping the coin, you assign P a ranged credence of $(0, 1)$, in recognition of your unspecific evidence concerning that proposition. That is, for every real number strictly between 0 and 1, there is a distribution in your representor assigning P that real number. Suppose also that every distribution in your representor initially assigns $\Pr(H) = 1/2$, and every distribution in your representor takes P and H to be probabilistically independent.

Now suppose that when you learn $P \equiv H$, each distribution in your representor updates by conditionalizing on this new evidence. After the update, what range of values does your representor associate with P? With H? (Hint: Try starting with a single distribution that assigns some particular real value to P, and see what happens when you update it. Then determine whether every other distribution will behave the same way.)

(c) Suppose that in answer to your question I had instead reported that the false side landed up. Assuming again that your distributions update by conditionalization, how would your ranges for P and H have changed in that case?

(d) Both Sturgeon (2010) and White (2010) see cases like this one as presenting *prima facie* counterexamples to the traditional Conditionalization-based method of updating ranged credences (and perhaps to the entire ranged-credence regime). Do the results you reported in parts (b) and (c) seem counterintuitive to you? Why or why not?

(e) Review your answers to Exercise 4.6—the "Speedy the racehorse" problem. Does this change your opinions about the updates from parts (b) and (c)?[69]

Problem 14.11. 🌶️🌶️

(a) In the representor described by Figure 14.1, what expected value does distribution Pr_x assign to a ticket that pays \$1 if A is true and nothing otherwise? What expected value does distribution Pr_y assign to such a ticket?

(b) Suppose that representor describes your doxastic state, and I offer to sell you a \$1 ticket on A for \$$x$. For what values of x will the minimax decision strategy described on page 526 permit you to buy the ticket?

(c) Now suppose I offer to sell you for \$$y$ a ticket that pays \$1 if A is *false* and nothing otherwise. For what values of y will minimax permit you to buy that ticket?

(d) Are there any values for x and y such that the minimax strategy will permit you to purchase a pair of tickets that guarantees you a sure loss? Explain why or why not.

(e) Are there any values for x and y such that the minimax strategy will permit you to pass up a sure gain? Explain why or why not.

Problem 14.12. 🌶️🌶️🌶️ Each of the following is true of probabilistic credence distributions. For each statement, replace all of the "cr"s with "Bel"s, then investigate whether the resulting statement is still true. If it is, prove it. If it's not, provide a counterexample by supplying a Bel function and its corresponding mass function m.

(a) For any cr and any proposition P, $\text{cr}(\sim P) = 1 - \text{cr}(P)$.

(b) For any cr and propositions P, Q, if $P \vDash Q$ then $\text{cr}(P) \leq \text{cr}(Q)$.

(c) For any cr, specifying the cr-values of the state-descriptions in a propositional language (or Boolean algebra) suffices to determine the cr-values of all other propositions in the language(/algebra).

Problem 14.13. 🌶️🌶️🌶️ Prove that a belief function is additive if and only if its associated mass function assigns positive value only to state-descriptions.

Problem 14.14. 🌶 If we start with two mass functions m_1 and m_2 defined over the same Boolean algebra, Dempster's rule of combination provides a four-step process for generating a new, combined mass function $m_3 = m_1 \oplus m_2$ over that algebra:

1. Start by assigning $m_3(X) = 0$ for every proposition X in the Boolean algebra.
2. Now select propositions X and Y in $\mathcal{L}$ such that $m_1(X) > 0$ and $m_2(Y) > 0$. Find the proposition in the algebra logically equivalent to the conjunction of X and Y, which we'll call Z. Then take the product $m_1(X) \cdot m_2(Y)$ and add it to the current value of $m_3(Z)$.
3. Repeat step 2 once for each pair of propositions in the algebra that receive nonzero m_1- and m_2-values. Notice that in some cases X and Y may be the same proposition, and in some cases the conjunction Z may be a contradiction.
4. A mass function may never assign a positive value to a contradiction. So set $m_3(F) = 0$. Then normalize the remaining nonzero values of m_3: multiply them all by the same constant so that they sum to 1.

Let's practice applying this combination algorithm!

(a) Let m_1 be the mass function described on page 528 for the Urdu urn testimony. Let m_2 be the mass function described in Equation (14.20) for the Swahili testimony. Suppose B, Y, and R form a partition, and are all members of the Boolean algebra over which these mass functions are defined. Use Dempster's combination rule to calculate $m_1 \oplus m_2$. Then construct the belief function corresponding to your result. (Hint: You don't need to consider *all* of the propositions in the Boolean algebra— focus on those that receive positive mass- or belief-values.)
(b) Does this belief function strike you as capturing the intuitive upshot of the combined Urdu and Swahili evidence?

Problem 14.15. 🌶🌶 This exercise will work through the details of Peter Walley's example from page 531.

Suppose we have a language whose two atomic propositions are H_1 and H_2, representing respectively that each of two fair coin flips came up heads. Yet the agent lacks any evidence as to whether the outcomes of these coin flips are correlated. So we construct a representor, such that all of its distributions are probabilistic, and all of its distributions assign $Pr(H_1) = Pr(H_2) = 1/2$.

(a) Imagine that the representor contains distributions representing all of the possible ways in which the two coin flips might be correlated. Of these, three distributions will be of particular interest. Let Pr_{ind} be the distribution on which the coin flips are probabilistically independent, Pr_{pos} be the distribution assigning $H_1 \equiv H_2$ a value of 1, and Pr_{neg} be the distribution assigning $H_1 \equiv H_2$ a value of 0.

Make a probability table representing the four state-descriptions of our language, and the values these three distributions assign to each one.

(b) Explain why Pr_{pos} will assign the lowest value out of all the infinitely many distributions in the representor to $H_1 \vee H_2$.

(c) Find the lower probability value associated with this representor for each of the following propositions: H_1 & H_2, H_1 & $\sim H_2$, $\sim H_1$ & H_2, $\sim H_1$ & $\sim H_2$, H_1, H_2, and $H_1 \vee H_2$.

(d) Suppose we attempted to construct a belief function such that for each proposition X, $Bel(X)$ equalled X's lower probability on this representor. Explain why this construction would be impossible. (<u>Hint</u>: Investigate the values of the associated mass function.)

(e) Does this impossibility result seem to you to present a problem for Dempster-Shafer theory? Can you construct a belief function that is intuitively faithful to the agent's evidence in this case (even if it doesn't match the lower probabilities of this representor)?

Problem 14.16. ✐ Of the formalisms you've studied in this book, which do you think does the best job of representing agents' doxastic attitudes and what rationality requires of them?

14.5 Further reading

INTRODUCTIONS AND OVERVIEWS

Peter C. Fishburn (1986). The Axioms of Subjective Probability. *Statistical Science* 1, pp. 335–45

Excellent overview of the various axiom systems for comparative confidence, and their intuitive strengths and weaknesses. (Note that Fishburn refers to what we've called the Scott Axiom as "strong additivity".)

Joseph Y. Halpern (2003). *Reasoning about Uncertainty*. Cambridge, MA: MIT Press

Halpern's Chapter 2 covers all of the formalisms discussed in this chapter, plus a number of other formalisms for representing uncertainty. The coverage is extensive and can get highly technical, but also is remarkably accessible at times.

Richard Pettigrew and Jonathan Weisberg, eds. (2019). *The Open Handbook of Formal Epistemology*. Published open access online by: PhilPapers. URL: https://philpapers.org/rec/PETTOH-2

Contains excellent full-length survey articles on imprecise probabilities, comparative probabilities, belief revision theory, ranking theory, doxastic logic, and other related topics.

Liping Liu and Ronald R. Yager (2008). Classic Works of the Dempster-Shafer Theory of Belief Functions: An Introduction. In: *Classic Works of the Dempster-Shafer Theory of Belief Functions*. Ed. by Ronald R. Yager and Liping Liu. Vol. 219. Studies in Fuzziness and Soft Computing. Berlin: Springer, pp. 1–34

This volume introduction provides a useful overview of Dempster-Shafer theory and its challenges, plus a history of the development of the theory.

CLASSIC TEXTS

Charles H. Kraft, John W. Pratt, and A. Seidenberg (1959). Intuitive Probability on Finite Sets. *The Annals of Mathematical Statistics* 30, pp. 408–19

Paper that disproved de Finetti's putative sufficient conditions for probabilistic representability of a comparative confidence ranking, then provided correct sufficient conditions to take their place.

Peter Walley (1991). *Statistical Reasoning with Imprecise Probabilities*. London: Chapman and Hall

Classic treatment of the mathematics of ranged credences.

Teddy Seidenfeld and Larry Wasserman (1993). Dilation for Sets of Probabilities. *The Annals of Statistics* 21, pp. 1139–54

First systematic mathematical analysis of the conditions under which ranged credences dilate. Also includes references to early discussions of dilation in the Bayesian literature.

> Arthur P. Dempster (1966). New Methods for Reasoning Towards Posterior Distributions Based on Sample Data. *Annals of Mathematical Statistics* 37, pp. 355–74

Paper in which Dempster first introduced what came to be known as belief and plausibility functions. (Though as I pointed out in note 65, he somewhat confusingly referred to them as "lower probabilities" and "upper probabilities" at the time.)

> Glenn Shafer (1976). *A Mathematical Theory of Evidence.* Princeton, NJ: Princeton University Press

Greatly developed Dempster's formalism and its interpretation, and made the framework a go-to for a myriad of applications.

EXTENDED DISCUSSION

> Terrence L. Fine (1973). *Theories of Probability: An Examination of Foundations.* New York, London: Academic Press

Comprehensive, insightful discussion of every aspect of comparative confidence, with many original results along the way.

> sipta.org

Website of the Society for Imprecise Probability: Theories and Applications. Contains resources and tutorials related to sets of probability distributions, including references to new works and announcements of upcoming events.

Notes

1. We have now repurposed the ">" and "~" symbols that were used in Chapter 7 to indicate an agent's preferences among acts. It would be nice to have different symbols for the act preference relation and the comparative confidence relation, but these symbols

are used almost universally in the literature for both purposes. So I've chosen to follow the literature here.

2. For recent discussions of this possibility, see Stefánsson (2017) and Haverkamp and Schulz (2012, p. 395).

3. Most authors who formalize comparative confidence don't mention a Comparative Equivalence rule. This isn't because those authors are fine with an agent's being more confident in one logical equivalent than another! It's because most authors define their rankings over a Boolean algebra rather than a full language of propositions $\mathcal{L}$. As we'll discuss in Section 14.3, a Boolean algebra is a proper subset of $\mathcal{L}$ in which no two distinct propositions are logically equivalent. Since a Boolean algebra contains no distinct, nonequivalent propositions, authors working with Boolean algebras have no need for a rule about comparing such propositions. But since we're working with a full propositional language $\mathcal{L}$, I will include Comparative Equivalence among all the axiom systems for comparative confidence we consider.

4. If you've run into technical talk about "orders" or "orderings", you might wonder what the difference is between a total preorder and a total ordering. For $\succeq$ to be an **ordering**, it would have to satisfy an antisymmetry condition: $P \sim Q$ only if $P \; \mathrel{=\!\mid\!\mid=} \; Q$. If we made $\succeq$ antisymmetric but not complete, it would be a "partial ordering"; adding completeness to a partial ordering would make it a "**total ordering**". But for many of the applications in which we're interested, being equally confident in two logically nonequivalent propositions is perfectly rational—consider the proposition that a fair die roll came up three and the proposition that it came up four. So we will not treat antisymmetry as a rational requirement, and will continue to assume only that $\succeq$ is a preorder.

   (Thanks to Catrin Campbell-Moore, Kenny Easwaran, Jason Konek, Ben Levinstein, and David McCarthy for helping me sort all this out and explain it clearly.)

5. Savage explains the intuition in a betting context. Suppose there is a partition of events $\{A, B, C, D, \ldots\}$, and an agent gets to choose whether the occurrence of B or C will win him a huge prize. Presumably the agent will choose B over C only if $B \succeq C$. Savage writes, "It may be helpful to remark that [Comparative Additivity] says, in effect, that it will not affect the person's guess to offer him a consolation prize in case neither B nor C obtains, but D happens to" (1954, §3.2). In other words, the agent should assign $B \succeq C$ just in case he assigns $B \vee D \succeq C \vee D$.

6. I realize that Comparative Additivity was framed in terms of $\succeq$, while this is an example of being strictly more confident in one destination than another. Yet given our definition of $\succ$ in terms of $\succeq$, you should be able to prove from Comparative Additivity a similar condition in which the "$\succeq$"s have all been replaced by "$\succ$"s.

7. Notice that given our definition of $\succ$ in terms of $\succeq$, Comparative Matching implies that $P \succ Q$ just in case $cr(P) > cr(Q)$. This will matter in examples later on.

8. Perhaps for this reason, Savage (1954, §3.2) referred to any ranking satisfying those six conditions as a "**qualitative probability**". While this label has lingered in the literature, we'll see in a moment that it's somewhat unfortunate.

9. Here I follow the presentation in Fine (1973, p. 22). For explanations of how Kraft et al. constructed their ingenious counterexample and how it works, see Krantz et al. (1971, §5.2.2) and Suppes (2002, pp. 228–9).

10. For more pragmatic arguments supporting various comparative constraints as rational requirements, see Savage (1954), Fishburn (1986), Halpern (2003), and Icard III (2016). For accuracy-based arguments for and against comparative constraints, see Fitelson and McCarthy (ms).

11. Just for illustration, here's a quick argument for Comparative Transitivity from the Scott Axiom, adapted from Suppes (2002, p. 230). Suppose our A-sequence is $\{P, Q, R\}$, and our B-sequence is $\{Q, R, P\}$. Since these equinumerous sequences contain the same propositions, they have the same number of truths in each possible world. We pair the members of the sequences in the order in which I've listed them. Now suppose the agent assigns $P \succeq Q$, $Q \succeq R$, and $R \succ P$. This ranking is forbidden by the Scott Axiom, since the agent is strictly more confident of the last member of the A-sequence than its paired member of the B-sequence, but is not strictly more confident of any member of the B-sequence than its paired member of the A-sequence. So according to the Scott Axiom, this kind of ranking is forbidden—if $P \succeq Q$ and $Q \succeq R$, we can't have $R \succ P$. Given Comparative Completeness, it follows that $P \succeq R$, which is what Comparative Transitivity requires in such a case.

12. I've endeavored to make it even easier to understand by rewriting the *then* part of the Scott Axiom. In the context of Comparative Completeness, the *then* condition I've presented is the contrapositive of the condition in Scott's original formulation.

13. Suppes proves this highly nonobvious fact immediately after introducing the Continuity condition at Suppes (1969, p. 6).

14. Much like Suppes's Continuity condition, these confidence conditions over infinite languages work by stipulating that the agent has—or lacks—propositions available with particular comparative profiles. For example, suppose the agent entertains propositions about fair lotteries of every finite size. Then we can pin down the first digit of her credence in any proposition P by establishing, say, that she is more confident in P than in the proposition that one of the first *three* tickets in a fair ten-ticket lottery will win, but less confident in P than the proposition that one of the first *four* tickets will win. If we want to establish the first two digits of her credence in P, we can use a fair 100-ticket lottery; and so on. Given the agent's rich set of confidence comparisons, this method determines a numerical credence matching the agent's confidence in P to as many digits as we like.

   On the other hand, moving to an infinite language introduces new problems. Depending on how one manages various issues around Countable Additivity (Section 5.4), it might be rational for an agent to be more confident in some contingent proposition than in a contradiction, yet still assign each of them credence 0. This would interfere with Comparative Matching. (Compare Savage 1954, §3.3.)

15. The Comparative Matching principle I've presented is a biconditional. André Neiva points out to me that a credence distribution (partial or full) could model the present example of incommensurability if we weakened Comparative Matching to "$P \succeq Q$ only if $cr(P) \geq cr(Q)$." Essentially, we would be saying that while all aspects of the confidence ranking must be recognized in the credence distribution, not all aspects of the credence distribution must be realized in the confidence ranking. (Presumably some extra formal feature would indicate which inequalities among the credences were to be treated as "real".)

Another Bayesian approach to modeling incommensurability replaces individual credence distributions with *sets* of distributions; we'll consider that option in Section 14.2.

16. This approach parallels the approach to rational assessment of partial credence distributions we saw in Section 12.2.1.

17. Compare Hawthorne (2016, p. 289, emphases in original): "Any [comparative] relation for which there cannot possibly be a *complete extension* must already contain a kind of looming syntactic inconsistency, owing to the forms of its *definite*... comparisons. It manages to stave off explicit formal inconsistency only by forcing at least some... pairs to remain incomparable."

18. For an interesting discussion of representing confidence-ratio claims in the framework of unconditional comparative confidence rankings, see Stefánsson (2018).

19. This approach can be compared to the conditional-credences-first approach to probabilistic credence that I described in Section 5.4.

20. Though see Fine (1973, §IIF) for the suggestion that independence might be underdetermined by confidence comparisons, and so might have to be formally represented by an additional relation grafted on top of the agent's confidence ranking.

21. Again, the comparison to utilities is instructive: Unlike rational credences, rational utilities need have no "top" or "bottom" outcome. If you like, a rational agent's utility dial need not have a maximum (or minimum). And even if we found two agents whose utility dials each had a maximum, there's no reason to suppose those maxima should occur in the same place on the dials.

22. Ellsberg is perhaps more widely known to the general public for being the individual who released the Pentagon Papers.

23. The Ellsberg Paradox is most frequently cited as a risk-aversion counterexample to Savage's decision theory, along the lines of the Allais Paradox (Section 7.2.4). But while Ellsberg does make the connection to decision theory, his (1961) is centered around the assessment of conditions for rational comparative confidence.

24. Notice that the Zynda scale yields distributions that violate Finite Additivity. For example, if we take the credence distribution from Equation (14.14) and apply the relevant affine transformation, B receives a value of 7 and N gets a 4. These values add up to more than the value of a tautology (10) on Zynda's scale, so (given that we defined B and N to be mutually exclusive) Zynda's scale is subadditive.

25. The details get a bit subtle here. Given the Revised Representation Theorem of Section 8.3.2, credence distributions that are positive affine but not scalar transformations of each other may yield different preferences when inputted with the same utility distribution into a traditional decision-theoretic valuation function. This returns us to the questions about confidence rankings and confidence ratios that I raised in the discussion of conditional comparisons above.

26. For another example, return to our old friends Mr. Prob, Mr. Bold, and Mr. Weak, who have the same comparative confidence rankings but differ substantially on relative credence-gap sizes.

27. Harrison-Trainor, Holliday, and Icard III (2016) examine the axioms a comparative confidence ranking must satisfy to be representable by a set of probability distributions.

28. To whittle the candidate distributions for the representor down to only these two would take far more information than just the fact that I'm more confident in A than $\sim A$

and R than $\sim R$. Moreover, we'll see later that some theorists think it's impossible for a rational representor to contain *exactly* two distributions. The representor in Figure 14.1 is offered as a simple early example to help us get the basics down—no realistic context would give rise to a representor this sparse.

29. A few decades back it was customary to refer to non-point-valued ranged credences as "numerically indeterminate credences". This locution indicated that no single number could properly be associated with the doxastic attitude in question. Nowadays, however, the terminology is often shortened to "indeterminate credence" or "**imprecise credence**". I avoid these labels because they suggest that the doxastic attitude under discussion is itself imprecise or indeterminate.

30. As a historical note, representors' ability to convey probabilistic relevance judgments was not an advantage Levi touted over mere numerical ranges. Using representors in that way would have conflicted with Levi's position on Convexity, which we'll discuss in Section 14.2.2. Most of the advantages Levi described in his (1974) concerned the interaction of representors with decision theory.

31. See, for example, the discussion at Weatherson (2015, pp. 535ff.)

32. Compare Zynda (2000, n. 10).

33. Walley (1991, pp. 84ff.) offers a comprehensive list of mathematical properties of upper and lower probabilities that follow from requiring the individual distributions in a representor to be probabilistic.

34. Could we capture these four conditions with a comparative confidence ranking? To get the incommensurability we'd have to use a partial preorder, and to represent the independence condition we'd need conditional comparisons. But at a first pass, it seems doable. Yet there are other ways in which representors outstrip comparative confidence rankings as a representational tool. As we saw in Section 14.1.3, comparative rankings typically underdetermine relative gap sizes between propositions, unless the ranking satisfies a strong condition like Suppes Continuity. And gap size information is important for such applications as decision theory. A representor, on the other hand, may easily convey precise gap-size information (when the distributions in the representor agree on various gap size facts).

35. Scott Sturgeon (2010, p. 133) discusses the possibility of ranged credences with imprecise, "fuzzy" boundaries. But if we go that route, presumably whatever tools we use to understand the fuzzy upper and lower probabilities of a ranged credence could equally be applied to generate "fuzzy" point-valued credences. So Sturgeon's suggestion confers no particular advantage on ranged credences over point-valued.

36. Compare also Jeffrey (1983, §12.5).

37. Kaplan (1996, pp. 27–8, n. 27) criticizes this point-valued approach to representing weight on the grounds that facts about the weight of an agent's evidence concerning a proposition should be reflected in the agent's attitude *toward that very proposition*. Suppose I go from having a 5-5 sample of the balls in the urn to having a 500-500 sample. Perhaps this change in the weight of my evidence should be reflected not just in the credences I assign to various chance propositions, but also in the attitude I assign directly to the proposition that Raj's draw will come up black. If that's right, then point-valued distributions lack the representational capacity to properly reflect evidential weight, and we need a formal tool like a representor to do the job.

38. Of course, we can *also* address these evidential features with ranged credences. For example, when an agent has scant evidence, her representor might contain a variety of flattish distributions over the available chance values. When her evidence increases, this added weight will be reflected in representor distributions that are more peaked, and therefore more resilient. The point is just that while balance and weight may be reflected in either point-valued credences or ranged, we have yet to see a feature of evidence that *demands* ranged representation.

39. While the risk/uncertainty distinction was introduced by Knight, it is often associated with Keynes, and was indeed popularized by him (for instance in his 1937).

40. Just to emphasize that the weight of evidence is distinct from its specificity: Compare the evidence you have *after* the stranger pulls the three objects out of his bag to the evidence you had *before* he showed you any of its contents. The weight of your evidence about whether the fourth object will be toothpaste clearly increases once you see the first three items. Yet it's not clear that your evidence becomes any more specific.

41. Joyce (2010), for instance, embraces this Objectivist approach.

42. Thanks to Jason Konek for helping me understand many of the issues in this section.

43. Here's why: Suppose we have a representor containing two distributions Pr_x and Pr_y, assigning proposition P values x and y respectively. Now let z be the average of x and y. The Interval Requirement demands that that representor also include some distribution Pr_z with $Pr_z(P) = z$. Yet the Interval Requirement doesn't dictate what values Pr_z assigns to other propositions. Convexity, on the other hand, requires that the representor contain a single distribution Pr_z assigning *every proposition* in $\mathcal{L}$ the average of the values assigned by Pr_x and Pr_y. (That is, a distribution with $\alpha = 0.5$.) There are many ways to satisfy the Interval Requirement without including such a distribution in the representor.

44. Does the Interval Requirement have a similar problem? Is it possible in every case to supply a set of probability distributions that agree on the independence of two propositions while also satisfying the Interval Requirement? While I don't know any formal results on this question, I think the answer depends on what additional constraints on the representor we allow such cases to stipulate. For instance, Jeffrey's example (described in Exercise 14.8) supplies two distributions Pr_q and Pr_r that agree on the independence of propositions A and B, but also agree on the probability of $A\&{\sim}B$ and on the probability of ${\sim}A\&B$. If we demand that every distribution in the representor agree with Pr_q and Pr_r on those two values, while also making A and B independent, then it's not going to be possible to satisfy the Interval Requirement. (For instance, there's no probabilistic distribution that satisfies those requirements while assigning a probability to $A \& B$ halfway between $Pr_q(A \& B)$ and $Pr_r(A \& B)$.)

45. The way I've been approaching things, the range associated with a representor for a proposition is the tightest interval containing every real number assigned by some distribution in that representor to that proposition. Then I've explained the lower probability as the first of the two numbers specifying that interval, and the upper probability as the second number. It follows that the ranges $(0, 1)$ and $[0, 1]$ have the same upper probability, 1. In each case, 1 is the lowest real number such that every distribution in the representor assigns a value equal to or below it. (Similarly, each range has the same lower probability of 0.) It's convenient to think of the upper probability

of a proposition as the highest value any distribution in the representor assigns to that proposition. But that isn't quite right; a representor that assigns black the range $(0, 1)$ has an upper probability of 1 for black despite the fact that no distribution in that representor assigns black that value. (In mathematical terms, an upper probability is a supremum rather than a maximum, while a lower probability is an infimum rather than a minimum.)

46. Note that we aren't dealing here with the kind of infinitistic situation that moved us to reconsider the meanings of extreme probability values in Section 5.4.

47. This objection to the closed interval would not bother Kaplan, because he denies (on independent grounds) that one should always update by Conditionalization.

48. The somewhat herky-jerky structure of this example—in which your range narrows from $(0, 1)$ to $(1/100, 99/100)$ and then stays at that value—occurs because you entertain a finite number of hypotheses about the possible contents of the urn. If in the example you didn't know how many balls were in the urn, and entertained literally every positive integer as a possibility for the total count, then your range would stay at $(0, 1)$ throughout. A similar effect often occurs when hypotheses concern the possible settings of a continuous parameter, as with a coin of unknown bias. Thus one finds many examples in the literature of the range $(0, 1)$ remaining "sticky" across conditionalizations.

49. Pedersen and Wheeler (2014) say that proposition Q is "epistemically irrelevant" to proposition P for an agent when conditionalizing on Q changes neither the agent's upper probability nor her lower probability for P. They say that Q is "stochastically independent" of P for an agent when every distribution in the agent's representor treats P and Q as probabilistically independent. In those terms, the point of the present example is that two propositions can be epistemically irrelevant to each other without being stochastically independent.

    (For what it's worth, while I think Pedersen and Wheeler's distinction here is important, I'm not thrilled with the "epistemically irrelevant" terminology. That makes it sound like any information that doesn't change an agent's upper or lower probabilities doesn't change anything about her doxastic state. But as I've been emphasizing, upper and lower probabilities don't capture all there is to say about an agent's doxastic state.)

50. Here's a somewhat silly analogy: Suppose you have a giant number line with at least one person standing on each of the integers. I want the people bunched up close to 0, so I tell you to have each of them cut the distance between themselves and 0 in half. Then I go away for a while while you arrange things. When I return, I complain that there are still people arbitrarily far away from 0! My complaint is ill-lodged; the trouble isn't with your execution of my order; it's with what I expected that order to produce.

51. Some authors define dilation as applying strictly to cases in which the prior range is a proper subset of the posterior range. While that makes a difference to the mathematical conditions required for dilation to occur, I don't see how it makes a difference to the philosophical significance or plausibility of the phenomenon. So I have adopted a more inclusive definition here.

52. For example, while setting up his critique of dilation Sturgeon at one point describes a batch of evidence as meriting a ranged credence because it's "rough", then later in the same sentence says the evidence requires a range because it's "meagre". These are distinct

evidential features: evidence may be rough and copious, or evidence may be sharp and meagre (Sturgeon 2010, p. 131).

53. Compare Joyce (2010, p. 300).

54. Here I'm ignoring various complications we discussed in Chapter 7, such as risk aversion and the distinction between evidential and causal decision theory. Generalizing decision theory to ranged credences is complicated enough without those extra wrinkles.

55. Quick proof: If *every* probabilistic distribution in your representor finds all of the bets in a set acceptable, then *at least one* probabilistic distribution finds all of those bets acceptable. By the Converse Dutch Book Theorems described in Section 9.2, any set of bets found acceptable by a probabilistic distribution cannot constitute a Dutch Book.

56. This possibility arises as soon as we go from a formalism in which each agent assigns exactly one credence distribution to a formalism that represents an agent's doxastic state with multiple distributions. But the problem also arises if you go from a formalism in which each agent assigns exactly one *utility* distribution to a formalism on which sets of utility distributions are allowed. Many authors have been concerned about that version of the problem as well.

57. Note that I'm using the term "global strategy" in a different way than Elga does in his typology of ranged-credence decision theories.

58. Notice that Elga's pair of bets is subtly different from the sort of "Czech Book" described in Section 9.2. Elga's bets are offered sequentially, at different times. If they were offered at once, as a package, then every distribution in your representor would evaluate that package positively. So every decision theory for ranged credences we're considering (global, local, or otherwise) would direct you to purchase the package and take advantage of the guaranteed gain.

59. While this section touches on the interaction between Dutch Books and ranged credences, you might wonder how ranged credences interact with representation theorems and accuracy arguments. The classic representation theorem for ranged credences appears in Jeffrey (1965); Jeffrey attributes the crucial mathematics to Ethan Bolker. As for the proper way to measure the accuracy of ranged credences, this is also an unsettled matter in the current literature. Miriam Schoenfield (2017) has even argued that on any plausible measure of accuracy, adopting a ranged credence rather than a point-valued credence can never be justified on accuracy grounds!

60. Strictly speaking, the mass function will assign 0.8 to some proposition in the Boolean algebra *equivalent* to $B \vee Y$; $B \vee Y$ itself may not be in the algebra. I'll take this caveat as understood in what follows.

61. Throughout this section we will be working only with examples in which the agent entertains a finite number of doxastic possibilities (possible worlds). In that context, Bel-functions may simply be defined in terms of m-functions, as I did in the previous paragraph. For contexts in which the agent entertains an infinite number of possibilities, we need a more general definition of belief functions, and it turns out that some functions satisfying this definition cannot be generated from any mass function. For more details, see Halpern (2003, pp. 33ff.)

62. In this discussion I always start with a mass function and generate a belief function from it—largely because I find that direction most intuitive. If one wanted to work the

other way, one could think of the mass assigned to a particular proposition X as the amount of belief assigned to X that isn't inherited from belief in any proposition that entails it. In the present Swahili example, $B \vee Y$ inherits some of its belief-value from B (which has a Bel of 0.25). Yet there's an extra 0.25 of Bel assigned to $B \vee Y$ that doesn't come from any proposition that entails it, so m($B \vee Y$) = 0.25.

63. This is a good example of a mass function that generates an additive belief function.
64. If you're wondering how belief functions relate to comparative confidence rankings, Wong et al. (1991) analyzes the conditions under which a confidence preorder is representable by a belief function. (See also Capotorti and Vantaggi 2000.)
65. In Dempster's early work, he referred to belief functions as "lower probabilities" and plausibility functions as "upper probabilities". But that terminology eventually became associated most closely with the ranged credence view, and since the lower and upper probabilities associated with representors can't always be mapped to belief and plausibility functions, Dempster-Shafer theorists agreed to employ other terminology.
66. How could two fair coin flips have their outcomes correlated (or anti-correlated)? Well, suppose you took two fair coins, laid them edge-to-edge so that their heads sides were both up, then glued them together. Flipping the joined coins would then yield two coin flip outcomes, each fair, but perfectly correlated. (Gluing them together with one heads up and one tails up would yield perfect anti-correlation.)
67. This problem is adapted from Krantz et al. (1971, p. 244).
68. I learned about this theorem from a manuscript by Richard Pettigrew, which says the result is "in the background" in Laddaga (1977) and Lehrer and Wagner (1983).
69. For more on Speedy and the White/Sturgeon objections, see Hart and Titelbaum (2015).

Glossary for Volumes 1 & 2

accuracy A doxastic attitude adopted toward a particular proposition is accurate to the extent that it appropriately reflects the truth-value of that proposition. 338

accuracy domination Given distributions cr and cr' over the same set of propositions, cr' accuracy-dominates cr just in case cr' is less inaccurate than cr in each and every logically possible world. 347

Accuracy Updating Theorem For any proper scoring rule, probabilistic distribution cr_i, and evidential partition in $\mathcal{L}$, an agent's t_i expected inaccuracy for updating by Conditionalization will be lower than that of any updating plan that diverges from it. 367

act In a decision problem, an agent must choose exactly one of the available acts. Depending on the state of the world, that act will produce one of a number of outcomes, to which the agent may assign varying utilities. 251

actual world The possible world in which we live. Events that actually happen happen in the actual world. 26

admissible evidence Evidence that, if it has any effect on an agent's credence in an outcome of an event, does so by way of affecting the agent's credences about the outcome's objective chance. 135

affine transformation Two measurement scales are related by an affine transformation when values on one scale can be obtained by multiplying values on the other scale by a particular constant, then adding another specified constant. The Fahrenheit and Celsius scales for temperature provide one example. 294

Allais' Paradox A set of gambles for which subjects' intuitions often fail to satisfy the Sure-Thing Principle; proposed by Maurice Allais as a counterexample to standard decision theory. 262

analogical effects A cluster of effects involving analogical reasoning, such as: the degree to which evidence that one object has a property confirms that another object has that property should increase in light of information that the objects have other properties in common. 214

analyst expert Expert to whom one defers because of her skill at forming attitudes on the basis of evidence. 142

antecedent In a conditional of the form "If P, then Q," P is the antecedent. 26

atomic proposition A proposition in language $\mathcal{L}$ that does not contain any connectives or quantifiers. An atomic proposition is usually represented either as a single capital letter (P, Q, R, etc.) or as a predicate applied to some constants (Fa, Lab, etc.). 26

balance The balance of a body of evidence relative to a particular proposition indicates which way that evidence leans on that proposition, and how far it leans. 513

Base Rate Fallacy Assigning a posterior credence to a hypothesis that overemphasizes the likelihoods associated with one's evidence and underemphasizes one's prior in the hypothesis. 97

Bayes factor For a given piece of evidence, the ratio of the likelihood of the hypothesis to the likelihood of the catchall. An update by Conditionalization multiplies your odds for the hypothesis by the Bayes factor of your evidence. 98

Bayes Net A diagram of causal relations among variables developed from information about probabilistic dependencies among them. 75

Bayes's Theorem For any H and E in $\mathcal{L}$, $\text{cr}(H\,|\,E) = \text{cr}(E\,|\,H) \cdot \text{cr}(H)/\text{cr}(E)$. 61

Belief Closure If some subset of the propositions an agent believes entails a further proposition, rationality requires the agent to believe that further proposition as well. 6

Belief Consistency Rationality requires the set of propositions an agent believes to be logically consistent. 6

belief function Distribution used in Dempster-Shafer theory to represent a doxastic state. 528

Bertrand's Paradox When asked how probable it is that a chord of a circle is longer than the side of an inscribed equilateral triangle, the Principle of Indifference produces different answers depending on how the chord is specified. 147

Boolean algebra For purposes of this book, a set containing 2^{2^n} logically nonequivalent propositions constructed from n atomic propositions using propositional connectives. (The term has a broader, more abstract meaning in general mathematics). 528

Brier score A scoring rule that measures the inaccuracy of a distribution by its Euclidean distance from the truth. 345

calibration A credence distribution over a finite set of propositions is perfectly calibrated when, for any real x, out of all the propositions to which the distribution assigns x, the fraction that turn out to be true is x. 340

catchall The proposition that the hypothesis H under consideration is false (in other words, the proposition $\sim H$). 64

Causal Decision Theory Decision theory in which expected utility depends on an act's causal tendency to promote various outcomes. 267

center Picks out a particular time, location, and individual within an uncentered possible world. 394

centered possible world An ordered pair of a traditional (uncentered) possible world and a center within it. 394

centered proposition A proposition whose associated set contains centered possible worlds. Centered propositions are capable of providing self-locating information. 394

classical probability The number of favorable event outcomes divided by the total number of outcomes possible. 124

classificatory concept Places an entity in one of a small number of kinds. 4

closed interval An interval that includes its endpoints. For instance, if the range of values assigned to a proposition by a particular representor is a closed interval, then there will be a distribution in the representor that assigns that proposition its upper probability, and a distribution that assigns the proposition its lower probability. 519

Clutter Avoidance principle Gilbert Harman's famous principle that "one should not clutter one's mind with trivialities". 429

coherent de Finetti (1937/1964) defined a coherent credence distribution as one not susceptible to Dutch Books. Other Bayesians define coherence as satisfying Kolmogorov's probability axioms. And others simply take coherence to be synonymous with rational consistency, whatever that requires. 335

common cause A single event that causally influences at least two other events. 73

commutative Updating by Conditionalization is commutative in the sense that updating first on E then on E' has the same effect as updating in the opposite order. 94

Comparative Additivity For any propositions P, Q, and R in $\mathcal{L}$, if P and Q are each mutually exclusive with R, then the agent is at least as confident in P as Q just in case the agent is at least as confident in $P \vee R$ as $Q \vee R$. 487

Comparative Completeness For any propositions P and Q in $\mathcal{L}$, the agent is at least as confident in P as Q or at least as confident in Q as P (or both!). 487

comparative concept Places one entity in order with respect to another. 4

Comparative Entailment For propositions P and Q, if $P \vDash Q$ then rationality requires an agent to be at least as confident of Q as P. 11

Comparative Equivalence For any propositions P and Q in $\mathcal{L}$, if $P \; \dashv\vDash \; Q$, then $P \sim Q$. 486

Comparative Matching An agent is at least as confident in P as Q just in case her credence in P is at least as great as her credence in Q. That is, for any P and Q in $\mathcal{L}$, $P \succeq Q$ just in case $\mathrm{cr}(P) \geq \mathrm{cr}(Q)$. 489

Comparative Non-Negativity For any proposition P and contradiction F in $\mathcal{L}$, the agent is at least as confident of P as F. 487

Comparative Non-Triviality For any tautology T and contradiction F in $\mathcal{L}$, the agent is more confident in T than F. 487

Comparative Transitivity For any propositions P, Q, and R in $\mathcal{L}$, if the agent is at least as confident in P as Q, and at least as confident in Q as R, then the agent is at least as confident in P as R. 486

condition In a conditional credence, the proposition the agent supposes. 56

conditional bet A conditional bet on P given Q wins or loses money for the agent only if Q is true; if Q is false the bet is called off. An agent's fair betting price for a conditional bet that pays \$1 on P (given Q) is typically $cr(P\,|\,Q)$. 316

Conditional Comparative Matching For any propositions P, Q, R, and S in $\mathcal{L}$, $P\,|\,R \geq Q\,|\,S$ just in case $cr(P\,|\,R) \geq cr(Q\,|\,S)$. 500

conditional credence A degree of belief assigned to an ordered pair of propositions, indicating how confident the agent is that the first proposition is true on the supposition that the second is. 56

conditional independence When $cr(Q\,\&\,R) > 0$, P is probabilistically independent of Q conditional on R just in case $cr(P\,|\,Q\,\&\,R) = cr(P\,|\,R)$. 68

Conditionalization For any time t_i and later time t_j, if proposition E in $\mathcal{L}$ represents everything the agent learns between t_i and t_j, and $cr_i(E) > 0$, then for any H in $\mathcal{L}$, $cr_j(H) = cr_i(H\,|\,E)$ (Bayesians' traditional updating rule). 91

confirmation Evidence confirms a hypothesis just in case the evidence supports that hypothesis (to any degree). 195

Confirmation and Increase principle Evidence E confirms hypothesis H for a rational agent if and only if any experience that increased her confidence in E would also increase her confidence in H. 424

Confirmation and Learning principle Evidence E confirms hypothesis H for a rational agent if and only if learning E would increase the agent's confidence in H. 414

Confirmation and Probability principle Evidence E confirms hypothesis H relative to probability distribution Pr if and only if $Pr(H\,|\,E) > Pr(H)$. 414

confirmation measure A numerical measure of the degree to which evidence E confirms hypothesis H relative to probability distribution Pr. 225

Confirmation Transitivity For any A, B, C, and K in $\mathcal{L}$, if A confirms B relative to K and B confirms C relative to K, then A confirms C relative to K. 201

Conglomerability For each proposition P and partition $\{Q_1, Q_2, Q_3, \ldots\}$ in $\mathcal{L}$, $cr(P)$ is no greater than the largest $cr(P\,|\,Q_i)$ and no less than the least $cr(P\,|\,Q_i)$. 153

conjunction $P\,\&\,Q$ is a conjunction; P and Q are its conjuncts. 26

Conjunction Fallacy Being more confident in a conjunction than you are in one of its conjuncts. 39

Consequence Condition If E in $\mathcal{L}$ confirms every member of a set of propositions relative to K and that set jointly entails H' relative to K, then E confirms H' relative to K. 203

consequent In a conditional of the form "If P, then Q," Q is the consequent. 26

Consistency Condition For any E and K in $\mathcal{L}$, the set of all hypotheses confirmed by E relative to K is logically consistent with E & K. 205

consistent The propositions in a set are consistent when at least one possible world makes all the propositions true. 28

constant A lower-case letter in language $\mathcal{L}$ representing an object in the universe of discourse. 30

constant act A decision-theoretic act that produces the same outcome for an agent regardless of which state of the world obtains. 288

contingent A proposition that is neither a tautology nor a contradiction. 28

contradiction A proposition that is false in every possible world. 28

Contradiction For any contradiction F in $\mathcal{L}$, $cr(\mathsf{F}) = 0$. 34

Converse Consequence Condition For any E, H, H', and K in $\mathcal{L}$ (with H' consistent with K), if E confirms H relative to K and H' & $K \vDash H$, then E confirms H' relative to K. 205

Converse Dutch Book Theorem A theorem showing that if an agent satisfies particular constraints on her credences, she will not be susceptible to a particular kind of Dutch Book. 319

Converse Entailment Condition For any consistent E, H, and K in $\mathcal{L}$, if H & $K \vDash E$ but $K \nvDash E$, then E confirms H relative to K. 206

Convexity A convex representor that contains two distributions will also contain every linear combination of those distributions. 518

Countable Additivity For any countable set $\{Q_1, Q_2, Q_3, \ldots\}$ of mutually exclusive propositions in $\mathcal{L}$, $cr(Q_1 \vee Q_2 \vee Q_3 \vee \ldots) = cr(Q_1) + cr(Q_2) + cr(Q_3) + \ldots$ 152

credence Degree of belief. 3

credence elicitation Structuring incentives so that rational agents will report the truth about the credence values they assign. 342

cumulative Updating by Conditionalization is cumulative in the sense that updating first on evidence E and then on evidence E' has the same net effect as updating once, on the conjunction E & E'. 94

Czech Book A set of bets, each placed with an agent at her fair betting price (or better), that together guarantee her a sure gain come what may. 319

database expert Expert to whom one defers because her evidence includes one's own, and more. 142

de dicto proposition Uncentered proposition. 410

de se proposition Centered proposition. 410

decision problem A situation in which an agent must choose exactly one out of a partition of available acts, in hopes of attaining particular outcomes. Decision problems are the targets of analysis in decision theory. 251

decision theory Searches for rational principles to evaluate the acts available to an agent in a decision problem. 246

Decomposition For any propositions P and Q in $\mathcal{L}$, $\text{cr}(P) = \text{cr}(P \& Q) + \text{cr}(P \& \sim Q)$. 34

decreasing marginal utility When a quantity has decreasing marginal utility, less utility is derived from each additional unit of that quantity the more units you already have. Economists often suggest that money has decreasing marginal utility for the typical agent. 251

defeat in expectation Given distributions cr and cr′ over the same set of propositions, cr′ defeats cr in expectation if cr calculates a lower expected inaccuracy for cr′ than it does for cr. 355

deference principle Any principle directing an agent to align her current credences with some other distribution (such as objective chances, credences of an expert, or credences of her future self). 141

Dempster-Shafer function Another name for a belief function. 527

dilation A ranged credence dilates when conditionalizing it on new evidence widens the gap between its upper and lower probabilities. 522

direct inference Determining how likely one is to obtain a particular experimental result from probabilistic hypotheses about the setup. 61

Disconfirmation Duality For any E, H, and K in $\mathcal{L}$, E confirms H relative to K just in case E disconfirms $\sim H$ relative to K. 207

disjunction $P \vee Q$ is a disjunction; P and Q are its disjuncts. 26

disjunctive normal form The disjunctive normal form of a non-contradictory proposition is the disjunction of state-descriptions that is equivalent to that proposition. 30

distribution An assignment of real numbers to each proposition in language $\mathcal{L}$. 31

Dominance Principle If act A produces a higher-utility outcome than act B in each possible state of the world, then A is preferred to B. 256

doxastic attitude A belief-like representational propositional attitude. 3

doxastically possible worlds The subset of possible worlds that a given agent entertains. 37

Dutch Book A set of bets, each placed with an agent at her fair betting price (or better), that together guarantee her a sure loss come what may. 313

Dutch Book Theorem If an agent's credence distribution violates at least one of the probability axioms (Non-Negativity, Normality, or Finite Additivity), then a Dutch Book can be constructed against her. 314

Dutch Strategy A strategy for placing different sets of bets with an agent over a period of time, depending on what the agent learns during that period of time. If the strategy is implemented correctly, the bets placed will guarantee the agent a sure loss come what may. 318

Ellsberg Paradox A set of gambles proposed by Daniel Ellsberg, in which risk-averse agents' behavior seems to violate both standard constraints on confidence comparisons and standard principles of decision theory. 503

entailment P entails Q ($P \vDash Q$) just in case there is no possible world in which P is true and Q is false. On a Venn diagram, the P-region is wholly contained in the Q-region. 28

Entailment Condition For any consistent E, H, and K in $\mathcal{L}$, if $E \& K \vDash H$ but $K \nvDash H$, then E confirms H relative to K. 201

Entailment For any propositions P and Q in $\mathcal{L}$, if $P \vDash Q$ then $cr(P) \leq cr(Q)$. 34

epistemic standards Applying an agent's ultimate epistemic standards to her total evidence at a given time yields her doxastic attitudes at that time. Bayesians represent ultimate epistemic standards as hypothetical priors. 107

epistemic utility A numerical measure of the epistemic value of a set of doxastic attitudes. 346

Equivalence Condition For any H, H', E, E', K and K' in $\mathcal{L}$, suppose $H =\!\!\vDash H'$, $E =\!\!\vDash E'$, and $K =\!\!\vDash K'$. Then E confirms (/disconfirms) H relative to background K just in case E' confirms (/disconfirms) H' relative to background K'. 200

Equivalence For any propositions P and Q in $\mathcal{L}$, if $P =\!\!\vDash Q$ then $cr(P) = cr(Q)$. 34

equivalent Equivalent propositions are associated with the same set of possible worlds. 28

ethically neutral A proposition P is ethically neutral for an agent if the agent is indifferent between any two gambles whose outcomes differ only in replacing P with $\sim P$. 287

Evidential Decision Theory Decision theory in which expected utility is calculated using an agent's credences in states conditional on the available acts. 266

evidential probability The degree to which a body of evidence probabilifies a hypothesis, understood as independent of any particular agent's attitudes. 131

evidentialism The position that what attitudes are rationally permissible for an agent supervenes on her evidence. 130

exhaustive The propositions in a set are jointly exhaustive if each possible world makes at least one of the propositions in the set true. 28

expectation An agent's expectation for the value of a particular quantity is a weighted average of the values that quantity might take, with weights provided by the agent's credences across those possible values. 247

extendable A partial credence distribution is extendable to a probabilistic complete distribution just in case there exists at least one complete distribution satisfying the probability axioms that assigns the same value as the partial credence distribution to every proposition toward which the latter adopts an attitude. A non-total preorder is extendable to a total preorder satisfying some set of norms just in case there is some way to add comparative relations into the preorder so as to make it a total preorder satisfying those norms. 432

Extensional Equivalence If two betting arrangements have the same payoff as one another in each possible world, a rational agent will value them equally. 328

fair price An agent's break-even point for a bet or investment. She will pay anything up to that amount of money for the bet/investment. 248

falsification A piece of evidence falsifies a hypothesis if it refutes that hypothesis relative to one's background assumptions. 67

Finite Additivity For any mutually exclusive propositions P and Q in $\mathcal{L}$, $cr(P \vee Q) = cr(P) + cr(Q)$ (one of the three probability axioms). 32

Finite Additivity (Extended) For any finite set of mutually exclusive propositions $\{P_1, P_2, \ldots, P_n\}$, $cr(P_1 \vee P_2 \vee \ldots \vee P_n) = cr(P_1) + cr(P_2) + \ldots + cr(P_n)$. 34

firmness concept of confirmation E confirms H relative to K just in case a probability distribution built on background K makes the probability of H on E high. 208

frequency theory An interpretation of probability according to which the probability is x that event A will have outcome B just in case fraction x of events like A have outcomes like B. 125

frequentism A cluster of traditional statistical tools, promoted on the grounds that, were they repeatedly applied, they would yield verdicts with a particular desirable feature fairly frequently (or an undesirable feature fairly infrequently). 459

Gambler's Fallacy Expecting later outcomes of an experiment to "compensate" for unexpected previous results despite the probabilistic independence of future results from those in the past. 68

General Additivity For any propositions P and Q in $\mathcal{L}$, $cr(P \vee Q) = cr(P) + cr(Q) - cr(P \& Q)$. 34

Gradational Accuracy Theorem Given a credence distribution cr over a finite set of propositions $\{X_1, X_2, \ldots, X_n\}$, if we use the Brier score $I_{BR}(cr, \omega)$ to measure inaccuracy then: (1) If cr does *not* satisfy the probability axioms, there exists a probabilistic distribution cr' over the same propositions such that $I_{BR}(cr', \omega) < I_{BR}(cr, \omega)$ in *every* logically possible world ω; and (2) If cr *does* satisfy the probability axioms, no such cr' exists. 347

higher-order credences An agent's credences about her own current credences. Includes both her credences about what her current credence-values *are* and her credences about what those values *should be*. 171

HTM approach Formal system for updating self-locating credences inspired by the work of Joseph Y. Halpern, Mark Tuttle, and Christopher J.G. Meacham. 397

Humphreys's Paradox Difficulty for the propensity interpretation of probability that when the probability of E given H can be understood in terms of propensities it is often difficult to interpret the probability of H given E as a propensity as well. 168

Hypothesis Symmetry For all H and E in $\mathcal{L}$ and every probabilistic Pr, the degree to which E confirms H is the opposite of the degree to which E confirms $\sim H$. 229

hypothetical frequency theory Interpretation of probability that looks not at the proportion of actual events producing a particular outcome but instead at the proportion of such events that would produce that outcome in the limit. 127

hypothetical prior distribution A regular, probabilistic distribution used to represent an agent's ultimate epistemic standards. The agent's credence distribution at a given time can be recovered by conditionalizing her hypothetical prior on her total evidence at that time. 110

Hypothetical Priors Theorem Given any finite series of credence distributions $\{cr_1, cr_2, \ldots, cr_n\}$ each of which satisfies the probability axioms and Ratio Formula, let E_i be a conjunction of the agent's total evidence at t_i. If each cr_i is related to cr_{i+1} as specified by Conditionalization, then there exists at least one regular probability distribution Pr_H such that for all $1 \leq i \leq n$, $cr_i(\cdot) = Pr_H(\cdot \mid E_i)$. 110

Hypothetical Representability Given any finite series of credence distributions $\{cr_1, cr_2, \ldots, cr_n\}$ that the agent assigns over time, there exists at least one regular probability distribution Pr_H such that for all $1 \leq i \leq n$, $cr_i(\cdot) = Pr_H(\cdot \mid E_i)$ (where E_i is the conjunction of the agent's total evidence at t_i). 386

hypothetico-deductivism Theory of confirmation on which E confirms H relative to background corpus K just in case $H \& K \vDash E$ and $K \nvDash E$. 242

ideally rational agent An agent who perfectly satisfies all the requirements of rationality. 382

IID trials Independent, identically distributed probabilistic events. Trials are IID if the probabilities associated with a given trial are unaffected by the outcomes of other trials (independence), and if each trial has the same probability of producing a given outcome (identically distributed). 87

immodest An immodest credence distribution is not defeated in accuracy expectation by any distribution. 356

imprecise credence Alternative terminology for ranged credence. 546

inconsistent The propositions in a set are inconsistent when there is no possible world in which all of them are true. 28

increase in firmness concept of confirmation E confirms H relative to K just in case a probability distribution built on K makes the posterior of H on E higher than the prior of H. 208

independence When $cr(Q) > 0$, proposition P is probabilistically independent of proposition Q relative to cr just in case $cr(P \,|\, Q) = cr(P)$. 65

infinitesimal A number that is greater than zero but less than any positive real number. 155

initial prior distribution Credence distribution assigned by an agent before she possessed any contingent evidence. 106

interference effect Any effect of placing the initial bets in a Dutch Book that makes an agent unwilling to accept the remaining bets (which she otherwise would have regarded as fair). 326

interpretations of probability Philosophical theories about the nature of probability and the meanings of linguistic probability expressions. 124

Interval Requirement If a representor contains distributions assigning a particular proposition the distinct values x and y, then for any value z between x and y, the representor must also contain a distribution assigning that proposition the value z. 517

inverse inference Determining how likely a probabilistic hypothesis is on the basis of a particular run of experimental data. 62

irrelevant Probabilistically independent. 65

Jeffrey Conditionalization Proposed by Richard C. Jeffrey as an alternative updating rule to Conditionalization, holds that for any t_i and t_j with $i < j$, any A in $\mathcal{L}$, and a finite partition $\{B_1, B_2, \ldots, B_n\}$ in $\mathcal{L}$ whose members each have nonzero cr_i, $cr_j(A) = cr_i(A \mid B_1) \cdot cr_j(B_1) + cr_i(A \mid B_2) \cdot cr_j(B_2) + \ldots + cr_i(A \mid B_n) \cdot cr_j(B_n)$. 157

Judy Benjamin Problem An example proposed by Bas van Fraassen in which an agent's experience directly alters some of her conditional credence values. van Fraassen argued that this example could not be addressed by traditional Conditionalization or by Jeffrey Conditionalization. 161

just in case If and only if. 26

Kolmogorov's axioms The three axioms (Non-Negativity, Normality, and Finite Additivity) that provide necessary and sufficient conditions for a probability distribution. 32

language dependence A theory is language dependent when the properties it ascribes to propositions vary depending on the language in which those propositions are expressed. 218

law of large numbers Any one of a number of mathematical results indicating roughly the following: the probability is 1 that as the number of trials approaches the limit, the average value of a quantity will approach its expected value. 247

Law of Likelihood Evidence E favors hypothesis H_1 over H_2 just in case $\Pr(E \mid H_1) > \Pr(E \mid H_2)$. In that case, the degree to which E favors H_1 over H_2 is measured by the likelihood ratio $\Pr(E \mid H_1)/\Pr(E \mid H_2)$. 467

Law of Total Probability For any proposition P and finite partition $\{Q_1, Q_2, \ldots, Q_n\}$ in $\mathcal{L}$, $cr(P) = cr(P \mid Q_1) \cdot cr(Q_1) + cr(P \mid Q_2) \cdot cr(Q_2) + \ldots + cr(P \mid Q_n) \cdot cr(Q_n)$. 59

likelihood The probability of some particular piece of evidence on the supposition of a particular hypothesis—$cr(E \mid H)$. 62

likelihood principle Likehoodist principle explaining when two distinct experimental observations should be taken to have the same evidential significance. 482

likelihood ratio Ratio of the likelihood of the evidence on one hypothesis to the likelihood of that evidence on another. That is, $cr(E \mid H_1)/cr(E \mid H_2)$. 153

likelihoodism Position in statistics and philosophy of probability that endorses Hacking's Law of Likelihood as our best tool for understanding evidential favoring. 467

linear combination When applied to three numerical variables, z is a linear combination of x and y just in case there exist constants a and b such that $z = ax + by$. For instance, Finite Additivity requires $cr(X \vee Y) = cr(X) + cr(Y)$, making the credence in a disjunction a linear combination of the credences in its mutually exclusive disjuncts (with the two constants a and b each set to 1). When applied to probability distributions, distribution $\Pr_z$ over language $\mathcal{L}$ is a linear combination of distributions $\Pr_x$ and $\Pr_y$ over $\mathcal{L}$ just in case there exists some $0 \leq \alpha \leq 1$ such that for every proposition $P \in \mathcal{L}$, $\Pr_z(P) = \alpha \cdot \Pr_x(P) + (1 - \alpha) \cdot \Pr_y(P)$. 54

Lockean thesis Connects believing a proposition with having a degree of confidence in that proposition above a numerical threshold. 15

logical omniscience requirement Requirement that an agent be certain of all logical truths. 429

logical probability The degree to which a body of evidence probabilifies a hypothesis, understood as a logical relation similar to deductive entailment. 131

Logicality All entailments receive the same degree of confirmation, and have a higher degree of confirmation than any non-entailing confirmations. 229

logically possible A world is logically possible when it violates no laws of logic. So, for instance, a world with no gravity is physically impossible because it violates the laws of physics, but it is logically possible. 37

Lottery Paradox Paradox for requirements of logical belief consistency and closure involving a lottery with a large number of tickets. 8

lower probability The lower numerical bound of a ranged credence. 506

mass function In Dempster-Shafer theory, a non-negative real-valued distribution over a Boolean algebra that: (1) assigns 0 to the contradiction; and (2) has values summing to 1. Among other things, used to represent the significance of a particular piece of evidence for the propositions in the algebra. 528

material biconditional A material biconditional $P \equiv Q$ is true just in case P and Q are both true or P and Q are both false. 26

material conditional A material conditional $P \supset Q$ is false just in case its antecedent P is true and its consequent Q is false. 26

Maximality For any proposition P in $\mathcal{L}$, $cr(P) \leqslant 1$. 34

maximin rule Decision rule that prefers the act with the highest minimum payoff. 252

Maximum Entropy Principle Given any partition of the space of possibilities, and any set of constraints on allowable credence distributions over that partition, the Maximum Entropy Principle selects the allowable distribution with the highest entropy. 147

money pump A situation in which an agent's preferences endorse her making a series of decisions, the net effect of which is to cost her some utility but otherwise leave her exactly where she began. Money pumps are used to argue that preferences violating Preference Transitivity or Preference Asymmetry are irrational. 254

monotonicity If a property is monotonic, then whenever proposition E bears the property, every conjunction containing E as a conjunct bears the property as well. 101

Monty Hall Problem A famous probabilistic puzzle case, demonstrating the importance of taking an agent's total evidence into account. 102

Multiplication When P and Q have nonextreme cr-values, P and Q are probabilistically independent relative to cr if and only if $cr(P \& Q) = cr(P) \cdot cr(Q)$. 65

mutually exclusive The propositions in a set are mutually exclusive when there is no possible world in which more than one of the propositions is true. 28

negation ~P is the negation of P. 26

Negation For any proposition P in $\mathcal{L}$, $\text{cr}(\sim P) = 1 - \text{cr}(P)$. 33

negative instance Fa & ~Ga is a negative instance of the universal generalization $(\forall x)(Fx \supset Gx)$. 197

negative relevance When $\text{cr}(Q) > 0$, Q is negatively relevant to P relative to cr just in case $\text{cr}(P \mid Q) < \text{cr}(P)$. 66

Newcomb's Problem A puzzle that prompted the development of Causal Decision Theory. Introduced to philosophy by Robert Nozick, who attributed its construction to William Newcomb. 263

Nicod's Criterion For any predicates F and G and constant a of $\mathcal{L}$, $(\forall x)(Fx \supset Gx)$ is confirmed by Fa & Ga and disconfirmed by Fa & ~Ga. 196

Non-Negativity For any proposition P in $\mathcal{L}$, $\text{cr}(P) \geq 0$ (one of the three probability axioms). 32

nonmonotonicity Probabilistic relations are nonmonotonic in the sense that even if H is highly probable given E, H might be improbable given the conjunction of E with some E'. 102

Normality For any tautology T in $\mathcal{L}$, $\text{cr}(\mathsf{T}) = 1$ (one of the three probability axioms). 32

normalization factor In an update by Conditionalization, state-descriptions inconsistent with E (the evidence learned) have their unconditional credences sent to zero. The remaining state-descriptions all have their unconditional credences multiplied by the same normalization factor, equal to the reciprocal of E's prior. 119

normative distinction The normative distinction between Subjective and Objective Bayesians concerns the strength of rationality's requirements. Distinguished this way, Objective Bayesians hold that there is exactly one rationally permissible set of epistemic standards (/hypothetical priors), so that any body of total evidence gives rise to a unique rational attitude toward any particular proposition. Subjective Bayesians deny that rational requirements are strong enough to mandate a unique attitude in every case. 129

null hypothesis Hypothesis assessed by a significance test. We typically select as our null a hypothesis that explains the results of a given statistical trial (or the statistical variation from one trial to another) entirely in terms of the chanciness of the process generating the data. 460

objective chance A type of physical probability that can be applied to the single case. 128

observation selection effect Effect on the appropriate conclusions to draw from a piece of evidence introduced by the manner in which that evidence was obtained (for example, the method by which a sample was drawn). 102

odds If an agent's unconditional credence in P is $cr(P)$, her odds for P are $cr(P)$: $cr(\sim P)$, and her odds against P are $cr(\sim P)$: $cr(P)$. 45

open interval An interval that does not include its endpoints. For instance, if the range of values assigned to a proposition by a particular representor is an open interval, then there will be no distribution in the representor assigning that proposition its upper probability, and no distribution assigning its lower probability. 520

ordering An antisymmetric preorder. 543

outcome The result of an agent's performing a particular act with the world in a particular state. Agents assign utilities to outcomes. 254

p-value In a significance test, the p-value of a particular hypothesis reports how probable it would be, were that hypothesis true, that a sample would yield at least as extreme an outcome as the one that's been observed. 460

Package Principle A rational agent's value for a package of bets equals the sum of her values for the individual bets it contains. 328

Paradox of the Ravens Counterintuitive consequence of many formal theories of confirmation that the proposition that a particular object is a non-black non-raven confirms the hypothesis that all ravens are black. 197

partial distribution An assignment of real numbers to some—but not all—of the propositions in language $\mathcal{L}$. 430

partial preorder A preorder that does not satisfy Comparative Completeness. 497

partition A mutually exclusive, jointly exhaustive set of propositions. On a Venn diagram, the regions representing propositions in a partition combine to fill the entire rectangle without overlapping at any point. 29

Partition For any finite partition of propositions in $\mathcal{L}$, the sum of their unconditional cr-values is 1. 34

permissive case An example in which two agents with identical total evidence assign different credences without either agent's thereby being irrational. Objective Bayesians in the normative sense deny the existence of permissive cases. 131

plausibility function In Dempster-Shafer theory, a real-valued distribution over a Boolean algebra indicating how plausible each proposition is in light of the agent's evidence. 530

positive instance Fa & Ga is a positive instance of the universal generalization $(\forall x)(Fx \supset Gx)$. 197

positive relevance When cr(Q) > 0, Q is positively relevant to P relative to cr just in case cr($P \mid Q$) > cr(P). 66

possible worlds Different ways the world might have come out. Possible worlds are maximally specified—for any event and any possible world that event either does or does not occur in that world—and the possible worlds are plentiful enough such that for any combination of events that *could* happen, there is a possible world in which that combination of events *does* happen. 26

posterior The probability of some hypothesis on the supposition of a particular piece of evidence—$P(H \mid E)$. 63

practical rationality Concerns the connections between attitudes and actions. 7

predicate A capital letter representing a property or relation in language $\mathcal{L}$. 30

Preface Paradox Paradox for requirements of logical belief consistency and closure in which the preface to a nonfiction book asserts that at least one of the claims in the book is false. 8

Preference Asymmetry There do not exist acts A and B such that the agent both prefers A to B and prefers B to A. 253

preference axioms Formal constraints we assume a rational agent's preferences satisfy in order to apply a representation theorem. 292

Preference Completeness For any acts A and B, exactly one of the following is true: the agent prefers A to B, the agent prefers B to A, or the agent is indifferent between the two. 253

Preference Transitivity For any acts A, B, and C, if the agent prefers A to B and B to C, then the agent prefers A to C. 252

preorder $\geq$ introduces a preorder on the members of $\mathcal{L}$ if it satisfies Comparative Equivalence and Transitivity. 486

Principal Principle David Lewis's proposal for how rational credences concerning an event incorporate suppositions about the objective chances of that event's possible outcomes. 136

Principle of Indifference If an agent has no evidence favoring any proposition in a partition over any other, she should spread her credence equally over the members of the partition. 146

Principle of the Common Cause When event outcomes are probabilistically correlated, either one causes the other or they have a common cause. 73

Principle of Total Evidence A rational agent's credence distribution takes into account all of the evidence she possesses. 102

prior An unconditional probability; the probability of a proposition before anything has been supposed. For example, an agent's prior credence in a particular hypothesis H is cr(H). 63

probabilism The thesis that rationality requires an agent's credences to satisfy the probability axioms. 33

probabilistic representability A total preorder $\geq$ over language $\mathcal{L}$ is probabilistically representable just in case there exists a probabilistic distribution Pr over $\mathcal{L}$ such that for any propositions P and Q in $\mathcal{L}$, $P \geq Q$ if and only if $\Pr(P) \geq \Pr(Q)$. 489

probabilistically independent When cr(Q) > 0, P is probabilistically independent of Q relative to cr just in case cr($P \mid Q$) = cr(P). 65

probability axioms Kolmogorov's axioms. 32

probability distribution Any distribution satisfying Kolmogorov's probability axioms. 32

probability kinematics What Richard C. Jeffrey, its inventor, called the updating rule now generally known as "Jeffrey Conditionalization". 157

probability table A table that assigns unconditional credences to each member in a partition. To satisfy the probability axioms, the values in each row must be non-negative and all the values must sum to 1. When the partition members are state-descriptions of a language $\mathcal{L}$, the values in the probability table suffice to specify all of the agent's credences over $\mathcal{L}$. 41

problem of irrelevant conjunction Counterintuitive consequence of many formal theories of confirmation that whenever evidence E confirms hypothesis H it will also confirm $H \& X$ for various Xs irrelevant to E and H. 206

problem of new theories Traditional Bayesian updating schemes seem unable to account for how an agent's credences change when she comes to entertain previously unconsidered possibilities. 418

Problem of Old Evidence On the standard Bayesian account of confirmation, evidence already part of an agent's background corpus cannot confirm any hypothesis for that agent. 416

Problem of the Priors Problem for Bayesian epistemology of how agents are to set the priors that influence the outcomes of a Bayesian update. 447

problem of the single case The challenge of interpreting probability such that single (and perhaps non-repeatable) events may receive nonextreme probabilities. 127

propensity theory Interpretation of probability identifying probability with a physical arrangement's quantifiable tendency to produce outcomes of a particular kind. 127

Proper Scoring Rule A scoring rule is proper just in case any agent with a probabilistic credence distribution who uses that rule assigns her own credences a lower expected inaccuracy than any other distribution over the same set of propositions. 358

proposition An abstract entity expressible by a declarative sentence and capable of having a truth-value. 3

propositional attitude An attitude adopted by an agent toward a proposition or set of propositions. 3

propositional connective One of five truth-functional symbols ($\sim$, &, $\vee$, $\supset$, $\equiv$) used to construct larger propositions from atomic propositions. 26

qualitative probability Any comparative confidence ranking $\geq$ that satisfies the Comparative Equivalence, Transitivity, Completeness, Non-Negativity, Non-Triviality, and Additivity conditions. Note that being a qualitative probability does *not* suffice for probabilistic representability. 543

quantitative concept Characterizes an entity by ascribing it a numerical value. 4

ranged credence A doxastic attitude better represented by a range of real numbers than by a single point value. 506

ratifiability Decision-theoretic requirement that an act is rationally permissible only if the agent assigns it the highest expected utility conditional on the supposition that she chooses to perform it. 271

Ratio Formula For any P and Q in $\mathcal{L}$, if $cr(Q) > 0$ then $cr(P \mid Q) = cr(P \& Q)/cr(Q)$. The Bayesian rational constraint relating an agent's conditional credences to her unconditional credences. 57

reference class problem When considering a particular event and one of its possible outcomes, the frequency with which this type of event produces that type of outcome depends on which reference class (event-type) we choose out of the many to which the event belongs. 126

Reflection Principle For any proposition A in $\mathcal{L}$, real number x, and times t_i and t_j with $j > i$, rationality requires $cr_i(A \mid cr_j(A) = x) = x$. 143

refutation P refutes Q just in case P entails $\sim Q$. When P refutes Q, every world that makes P true makes Q false. 28

regular A distribution that does not assign the value 0 to any logically contingent propositions. 99

Regularity Principle In a rational credence distribution, no logically contingent proposition receives unconditional credence 0. 99

relevance measure A confirmation measure that indicates confirmation just in case E is positively relevant to H on Pr; disconfirmation just in case E is negatively relevant to H on Pr; and neither just in case E in independent of H on Pr. 227

Relevance-Limiting Thesis If an update does not eliminate any uncentered possible worlds, then it does not change the agent's credence distribution over uncentered propositions. 400

relevant Not probabilistically independent. 66

Representation Theorem If an agent's preferences satisfy certain constraints, then there exists a unique probabilistic credence distribution and unique utility distribution (up to positive affine transformation) that yield those preferences when the agent maximizes expected utility. 291

representor A set of probability distributions used to represent an agent's doxastic state, especially a state containing ranged credences. 507

resilience A measure of how much an agent's credence in a particular proposition is apt to move in light of future courses of evidence. 515

Rigidity condition For any A in $\mathcal{L}$ and any B_m in the finite partition $\{B_1, B_2, \ldots, B_n\}$, $cr_j(A \mid B_m) = cr_i(A \mid B_m)$. This condition obtains between t_i and t_j just in case the agent Jeffrey Conditionalizes across $\{B_1, B_2, \ldots, B_n\}$. 159

risk aversion Preferring an act with lesser expected value because it offers a surer payout. 260

rule of succession Laplace's rule directing an agent who has witnessed h heads on n independent flips of a coin to set credence $(h+1)/(n+2)$ that the next flip will come up heads. 169

scalar transformation Two measurement scales are related by a scalar transformation when values on one scale can be converted to values on the other by multiplying by a specified constant. The pound and kilogram scales for mass provide one example. 293

scoring rule A quantitative measure of the accuracy (or inaccuracy) of distributions. 341

Scott Axiom For any two equinumerous, finite sequences of propositions drawn from $\mathcal{L}, \{A_1, A_2, \ldots, A_n\}$ and $\{B_1, B_2, \ldots, B_n\}$: *if* in each of the agent's doxastically possible worlds, the A-sequence contains the same number of truths as the B-sequence; *then* if there exists some i such that the agent is strictly more confident of A_i than B_i, there must also exist some j such that the agent is strictly more confident of B_j than A_j. 492

screening off R screens off P from Q when P is unconditionally relevant to Q but not relevant to Q conditional on either R or $\sim R$. 68

self-locating information Information about what time it is, or where the agent is located spatially, or who that agent is. 393

semantic distinction When classified according to the semantic distinction, Subjective Bayesians take "probability" talk to reveal the credences of agents, while Objective Bayesians assign "probability" assertions truth-conditions independent of the attitudes of particular agents or groups of agents. 129

separable A separable scoring rule measures how far a distribution is from the truth one proposition at a time, then sums the results. 345

sigma algebra A set of sets closed under union, intersection, and complementation. A probability distribution can be assigned over a sigma algebra containing sets of possible worlds instead of over a language containing propositions. 52

significance level In a significance test, any p-value below the pre-selected significance level α suffices to reject a hypothesis. 460

Simple Binarist A made-up character who describes agents' doxastic propositional attitudes exclusively in terms of belief, disbelief, and suspension of judgment. 5

Simpson's Paradox Two propositions may be correlated conditional on each member of a partition yet anti-correlated unconditionally. 70

Special Consequence Condition For any E, H, H', and K in $\mathcal{L}$, if E confirms H relative to K and $H \& K \vDash H'$, then E confirms H' relative to K. 203

specificity Specific evidence about a proposition provides unambiguous information about that proposition's probability, rationalizing the assignment of a narrow ranged credence—or even a point-valued credence—to the proposition. 515

state In decision theory, an arrangement of the world (usually represented as a proposition). Which state obtains affects which outcome will be generated by the agent's performing a particular act. 254

state-description A conjunction of language $\mathcal{L}$ in which (1) each conjunct is either an atomic proposition of $\mathcal{L}$ or its negation; and (2) each atomic proposition of $\mathcal{L}$ appears exactly once. 29

straight rule Reichenbach's name for the norm setting an agent's credence that the next event of type A will produce an outcome of type B exactly equal to the observed frequency of B-outcomes in past A-events. 169

strict Conditionalization Another name for the Conditionalization updating rule. The "strict" is usually used to emphasize a contrast with Jeffrey Conditionalization. 160

structure-description Given a specific language, a structure-description says how many objects possess each of the available property profiles, but doesn't say which particular objects have which profiles. 213

subadditive In a subadditive distribution, there exist mutually exclusive P and Q in $\mathcal{L}$ such that $cr(P \vee Q) < cr(P) + cr(Q)$. 51

substitution instance A substitution instance of a quantified sentence is produced by removing the quantifier and replacing its variable throughout what remains with the same constant. 31

superadditive In a superadditive distribution, there exist mutually exclusive P and Q in $\mathcal{L}$ such that $cr(P \vee Q) > cr(P) + cr(Q)$. 47

supervenience *A*-properties supervene on *B*-properties just in case any two objects that differ in their *A*-properties also differ in their *B*-properties. For example, one's score on a test supervenes on the answers one provides; if two students got different scores on the same test, their answers must have differed. 58

Suppes Continuity For any propositions *P* and *Q* in $\mathcal{L}$ such that the agent is at least as confident of *P* as *Q*, there exists a proposition *R* in $\mathcal{L}$ such that the agent is equally confident in *P* as she is in the disjunction of *Q* and *R*. 495

Suppositional Consistency Stipulate that $cr_X(Y)$ represents the credence in *Y* that a particular set of epistemic standards dictates relative to total evidence *X*, and $cr_X(Y\,|\,Z) = cr_X(Y \,\&\, Z)/cr_X(Z)$. Then Suppositional Consistency requires of a set of standards that for any propositions *A*, *B*, *C* in $\mathcal{L}$, $cr_{A\&C}(B) = cr_A(B\,|\,C)$. 390

Sure-Thing Principle If two acts yield the same outcome on a particular state, any preference between them remains the same if that outcome is changed. 261

tautological background A background corpus containing no contingent information, logically equivalent to a tautology T. 198

tautology A proposition that is true in every possible world. 28

theoretical rationality Evaluates representational attitudes in their capacity as representations, without considering how they influence action. 7

total ordering A complete ordering—that is, an ordering in which every pair of items in the set gets compared. For an ordering to be total, it must satisfy equivalence, antisymmetry, transitivity, and completeness. 543

total preorder $\geq$ introduces a total preorder on the members of $\mathcal{L}$ if it satisfies Comparative Equivalence, Transitivity, and Completeness. 487

Truth-Directedness If an inaccuracy score is truth-directed, altering a distribution by moving some of its values closer to the truth and none of its values farther from the truth will decrease that distribution's inaccuracy. 351

Truth-Functional Bayesianism A Bayesianism that requires certainty in truth-functional truths, but not other kinds of logical truths. 434

truth-functional truth A proposition that is true by virtue of its truth-functional form—by virtue of the arrangement of truth-functional connectives within it. 434

truth-value *True* and *false* are truth-values. We assume propositions are capable of having truth-values. 3

uncentered possible world A traditional possible world, which maximally specifies what events occur but is not centered on a particular individual or spatio-temporal location. 394

uncentered proposition A special type of centered proposition: For any uncentered proposition and any uncentered possible world, the set associated with the uncentered proposition either contains all of the centered worlds indexed to that uncentered world, or none of them. An uncentered proposition tells you something about what the world is like, without telling you anything about where you are in it. 394

unconditional credence An agent's degree of belief in a proposition, without making any suppositions beyond her current background information. 32

Uniqueness Thesis Given any proposition and body of total evidence, there is exactly one attitude it is rationally permissible for agents with that body of total evidence to adopt toward that proposition. 129

universe of discourse The set of objects under discussion. 30

upper probability The upper numerical bound of a ranged credence. 506

ur-prior Alternate name for a hypothetical prior distribution. 106

util A single unit of utility. 250

utility A numerical measure of the degree to which an agent values a particular proposition's being true. 250

valuation function In a decision problem, the agent's valuation function combines her credences and utilities to assign each available act a numerical score. The agent then prefers the act with the highest score. 252

Venn Diagram Diagram in which an agent's doxastically possible worlds are represented as points in a rectangle. Propositions are represented by regions containing those points, with the area of a region often representing the agent's credence in an associated proposition. 26

washing out of priors Bayesian convergence results guarantee that as evidence accumulates, conditionalizing agents' credences will be more and more dictated by that evidence and less influenced by the particular values of their priors. 455

weight The weight of a body of evidence relative to a particular proposition captures how much information about that proposition the evidence provides. 513

Bibliography of Volumes 1 & 2

Achinstein, Peter (1963). Variety and Analogy in Confirmation Theory. *Philosophy of Science* 3, pp. 207–21.

Adams, Ernest (1962). On Rational Betting Systems. *Archiv für mathematische Logik und Grundlagenforschung* 6, pp. 7–29.

Adams, Ernest (1965). The Logic of Conditionals. *Inquiry* 8, pp. 166–97.

Alchourrón, Carlos E., Peter Gärdenfors, and David Makinson (1985). On the Logic of Theory Change: Partial Meet Contraction and Revision Functions. *The Journal of Symbolic Logic* 50, pp. 510–30.

Allais, Maurice (1953). Le comportement de l'homme rationnel devant le risque: Critique des postulates et axiomes de l'école Américaine. *Econometrica* 21, pp. 503–46.

Amrhein, Valentin, Sander Greenland, and Blake McShane (2019). Scientists Rise up against Statistical Significance. *Nature* 567, pp. 305–7.

Armendt, Brad (1980). Is There a Dutch Book Argument for Probability Kinematics? *Philosophy of Science* 47, pp. 583–8.

Armendt, Brad (1992). Dutch Strategies for Diachronic Rules: When Believers See the Sure Loss Coming. *PSA: Proceedings of the Biennial Meeting of the Philosophy of Science Association* 1, pp. 217–29.

Arntzenius, Frank (1993). The Common Cause Principle. *PSA: Proceedings of the Biennial Meeting of the Philosophy of Science Association* 2, pp. 227–37.

Arntzenius, Frank (2003). Some Problems for Conditionalization and Reflection. *The Journal of Philosophy* 100, pp. 356–70.

Arrow, Kenneth J. (1951). *Social Choice and Individual Values*. New York: John Wiley and Sons.

Bandyopadhyay, Prasanta S. and Malcolm R. Forster (2011). *Philosophy of Statistics*. Vol. 7. Handbook of the Philosophy of Science. Amsterdam: Elsevier.

Barnett, Jeffrey A. (1981). Computational Methods for a Mathematical Theory of Evidence. In: *Proceedings of the 7th International Joint Conference on AI*, pp. 868–75.

Bartha, Paul and Christopher R. Hitchcock (1999). No One Knows the Date or the Hour: An Unorthodox Application of Rev. Bayes's Theorem. *Philosophy of Science* 66, S339–53.

Berger, James O. and Donald A. Berry (1988). Statistical Analysis and the Illusion of Objectivity. *American Scientist* 76, pp. 159–65.

Bergmann, Merrie, James Moor, and Jack Nelson (2013). *The Logic Book*. 6th edition. New York: McGraw Hill.

Berker, Selim (2013). Epistemic Teleology and the Separateness of Propositions. *Philosophical Review* 122, pp. 337–93.

Bernoulli, Daniel (1738/1954). Exposition of a New Theory on the Measurement of Risk. *Econometrica* 22, pp. 23–36.

Bernoulli, Jacob (1713). *Ars Conjectandi*. Basiliae.

Bertrand, Joseph (1888/1972). *Calcul des probabilités*. 2nd edition. New York: Chelsea Publishing Company.

Bickel, P.J., E.A. Hammel, and J.W. O'Connell (1975). Sex Bias in Graduate Admissions: Data from Berkeley. *Science* 187, pp. 398–404.

Birnbaum, Allan (1962). On the Foundations of Statistical Inference. *Journal of the American Statistical Association* 57, pp. 269–306.

Bolzano, Bernard (1837/1973). *Wissenschaftslehre*. Translated by Jan Berg under the title *Theory of Science*. Dordrecht: Reidel.

Bovens, Luc and Stephan Hartmann (2003). *Bayesian Epistemology*. Oxford: Oxford University Press.

Bradley, Darren (2010). Conditionalization and Belief *De Se*. *Dialectica* 64, pp. 247–50.

Bradley, Darren (2011). Self-location Is No Problem for Conditionalization. *Synthese* 182, pp. 393–411.

Bradley, Darren (2015). *A Criticial Introduction to Formal Epistemology*. London: Bloomsbury.

Bratman, Michael E. (1987). *Intention, Plans, and Practical Reason*. Cambridge, MA: Harvard University Press.

Brier, George (1950). Verification of Forecasts Expressed in Terms of Probability. *Monthly Weather Review* 78, pp. 1–3.

Briggs, R.A. (2010). Putting a Value on Beauty. In: *Oxford Studies in Epistemology*. Ed. by Tamar Szabó Gendler and John Hawthorne. Vol. 3. Oxford University Press, pp. 3–34.

Briggs, R.A. and Richard Pettigrew (2020). An Accuracy-Dominance Argument for Conditionalization. *Noûs* 54, pp. 162–81.

Broome, John (1999). Normative Requirements. *Ratio* 12, pp. 398–419.

Buchak, Lara (2013). *Risk and Rationality*. Oxford: Oxford University Press.

Campbell-Moore, Catrin and Benjamin A. Levinstein (2021). Strict Propriety Is Weak. *Analysis* 81.1, pp. 8–13.

Capotorti, Andrea and Barbara Vantaggi (2000). Axiomatic Characterization of Partial Ordinal Relations. *International Journal of Approximate Reasoning* 24, pp. 207–19.

Carnap, Rudolf (1945). On Inductive Logic. *Philosophy of Science* 12, pp. 72–97.

Carnap, Rudolf (1947). On the Application of Inductive Logic. *Philosophy and Phenomenological Research* 8, pp. 133–48.

Carnap, Rudolf (1950). *Logical Foundations of Probability*. Chicago: University of Chicago Press.

Carnap, Rudolf (1955/1989). Statistical and Inductive Probability. In: *Readings in the Philosophy of Science*. Ed. by Baruch A. Brody and Richard E. Grandy. 2nd edition. Hoboken: Prentice-Hall, pp. 279–87.

Carnap, Rudolf (1962a). *Logical Foundations of Probability*. 2nd edition. Chicago: University of Chicago Press.

Carnap, Rudolf (1962b). The Aim of Inductive Logic. In: *Logic, Methodology, and the Philosophy of Science*. Ed. by P. Suppes, E. Nagel, and A. Tarski. Stanford University: Stanford University Press, pp. 303–18.

Carnap, Rudolf (1971). A Basic System of Inductive Logic, Part 1. In: *Studies in Inductive Logic and Probability*. Ed. by Rudolf Carnap and Richard C. Jeffrey. Vol. I. Berkeley: University of California Press, pp. 33–166.

Carnap, Rudolf (1980). A Basic System of Inductive Logic, Part 2. In: *Studies in Inductive Logic and Probability*. Ed. by Richard C. Jeffrey. Vol. II. Berkeley: University of California Press, pp. 7–156.

Carr, Jennifer (2017). Epistemic Utility Theory and the Aim of Belief. *Philosophy and Phenomenological Research* 95, pp. 511–34.

Cartwright, Nancy (1979). Causal Laws and Effective Strategies. *Noûs* 13, pp. 419–37.

Carver, Ronald P. (1978). The Case against Statistical Significance Testing. *Harvard Educational Review* 48, pp. 378–99.

Chandler, Jake (2013). Contrastive Confirmation: Some Competing Accounts. *Synthese* 190, pp. 129–38.

Cherniak, Christopher (1986). *Minimal Rationality.* Cambridge, MA: The MIT Press.

Chihara, C. (1981). Quine and the Confirmational Paradoxes. In: *Midwest Studies in Philosophy 6: Foundations of Analytic Philosophy.* Ed. by P. French, H. Wettstein, and T. Uehling. Minneapolis: University of Minnesota Press, pp. 425–52.

Christensen, David (1991). Clever Bookies and Coherent Beliefs. *The Philosophical Review* 100, pp. 229–47.

Christensen, David (1999). Measuring Confirmation. *The Journal of Philosophy* 96, pp. 437–61.

Christensen, David (2001). Preference-Based Arguments for Probabilism. *Philosophy of Science* 68, pp. 356–76.

Christensen, David (2004). *Putting Logic in its Place.* Oxford: Oxford University Press.

Colyvan, Mark (2004). The Philosophical Significance of Cox's Theorem. *International Journal of Approximate Reasoning* 37, pp. 71–85.

Cox, Richard T. (1946). Probability, Frequency and Reasonable Expectation. *American Journal of Physics* 14, pp. 1–13.

Cox, Richard T. (1961). *The Algebra of Probable Inference.* Baltimore, MD: The Johns Hopkins Press.

Crupi, Vincenzo, Branden Fitelson, and Katya Tentori (2008). Probability, Confirmation, and the Conjunction Fallacy. *Thinking & Reasoning* 14, pp. 182–99.

Crupi, Vincenzo, Katya Tentori, and Michel Gonzalez (2007). On Bayesian Measures of Evidential Support: Theoretical and Empirical Issues. *Philosophy of Science* 74, pp. 229–52.

Davidson, Donald (1984). *Inquiries into Truth and Interpretation.* Oxford: Clarendon Press.

Davidson, Donald, J.C.C. McKinsey, and Patrick Suppes (1955). Outlines of a Formal Theory of Value, I. *Philosophy of Science* 22, pp. 14–60.

de Finetti, Bruno (1931/1989). Probabilism: A Critical Essay on the Theory of Probability and the Value of Science. *Erkenntnis* 31, pp. 169–223. Translation of B. de Finetti, *Probabilismo,* Logos 14, pp. 163–219.

de Finetti, Bruno (1937/1964). Foresight: Its Logical Laws, its Subjective Sources. In: *Studies in Subjective Probability.* Ed. by Henry E. Kyburg Jr and H.E. Smokler. New York: Wiley, pp. 94–158. Originally published as "La prévision; ses lois logiques, ses sources subjectives" in *Annales de l'Institut Henri Poincaré* 7, pp. 1–68.

de Finetti, Bruno (1949/1951). La 'Logica del Plausibile' Secondo la Conçezione di Polya. *Atti della XLII Riunione, Societa Italiana per il Progresso delle Scienze.* Presented in 1949, published in 1951, pp. 227–36.

de Finetti, Bruno (1974). *Theory of Probability.* Vol. 1. New York: Wiley.

de Finetti, Bruno (1995). *Filosofia della probabilità.* Ed. by Alberto Mura. Milan: Il Saggiatore.

DeGroot, Morris H. (1970). *Optimal Statistical Decisions.* Hoboken, New Jersey: Wiley.

Dempster, Arthur P. (1966). New Methods for Reasoning towards Posterior Distributions Based on Sample Data. *Annals of Mathematical Statistics* 37, pp. 355–74.

Dickson, Michael and Davis Baird (2011). Significance Testing. In: *Philosophy of Statistics.* Ed. by Prasanta S. Bandyopadhyay and Malcolm R. Forster. Vol. 7. Handbook of the Philosophy of Science. Amsterdam: Elsevier.

Domotor, Zoltan (1969). *Probabilistic Relational Structures and their Applications*. Technical Report 144. Stanford University: Institute for Mathematical Studies in the Social Sciences.

Earman, John (1992). *Bayes or Bust? A Critical Examination of Bayesian Confirmation Theory*. Cambridge, MA: The MIT Press.

Easwaran, Kenny (2013). Expected Accuracy Supports Conditionalization—and Conglomerability and Reflection. *Philosophy of Science* 80, pp. 119–42.

Easwaran, Kenny (2014a). Decision Theory without Representation Theorems. *Philosophers' Imprint* 14, pp. 1–30.

Easwaran, Kenny (2014b). Regularity and Hyperreal Credences. *Philosophical Review* 123, pp. 1–41.

Eddington, A. (1939). *The Philosophy of Physical Science*. Cambridge: Cambridge University Press.

Edwards, A.W.F. (1972). *Likelihood: An Account of the Statistical Concept of Likelihood and its Application to Scientific Inference*. Cambridge: Cambridge University Press.

Eells, Ellery (1982). *Rational Decision and Causality*. Cambridge Studies in Philosophy. Cambridge: Cambridge University Press.

Eells, Ellery (1985). Problems of Old Evidence. *Pacific Philosophical Quarterly* 66, pp. 283–302.

Eells, Ellery and Branden Fitelson (2000). Measuring Confirmation and Evidence. *Journal of Philosophy* 97, pp. 663–72.

Eells, Ellery and Branden Fitelson (2002). Symmetries and Asymmetries in Evidential Support. *Philosophical Studies* 107, pp. 129–42.

Efron, B. (1986). Why Isn't Everyone a Bayesian? *The American Statistician* 40, pp. 1–5.

Egan, Andy (2007). Some Counterexamples to Causal Decision Theory. *Philosophical Review* 116, pp. 93–114.

Egan, Andy and Michael G. Titelbaum (2022). Self-locating Belief. Forthcoming in *The Stanford Encyclopedia of Philosophy*.

Elga, Adam (2000). Self-locating Belief and the Sleeping Beauty Problem. *Analysis* 60, pp. 143–7.

Elga, Adam (2007). Reflection and Disagreement. *Noûs* 41, pp. 478–502.

Elga, Adam (2010). Subjective Probabilities Should Be Sharp. *Philosophers' Imprint* 10, pp. 1–11.

Ellenberg, Jordan (2014). *How Not to Be Wrong: The Power of Mathematical Thinking*. New York: Penguin Press.

Ellis, Robert Leslie (1849). On the Foundations of the Theory of Probabilities. *Transactions of the Cambridge Philosophical Society* VIII, pp. 1–6.

Ellsberg, Daniel (1961). Risk, Ambiguity, and the Savage Axioms. *The Quarterly Journal of Economics* 75, pp. 643–69.

Feldman, Richard (2007). Reasonable Religious Disagreements. In: *Philosophers without Gods: Meditations on Atheism and the Secular Life*. Ed. by Louise M. Antony. Oxford: Oxford University Press, pp. 194–214.

Feller, William (1968). *An Introduction to Probability Theory and its Applications*. 3rd edition. New York: Wiley.

Fermat, Pierre and Blaise Pascal (1654/1929). Fermat and Pascal on Probability. In: *A Source Book in Mathematics*. Ed. by D. Smith. Translated by Vera Sanford. New York: McGraw-Hill, pp. 546–65.

Fine, Terrence L. (1973). *Theories of Probability: An Examination of Foundations*. New York, London: Academic Press.

Fishburn, Peter C. (1981). Subjective Expected Utility: A Review of Normative Theories. *Theory and Decision* 13, pp. 129–99.

Fishburn, Peter C. (1986). The Axioms of Subjective Probability. *Statistical Science* 1, pp. 335–45.

Fisher, Ronald A. (1956). *Statistical Methods and Scientific Inference*. Edinburgh: Oliver and Boyd.

Fitelson, Branden (2006). Logical Foundations of Evidential Support. *Philosophy of Science* 73, pp. 500–12.

Fitelson, Branden (2007). Likelihoodism, Bayesianism, and Relational Confirmation. *Synthese* 156, pp. 473–89.

Fitelson, Branden (2008). A Decision Procedure for Probability Calculus with Applications. *The Review of Symbolic Logic* 1, pp. 111–25.

Fitelson, Branden (2012). Evidence of Evidence Is Not (Necessarily) Evidence. *Analysis* 72, pp. 85–8.

Fitelson, Branden (2015). The Strongest Possible Lewisian Triviality Result. *Thought* 4, pp. 69–74.

Fitelson, Branden and Alan Hájek (2014). Declarations of Independence. *Synthese* 194, pp. 3979–95.

Fitelson, Branden and James Hawthorne (2010a). How Bayesian Confirmation Theory Handles the Paradox of the Ravens. *The Place of Probability in Science*. Ed. by Ellery Eells and J. Fetzer. *Boston Studies in the Philosophy of Science* 284, pp. 247–75.

Fitelson, Branden and James Hawthorne (2010b). The Wason Task(s) and the Paradox of Confirmation. *Philosophical Perspectives* 24, pp. 207–41.

Fitelson, Branden and David McCarthy (ms). Accuracy and Comparative Likelihood. Unpublished manuscript.

Fitelson, Branden and Daniel Osherson (2015). Remarks on "Random Sequences". *Australasian Journal of Logic* 12, pp. 11–16.

Foley, Richard (1987). *The Theory of Epistemic Rationality*. Cambridge, MA: Harvard University Press.

Foley, Richard (1993). *Working without a Net*. Oxford: Oxford University Press.

Foley, Richard (2009). Beliefs, Degrees of Belief, and the Lockean Thesis. In: *Degrees of Belief*. Ed. by Franz Huber and Christoph Schmidt-Petri. Vol. 342. Synthese Library. Springer, pp. 37–48.

Frigerio, Roberta et al. (2005). Education and Occupations Preceding Parkinson's Disease. *Neurology* 65, pp. 1575–83.

Gaifman, H. and M. Snir (1982). Probabilities over Rich Languages. *Journal of Symbolic Logic* 47, pp. 495–548.

Gaifman, Haim (2004). Reasoning with Limited Resources and Assigning Probabilities to Arithmetical Statements. *Synthese* 140, pp. 97–119.

Galavotti, Maria Carla (2005). *Philosophical Introduction to Probability*. CSLI Lecture Notes 167. Stanford, CA: CSLI Publications.

Gallo, Valentina et al. (2018). Exploring Causality of the Association between Smoking and Parkinson's Disease. *International Journal of Epidemiology* 48, pp. 912–25.

Gandenberger, Greg (2014). *Titelbaum's Counterexample to the Law of Likelihood*. Blog post available at URL: https://gandenberger.org/2014/05/26/titelbaum-counterexample/.

Gandenberger, Greg (2015). A New Proof of the Likelihood Principle. *British Journal for the Philosophy of Science* 66, pp. 475–503.

Gandenberger, Greg (ms). New Responses to Three Purported Counterexamples to the Likelihood Principle. Unpublished manuscript.

Garber, Daniel (1983). Old Evidence and Logical Omniscience in Bayesian Confirmation Theory. In: *Testing Scientific Theories*. Ed. by John Earman. Vol. 10. Minnesota Studies in the Philosophy of Science. Minneapolis: University of Minnesota Press, pp. 99–132.

Gibbard, A. and W. Harper (1978/1981). Counterfactuals and Two Kinds of Expected Utility. In: *Ifs: Conditionals, Belief, Decision, Chance, and Time*. Ed. by W. Harper, Robert C. Stalnaker, and G. Pearce. Dordrecht: Reidel, pp. 153–90.

Gillies, Donald (2000). Varieties of Propensity. *British Journal for the Philosophy of Science* 51, pp. 807–35.

Glass, David H. and Mark McCartney (2015). A New Argument for the Likelihood Ratio Measure of Confirmation. *Acta Analytica* 30, pp. 59–65.

Glymour, Clark (1980). *Theory and Evidence*. Princeton, NJ: Princeton University Press.

Good, I.J. (1952). Rational Decisions. *Journal of the Royal Statistical Society, Series B* 14, pp. 107–14.

Good, I.J. (1962). Subjective Probability as the Measure of a Non-measurable Set. In: *Logic, Methodology, and the Philosophy of Science*. Ed. by P. Suppes, E. Nagel, and A. Tarski. Stanford, CA: Stanford University Press, pp. 319–29.

Good, I.J. (1967). The White Shoe Is a Red Herring. *British Journal for the Philosophy of Science* 17, p. 322.

Good, I.J. (1968). The White Shoe qua Herring Is Pink. *British Journal for the Philosophy of Science* 19, pp. 156–7.

Good, I.J. (1971). Letter to the Editor. *The American Statistician* 25, pp. 62–3.

Goodman, Nelson (1946). A Query on Confirmation. *The Journal of Philosophy* 43, pp. 383–5.

Goodman, Nelson (1955). *Fact, Fiction, and Forecast*. Cambridge, MA: Harvard University Press.

Greaves, Hilary (2013). Epistemic Decision Theory. *Mind* 122, pp. 915–52.

Greaves, Hilary and David Wallace (2006). Justifying Conditionalization: Conditionalization Maximizes Expected Epistemic Utility. *Mind* 115, pp. 607–32.

Hacking, Ian (1965). *The Logic of Statistical Inference*. Cambridge: Cambridge University Press.

Hacking, Ian (1967). Slightly More Realistic Personal Probability. *Philosophy of Science* 34, pp. 311–25.

Hacking, Ian (1971). The Leibniz-Carnap Program for Inductive Logic. *The Journal of Philosophy* 68, pp. 597–610.

Hacking, Ian (2001). *An Introduction to Probability and Inductive Logic*. Cambridge: Cambridge University Press.

Hájek, Alan (1996). 'Mises Redux'—Redux: Fifteen Arguments against Finite Frequentism. *Erkenntnis* 45, pp. 209–27.

Hájek, Alan (2003). What Conditional Probability Could Not Be. *Synthese* 137, pp. 273–323.

Hájek, Alan (2009a). Arguments for—or against—Probabilism? In: *Degrees of Belief*. Ed. by Franz Huber and Christoph Schmidt-Petri. Vol. 342. Synthese Library. Springer, pp. 229–51.

Hájek, Alan (2009b). Fifteen Arguments against Hypothetical Frequentism. *Erkenntnis* 70, pp. 211–35.

Hájek, Alan (2011a). Conditional Probability. In: *Philosophy of Statistics*. Ed. by Prasanta S. Bandyopadhyay and Malcolm R. Forster. Vol. 7. Handbook of the Philosophy of Science. Amsterdam: Elsevier, pp. 99–136.

Hájek, Alan (2011b). Triviality Pursuit. *Topoi* 30, pp. 3–15.

Hájek, Alan (2019). Interpretations of Probability. In: *The Stanford Encyclopedia of Philosophy*. Ed. by Edward N. Zalta. Fall 2019. URL: http://plato.stanford.edu/archives/fall2019/entries/probability-interpret/.

Hájek, Alan and James M. Joyce (2008). Confirmation. In: *The Routledge Companion to Philosophy of Science*. Ed. by Stathis Psillos and Martin Curd. New York: Routledge, pp. 115–28.

Hall, Ned (2004). Two Mistakes about Credence and Chance. *Australasian Journal of Philosophy* 82, pp. 93–111.

Halpern, Joseph Y. (1999). Cox's Theorem Revisited. *Journal of Artificial Intelligence Research* 11, pp. 429–35.

Halpern, Joseph Y. (2003). *Reasoning about Uncertainty*. Cambridge, MA: MIT Press.

Halpern, Joseph Y. (2004). Sleeping Beauty Reconsidered: Conditioning and Reflection in Asynchronous Systems. *Proceedings of the Twentieth Conference on Uncertainty in AI*, pp. 226–34.

Halpern, Joseph Y. (2005). Sleeping Beauty Reconsidered: Conditioning and Reflection in Asynchronous Systems. In: *Oxford Studies in Epistemology*. Ed. by Tamar Szabó Gendler and John Hawthorne. Vol. 1. Oxford: Oxford University Press, pp. 111–42.

Harman, Gilbert (1986). *Change in View*. Boston: The MIT Press.

Harrison-Trainor, Matthew, Wesley H. Holliday, and Thomas F. Icard III (2016). A Note on Cancellation Axioms for Comparative Probability. *Theory and Decision* 80, pp. 159–66.

Hart, Casey and Michael G. Titelbaum (2015). Intuitive Dilation? *Thought* 4, pp. 252–62.

Haverkamp, Nick and Moritz Schulz (2012). A Note on Comparative Probability. *Erkenntnis* 76, pp. 395–402.

Hawthorne, James (2014). Inductive Logic. In: *The Stanford Encyclopedia of Philosophy*. Ed. by Edward N. Zalta. Winter 2014. URL: http://plato.stanford.edu/archives/win2014/entries/logic-inductive/.

Hawthorne, James (2016). A Logic of Comparative Support: Qualitative Conditional Probability Relations Representable by Popper Functions. In: *The Oxford Handbook of Probability and Philosophy*. Ed. by Alan Hájek and Christopher R. Hitchcock. Oxford: Oxford University Press, pp. 277–95.

Hawthorne, James and Branden Fitelson (2004). Re-solving Irrelevant Conjunction with Probabilistic Independence. *Philosophy of Science* 71, pp. 505–14.

Hedden, Brian (2013). Incoherence without Exploitability. *Noûs* 47, pp. 482–95.

Hempel, Carl G. (1945a). Studies in the Logic of Confirmation (I). *Mind* 54, pp. 1–26.

Hempel, Carl G. (1945b). Studies in the Logic of Confirmation (II). *Mind* 54, pp. 97–121.

Hesse, Mary (1963). *Models and Analogies in Science*. London: Sheed & Ward.

Heukelom, Floris (2015). A History of the Allais Paradox. *The British Journal for the History of Science* 48, pp. 147–69.

Hintikka, Jaakko (1975). Impossible Possible Worlds Vindicated. *Journal of Philosophical Logic* 4, pp. 475–84.

Hitchcock, Christopher R. (2004). Beauty and the Bets. *Synthese* 139, pp. 405–20.

Hitchcock, Christopher R. (2021). Probabilistic Causation. In: *The Stanford Encyclopedia of Philosophy*. Ed. by Edward N. Zalta. Spring 2021. URL: https://plato.stanford.edu/archives/spr2021/entries/causation-probabilistic/.

Holton, Richard (2014). Intention as a Model for Belief. In: *Rational and Social Agency: The Philosophy of Michael Bratman*. Ed. by Manuel Vargas and Gideon Yaffe. Oxford: Oxford University Press, pp. 12–37.

Hooker, C.A. (1968). Goodman, 'Grue' and Hempel. *Philosophy of Science* 35, pp. 232–47.

Hosiasson-Lindenbaum, Janina (1940). On Confirmation. *Journal of Symbolic Logic* 5, pp. 133–48.

Howson, Colin (1992). Dutch Book Arguments and Consistency. *PSA: Proceedings of the Biennial Meeting of the Philosophy of Science Association* 2, pp. 161–8.

Howson, Colin (2014). Finite Additivity, Another Lottery Paradox and Conditionalisation. *Synthese* 191, pp. 989–1012.

Howson, Colin and Peter Urbach (2006). *Scientific Reasoning: The Bayesian Approach*. 3rd edition. Chicago: Open Court.

Huber, Franz (2016). Formal Representations of Belief. In: *The Stanford Encyclopedia of Philosophy*. Ed. by Edward N. Zalta. Spring 2016. URL: https://plato.stanford.edu/archives/spr2016/entries/formal-belief/.

Hume, David (1739–40/1978). *A Treatise of Human Nature*. Ed. by L.A. Selby-Bigge and Peter H. Nidditch. 2nd edition. Oxford: Oxford University Press.

Humphreys, Paul (1985). Why Propensities Cannot Be Probabilities. *Philosophical Review* 94, pp. 557–70.

Icard III, Thomas F. (2016). Pragmatic Considerations on Comparative Probability. *Philosophy of Science* 83, pp. 348–70.

Jackson, Elizabeth G. (2020). The Relationship between Belief and Credence. *Philosophy Compass* 15, pp. 1–13.

Jaynes, E.T. (1957a). Information Theory and Statistical Mechanics I. *Physical Review* 106, pp. 62–30.

Jaynes, E.T. (1957b). Information Theory and Statistical Mechanics II. *Physical Review* 108, pp. 171–90.

Jaynes, E.T. (1968). Prior Probabilities. *IEEE Transactions on Systems Science and Cybernetics* SEC-4, pp. 227–41.

Jeffrey, Richard C. (1965). *The Logic of Decision*. 1st edition. McGraw-Hill Series in Probability and Statistics. New York: McGraw-Hill.

Jeffrey, Richard C. (1970). Dracula Meets Wolfman: Acceptance vs. Partial Belief. In: *Induction, Acceptance, and Rational Belief*. Ed. by M. Swain. Dordrecht: Reidel, pp. 157–85.

Jeffrey, Richard C. (1983). *The Logic of Decision*. 2nd edition. Chicago: University of Chicago Press.

Jeffrey, Richard C. (1987). Indefinite Probability Judgment: A Reply to Levi. *Philosophy of Science* 54, pp. 586–91.

Jeffrey, Richard C. (1993). Causality and the Logic of Decision. *Philosophical Topics* 21, pp. 139–51.

Jeffrey, Richard C. (2004). *Subjective Probability: The Real Thing*. Cambridge: Cambridge University Press.

Johnson, W.E. (1932). Probability: The Deductive and Inductive Problems. *Mind* 41, pp. 409–23.

Joyce, James M. (1998). A Nonpragmatic Vindication of Probabilism. *Philosophy of Science* 65, pp. 575–603.

Joyce, James M. (1999). *The Foundations of Causal Decision Theory*. Cambridge: Cambridge University Press.

Joyce, James M. (2005). How Probabilities Reflect Evidence. *Philosophical Perspectives* 19, pp. 153–78.

Joyce, James M. (2009). Accuracy and Coherence: Prospects for an Alethic Epistemology of Partial Belief. In: *Degrees of Belief*. Ed. by Franz Huber and Christoph Schmidt-Petri. Vol. 342. Synthese Library. Springer, pp. 263–97.

Joyce, James M. (2010). A Defense of Imprecise Credences in Inference and Decision Making. *Philosophical Perspectives* 24, pp. 281–323.

Kahneman, Daniel and Amos Tversky (1979). Prospect Theory: An Analysis of Decision under Risk. *Econometrica* XLVII, pp. 263–91.

Kaplan, Mark (1996). *Decision Theory as Philosophy*. Cambridge: Cambridge University Press.

Kelley, Mikayla (ms). On Accuracy and Coherence with Infinite Opinion Sets. Unpublished manuscript.

Kemeny, John G. (1955). Fair Bets and Inductive Probabilities. *The Journal of Symbolic Logic* 20, pp. 263–73.

Kemeny, John G. and Paul Oppenheim (1952). Degree of Factual Support. *Philosophy of Science* 19, pp. 307–24.

Keynes, John Maynard (1921). *Treatise on Probability*. London: Macmillan and Co., Limited.

Keynes, John Maynard (1923). *A Tract on Monetary Reform*. Macmillan and Co., Limited.

Keynes, John Maynard (1937). The General Theory of Employment. *The Quarterly Journal of Economics* 51, pp. 209–23.

Kierland, Brian and Bradley Monton (2005). Minimizing Inaccuracy for Self-locating Beliefs. *Philosophy and Phenomenological Research* 70, pp. 384–95.

Kim, Jaegwon (1988). What Is "Naturalized Epistemology"? *Philosophical Perspectives* 2, pp. 381–405.

Kim, Namjoong (2009). Sleeping Beauty and Shifted Jeffrey Conditionalization. *Synthese* 168, pp. 295–312.

Knight, Frank Hyneman (1921). *Risk, Uncertainty and Profit*. Boston and New York: Houghton Mifflin.

Kolmogorov, A.N. (1933/1950). *Foundations of the Theory of Probability*. Translation edited by Nathan Morrison. New York: Chelsea Publishing Company.

Konek, Jason and Benjamin A. Levinstein (2019). The Foundations of Epistemic Decision Theory. *Mind* 128, pp. 69–107.

Kornblith, Hilary (1993). Epistemic Normativity. *Synthese* 94, pp. 357–76.

Kraft, Charles H., John W. Pratt, and A. Seidenberg (1959). Intuitive Probability on Finite Sets. *The Annals of Mathematical Statistics* 30, pp. 408–19.

Krantz, D.H. et al. (1971). *Foundations of Measurement, Vol 1: Additive and Polynomial Representations*. Cambridge: Academic Press.

Kuhn, Thomas S. (1957). *The Copernican Revolution: Planetary Astronomy in the Development of Western Thought*. New York: MJF Books.

Kyburg Jr, Henry E. (1961). *Probability and the Logic of Rational Belief*. Middletown: Wesleyan University Press.

Kyburg Jr, Henry E. (1970). Conjunctivitis. In: *Induction, Acceptance, and Rational Belief*. Ed. by M. Swain. Boston: Reidel, pp. 55–82.

Laddaga, R. (1977). Lehrer and the Consensus Proposal. *Synthese* 36, pp. 473–7.

Lance, Mark Norris (1995). Subjective Probability and Acceptance. *Philosophical Studies* 77, pp. 147–79.

Lange, Alexandra (2019). Can Data Be Human? The Work of Giorgia Lupi. *The New Yorker*. Published May 25, 2019.

Lange, Marc (2000). Is Jeffrey Conditionalization Defective by Virtue of Being Non-commutative? Remarks on the Sameness of Sensory Experience. *Synthese* 123, pp. 393–403.

Laplace, Pierre-Simon (1814/1995). *Philosophical Essay on Probabilities*. Translated from the French by Andrew Dale. New York: Springer.

Lehman, R. Sherman (1955). On Confirmation and Rational Betting. *Journal of Symbolic Logic* 20, pp. 251–62.

Lehrer, K. and Carl Wagner (1983). Probability Amalgamation and the Independence Issue: A Reply to Laddaga. *Synthese* 55, pp. 339–46.

Leitgeb, Hannes and Richard Pettigrew (2010a). An Objective Justification of Bayesianism I: Measuring Inaccuracy. *Philosophy of Science* 77, pp. 201–35.

Leitgeb, Hannes and Richard Pettigrew (2010b). An Objective Justification of Bayesianism II: The Consequences of Minimizing Inaccuracy. *Philosophy of Science* 77, pp. 236–72.

Lele, Subhash R. (2004). Evidence Functions and the Optimality of the Law of Likelihood. In: *The Nature of Scientific Evidence: Statistical, Philosophical, and Empirical Considerations*. Chicago: University of Chicago Press, pp. 191–216.

Levi, Isaac (1974). On Indeterminate Probabilities. *The Journal of Philosophy* 71, pp. 391–418.

Levi, Isaac (1980). *The Enterprise of Knowledge*. Boston: The MIT Press.

Levi, Isaac (1987). The Demons of Decision. *The Monist* 70, pp. 193–211.

Lewis, C.I. (1946). *An Analysis of Knowledge and Valuation*. La Salle, IL: Open Court.

Lewis, David (1971). Immodest Inductive Methods. *Philosophy of Science* 38, pp. 54–63.

Lewis, David (1976). Probabilities of Conditionals and Conditional Probabilities. *The Philosophical Review* 85, pp. 297–315.

Lewis, David (1979). Atittudes *de dicto* and *de se*. *The Philosophical Review* 88, pp. 513–43.

Lewis, David (1980). A Subjectivist's Guide to Objective Chance. In: *Studies in Inductive Logic and Probability*. Ed. by Richard C. Jeffrey. Vol. 2. Berkeley: University of California Press, pp. 263–94.

Lewis, David (1981a). Causal Decision Theory. *Australasian Journal of Philosophy* 59, pp. 5–30.

Lewis, David (1981b). 'Why Ain'cha Rich?' *Noûs* 15, pp. 377–80.

Lewis, David (1994). Humean Supervenience Debugged. *Mind* 103, pp. 473–90.

Lewis, David (1996). Desire as Belief II. *Mind* 105, pp. 303–13.

Lewis, David (2001). Sleeping Beauty: Reply to Elga. *Analysis* 61, pp. 171–6.

Lichtenstein, S., B. Fischoff, and L. Phillips (1982). Calibration of Probabilities: The State of the Art to 1980. In: *Judgment under Uncertainty: Heuristics and Biases*. Ed. by Daniel Kahneman, P. Slovic, and Amos Tversky. Cambridge: Cambridge University Press, pp. 306–34.

Lindley, Dennis V. (1982). Scoring Rules and the Inevitability of Probability. *International Statistical Review* 50, pp. 1–26.

Lindley, Dennis V. (1985). *Making Decisions*. 2nd edition. London: Wiley.

Liu, Liping and Ronald R. Yager (2008). Classic Works of the Dempster-Shafer Theory of Belief Functions: An Introduction. In: *Classic Works of the Dempster-Shafer Theory of Belief Functions*. Ed. by Ronald R. Yager and Liping Liu. Vol. 219. Studies in Fuzziness and Soft Computing. Berlin: Springer, pp. 1–34.

Locke, John (1689/1975). *An Essay Concerning Human Understanding*. Ed. by Peter H. Nidditch. Oxford: Oxford University Press.

MacFarlane, John (2005). Making Sense of Relative Truth. *Proceedings of the Aristotelian Society* 105, pp. 321–39.

Maher, Patrick (1993). *Betting on Theories*. Cambridge Studies in Probability, Induction, and Decision Theory. Cambridge: Cambridge University Press.

Maher, Patrick (1996). Subjective and Objective Confirmation. *Philosophy of Science* 63, pp. 149–74.

Maher, Patrick (2002). Joyce's Argument for Probabilism. *Philosophy of Science* 96, pp. 73–81.

Maher, Patrick (2010). Explication of Inductive Probability. *Journal of Philosophical Logic* 39, pp. 593–616.

Makinson, David C. (1965). The Paradox of the Preface. *Analysis* 25, pp. 205–7.

Makinson, David C. (2011). Conditional Probability in the Light of Qualitative Belief Change. *Journal of Philosophical Logic* 40, pp. 121–53.

Mayo, Deborah (2018). *Statistical Inference as Severe Testing: How to Get beyond the Statistics Wars*. Cambridge: Cambridge University Press.

Mazurkiewicz, Stefan (1932). Zur Axiomatik der Wahrscheinlichkeitsrechnung. *Comptes rendues des séances de la Société des Sciences et des Lettres de Varsovie* 25, pp. 1–4.

Meacham, Christopher J. G. (2008). Sleeping Beauty and the Dynamics of De Se Beliefs. *Philosophical Studies* 138, pp. 245–70.

Meacham, Christopher J. G. (2010a). Unravelling the Tangled Web: Continuity, Internalism, Uniqueness, and Self-locating Belief. In: *Oxford Studies in Epistemology*. Ed. by Tamar Szabó Gendler and John Hawthorne. Vol. 3. Oxford: Oxford University Press, pp. 86–125.

Meacham, Christopher J.G. (2010b). Two Mistakes Regarding the Principal Principle. *British Journal for the Philosophy of Science* 61, pp. 407–31.

Meacham, Christopher J. G. (2016). Ur-Priors, Conditionalization, and Ur-Prior Conditionalization. *Ergo* 3, pp. 444–92.

Meacham, Christopher J.G. and Jonathan Weisberg (2011). Representation Theorems and the Foundations of Decision Theory. *Australasian Journal of Philosophy* 89, pp. 641–63.

Mellor, D.H. (2013). Review of *Probability in the Philosophy of Religion*. *Analysis* 73, pp. 548–54.

Moore, David S., George P. McCabe, and Bruce A. Craig (2009). *Introduction to the Practice of Statistics*. 6th edition. New York: W.H. Freeman and Company.

Moore, G.E. (1939). Proof of an External World. *Proceedings of the British Academy* 25, pp. 273–300.

Moss, Sarah (2012). Updating as Communication. *Philosophy and Phenomenological Research* 85, pp. 225–48.

Moss, Sarah (2015). Credal Dilemmas. *Noûs* 49, pp. 665–83.

Moss, Sarah (2018). *Probabilistic Knowledge*. Oxford: Oxford University Press.

Murphy, Allan H. (1973). A New Vector Partition of the Probability Score. *Journal of Applied Meteorology* 12, pp. 595–600.

Murphy, Allan H. and Robert L. Winkler (1977). Reliability of Subjective Probability Forecasts of Precipitation and Temperature. *Journal of the Royal Statistical Society, Series C* 26, pp. 41–7.

Neal, Radford M. (2006). *Puzzles of Anthropic Reasoning Resolved Using Full Non-indexical Conditioning*. Tech. rep. 0607. Department of Statistics, University of Toronto.

Neyman, J. and Egon Pearson (1967). *Joint Statistical Papers*. Cambridge: Cambridge University Press.

Nicod, Jean (1930). *Foundations of Geometry and Induction*. Translated by Philip Wiener. New York: Harcourt, Brace and Company.

Nozick, Robert (1969). Newcomb's Problem and Two Principles of Choice. In: *Essays in Honor of Carl G. Hempel*. Synthese Library. Dordrecht: Reidel, pp. 114–15.

Open Science Collaboration (2015). Estimating the reproducibility of psychological science. *Science* 349. DOI: 10.1126/science.aac4716.

Papineau, David (2012). *Philosophical Devices: Proofs, Probabilities, Possibilities, and Sets*. Oxford: Oxford University Press.

Paris, J.B. (1994). *The Uncertain Reasoner's Companion*. Cambridge: Cambridge University Press.

Pascal, Blaise (1670/1910). *Pensées*. Translated by W.F. Trotter. London: Dent.

Pearson, K., A. Lee, and L. Bramley-Moore (1899). Genetic (Reproductive) Selection: Inheritance of Fertility in Man. *Philosophical Transactions of the Royal Society A* 73, pp. 534–39.

Pedersen, Arthur Paul and Gregory Wheeler (2014). Demystifying Dilation. *Erkenntnis* 79, pp. 1305–42.

Peirce, Charles Sanders (1910/1932). Notes on the Doctrine of Chances. In: *Collected Papers of Charles Sanders Peirce*. Ed. by Charles Hartshorne and Paul Weiss. Cambridge, MA: Harvard University Press, pp. 404–14.

Peterson, Martin (2009). *An Introduction to Decision Theory*. Cambridge Introductions to Philosophy. Cambridge: Cambridge University Press.

Pettigrew, Richard (2013a). A New Epistemic Utility Argument for the Principal Principle. *Episteme* 10, pp. 19–35.

Pettigrew, Richard (2013b). Epistemic Utility and Norms for Credences. *Philosophy Compass* 8, pp. 897–908.

Pettigrew, Richard (2014). Accuracy, Risk, and the Principle of Indifference. *Philosophy and Phenomenological Research* 92(1), pp. 35–59.

Pettigrew, Richard (2016). *Accuracy and the Laws of Credence*. Oxford: Oxford University Press.

Pettigrew, Richard (2021). On the Expected Utility Objection to the Dutch Book Argument for Probabilism. *Noûs* 55, pp. 23–38.

Pettigrew, Richard and Michael G. Titelbaum (2014). Deference Done Right. *Philosophers' Imprint* 14, pp. 1–19.

Pettigrew, Richard and Jonathan Weisberg, eds (2019). *The Open Handbook of Formal Epistemology*. Published open access online by PhilPapers. URL: https://philpapers.org/rec/PETTOH-2.

Piccione, Michele and Ariel Rubinstein (1997). On the Interpretation of Decision Problems with Imperfect Recall. *Games and Economic Behavior* 20, pp. 3–24.

Pollock, John L. (2001). Defeasible Reasoning with Variable Degrees of Justification. *Artificial Intelligence* 133, pp. 233–82.

Popper, Karl R. (1935/1959). *The Logic of Scientific Discovery*. London: Hutchinson & Co.

Popper, Karl R. (1938). A Set of Independent Axioms for Probability. *Mind* 47, pp. 275–9.

Popper, Karl R. (1954). Degree of Confirmation. *British Journal for the Philosophy of Science* 5, pp. 143–9.

Popper, Karl R. (1955). Two Autonomous Axiom Systems for the Calculus of Probabilities. *British Journal for the Philosophy of Science* 6, pp. 51–7.

Popper, Karl R. (1957). The Propensity Interpretation of the Calculus of Probability and the Quantum Theory. *The Colston Papers* 9. Ed. by S. Körner, pp. 65–70.

Predd, J. et al. (2009). Probabilistic Coherence and Proper Scoring Rules. *IEEE Transactions on Information Theory* 55, pp. 4786–92.

Pryor, James (2004). What's Wrong with Moore's Argument? *Philosophical Issues* 14, pp. 349–78.

Quinn, Warren S. (1990). The Puzzle of the Self-Torturer. *Philosophical Studies* 59, pp. 79–90.

Ramsey, Frank P. (1929/1990). General Propositions and Causality. In: *Philosophical Papers*. Ed. by D.H. Mellor. Cambridge: Cambridge University Press, pp. 145–63.

Ramsey, Frank P. (1931). Truth and Probability. In: *The Foundations of Mathematics and other Logic Essays*. Ed. by R.B. Braithwaite. New York: Harcourt, Brace and Company, pp. 156–98.

Rees, Martin (2000). *Just Six Numbers: The Deep Forces that Shape the Universe*. New York: Basic Books.

Reichenbach, Hans (1935/1949). *The Theory of Probability.* English expanded version of the German original. Berkeley: University of California Press.

Reichenbach, Hans (1938). *Experience and Prediction.* Chicago: University of Chicago Press.

Reichenbach, Hans (1956). The Principle of Common Cause. In: *The Direction of Time.* Berkeley: University of California Press, pp. 157–60.

Reiss, Julian and Jan Sprenger (2017). Scientific Objectivity. In: *The Stanford Encyclopedia of Philosophy.* Ed. by Edward N. Zalta. Winter 2017. Metaphysics Research Lab, Stanford University. URL: https://plato.stanford.edu/archives/win2017/entries/scientific-objectivity/.

Renyi, Alfred (1970). *Foundations of Probability.* San Francisco: Holden-Day.

Resnik, Michael D. (1987). *Choices: An Introduction to Decision Theory.* Minneapolis: University of Minnesota Press.

Roche, William (2014). Evidence of Evidence Is Evidence under Screening-off. *Episteme* 11, pp. 119–24.

Roeper, P. and H. Leblanc (1999). *Probability Theory and Probability Logic.* Toronto: University of Toronto Press.

Rosenkrantz, Roger (1981). *Foundations and Applications of Inductive Probability.* Atascadero, CA: Ridgeview Press.

Rosenthal, Jeffrey S. (2009). A Mathematical Analysis of the Sleeping Beauty Problem. *The Mathematical Intelligencer* 31, pp. 32–37.

Royall, Richard M. (1997). *Statistical Evidence: A Likelihood Paradigm.* New York: Chapman & Hall/CRC.

Salmon, Wesley (1966). *The Foundations of Scientific Inference.* Pittsburgh: University of Pittsburgh Press.

Salmon, Wesley (1975). Confirmation and Relevance. In: *Induction, Probability, and Confirmation.* Ed. by Grover Maxwell and Robert M. Jr. Anderson. Vol. VI. Minnesota Studies in the Philosophy of Science. Minneapolis: University of Minnesota Press, pp. 3–36.

Savage, Leonard J. (1954). *The Foundations of Statistics.* New York: Wiley.

Savage, Leonard J. (1967). Difficulties in the Theory of Personal Probability. *Philosophy of Science* 34, pp. 305–10.

Schervish, M.J., T. Seidenfeld, and J.B. Kadane (2004). Stopping to Reflect. *Journal of Philosophy* 101, pp. 315–22.

Schervish, M.J., T. Seidenfeld, and J.B. Kadane (2009). Proper Scoring Rules, Dominated Forecasts, and Coherence. *Decision Analysis* 6, pp. 202–21.

Schick, Frederic (1986). Dutch Bookies and Money Pumps. *The Journal of Philosophy* 83, pp. 112–19.

Schoenfield, Miriam (2017). The Accuracy and Rationality of Imprecise Credences. *Noûs* 51, pp. 667–85.

Schwarz, Wolfgang (2010). *Lewis on Updating and Self-location.* Blog post available at URL: https://www.umsu.de/wo/2010/563.

Schwarz, Wolfgang (2018). Subjunctive Conditional Probability. *Journal of Philosophical Logic* 47, pp. 47–66.

Scott, Dana (1964). Measurement Structures and Linear Inequalities. *Journal of Mathematical Psychology* 1, pp. 233–47.

Seelig, C. (1956). *Albert Einstein: A Documentary Biography.* London: Staples Press.

Seidenfeld, Teddy (1986). Entropy and Uncertainty. *Philosophy of Science* 53, pp. 467–91.

Seidenfeld, Teddy, M.J. Schervish, and J.B. Kadane (2017). Non-Conglomerability for Countably Additive Measures That Are Not κ-Additive. *The Review of Symbolic Logic* 10, pp. 284–300.

Seidenfeld, Teddy and Larry Wasserman (1993). Dilation for Sets of Probabilities. *The Annals of Statistics* 21, pp. 1139–54.

Selvin, Steve (1975). A Problem in Probability. *The American Statistician* 29. Published among the Letters to the Editor, p. 67.

Shafer, Glenn (1976). *A Mathematical Theory of Evidence.* Princeton, NJ: Princeton University Press.

Shafer, Glenn (1981). Constructive Probability. *Synthese* 48, pp. 1–60.

Shapiro, Amram, Louise Firth Campbell, and Rosalind Wright (2014). *The Book of Odds.* New York: Harper Collins.

Shimony, Abner (1955). Coherence and the Axioms of Confirmation. *Journal of Symbolic Logic* 20, pp. 1–28.

Shimony, Abner (1988). An Adamite Derivation of the Calculus of Probability. In: *Probability and Causality.* Ed. by J.H. Fetzer. Dordrecht: Reidel, pp. 151–61.

Shogenji, Tomoji (2003). A Condition for Transitivity in Probabilistic Support. *British Journal for the Philosophy of Science* 54, pp. 613–16.

Shogenji, Tomoji (2012). The Degree of Epistemic Justification and the Conjunction Fallacy. *Synthese* 184, pp. 29–48.

Shope, R.K. (1978). The Conditional Fallacy in Contemporary Philosophy. *Journal of Philosophy* 75, pp. 397–413.

Shortliffe, E. and B. Buchanan (1975). A Model of Inexact Reasoning in Medicine. *Mathematical Biosciences* 23, pp. 351–79.

Simpson, E.H. (1951). The Interpretation of Interaction in Contingency Tables. *Journal of the Royal Statistical Society, Series B* 13, pp. 238–41.

Skipper, Mattias and Jens Christian Bjerring (2022). Bayesianism for Nonideal Agents. *Erkenntnis* 87, pp. 93–115.

Skyrms, Brian (1980a). *Causal Necessity: A Pragmatic Investigation of the Necessity of Laws.* New Haven CT: Yale University Press.

Skyrms, Brian (1980b). Higher Order Degrees of Belief. In: *Prospects for Pragmatism.* Ed. by D.H. Mellor. Cambridge: Cambridge University Press, pp. 109–37.

Skyrms, Brian (1983). Three Ways to Give a Probability Function a Memory. In: *Testing Scientific Theories.* Ed. by John Earman. Vol. 10. Minnesota Studies in the Philosophy of Science. Minneapolis: University of Minnesota Press, pp. 157–61.

Skyrms, Brian (1987a). Coherence. In: *Scientific Inquiry in Philosophical Perspective.* Ed. by N. Rescher. Pittsburgh: University of Pittsburgh Press, pp. 225–42.

Skyrms, Brian (1987b). Dynamic Coherence and Probability Kinematics. *Philosophy of Science* 54, pp. 1–20.

Skyrms, Brian (2000). *Choice & Chance: An Introduction to Inductive Logic.* 4th edition. Stamford, CT: Wadsworth.

Smith, C.A.B. (1961). Consistency in Statistical Inference and Decision. *Journal of the Royal Statistical Society, Series B* 23, pp. 1–25.

Smithies, Declan (2015). Ideal Rationality and Logical Omniscience. *Synthese* 192, pp. 2769–93.

Sober, Elliott (2008). *Evidence and Evolution.* Cambridge: Cambridge University Press.

Spohn, Wolfgang (2012). *The Laws of Belief: Ranking Theory and its Philosophical Applications.* Oxford: Oxford University Press.

Staffel, Julia (2019). *Unsettled Thoughts: Reasoning and Uncertainty in Epistemology.* Oxford: Oxford University Press.

Stalnaker, Robert C. (1972/1981). Letter to David Lewis. In: *Ifs: Conditionals, Belief, Decision, Chance, and Time.* Ed. by W. Harper, Robert C. Stalnaker, and G. Pearce. Dordrecht: Reidel, pp. 151–2.

Stalnaker, Robert C. (2008). *Our Knowledge of the Internal World.* Oxford: Oxford University Press.

Stalnaker, Robert C. (2011). Responses to Stoljar, Weatherson and Boghossian. *Philosophical Studies* 155, pp. 467–79.

Stefánsson, H. Orri (2017). What Is "Real" in Probabilism? *Australasian Journal of Philosophy* 95, pp. 573–87.

Stefánsson, H. Orri (2018). On the Ratio Challenge for Comparativism. *Australasian Journal of Philosophy* 96, pp. 380–90.

Stephenson, Todd A. (2000). *An Introduction to Bayesian Network Theory and Usage.* Tech. rep. 03. IDIAP.

Sturgeon, Scott (2010). Confidence and Coarse-grained Attitudes. In: *Oxford Studies in Epistemology.* Ed. by Tamar Szabó Gendler and John Hawthorne. Vol. 3. Oxford: Oxford University Press, pp. 126–49.

Suppes, Patrick (1955). The Role of Subjective Probability and Utility in Decision-making. *Proceedings of the Third Berkeley Symposium on Mathematical Statistics and Probability,* pp. 61–73.

Suppes, Patrick (1969). *Studies in the Methodology and Foundations of Science.* Berlin: Springer.

Suppes, Patrick (1974). *Probabilistic Metaphysics.* Uppsala: University of Uppsala Press.

Suppes, Patrick (2002). *Representation and Invariance of Scientific Structures.* Stanford, CA: CSLI Publications.

Tal, Eyal and Juan Comesaña (2017). Is Evidence of Evidence Evidence? *Noûs* 51, pp. 95–112.

Talbott, William J. (1991). Two Principles of Bayesian Epistemology. *Philosophical Studies* 62, pp. 135–50.

Talbott, William J. (2005). Review of "Putting Logic in its Place: Formal Constraints on Rational Belief". *Notre Dame Philosophical Reviews.* URL: http://ndpr.nd.edu/.

Talbott, William J. (2016). Bayesian Epistemology. In: *The Stanford Encyclopedia of Philosophy.* Ed. by Edward N. Zalta. Winter 2016. Metaphysics Research Lab, Stanford University. URL: https://plato.stanford.edu/archives/win2016/entries/epistemology-bayesian/.

Teller, Paul (1973). Conditionalization and Observation. *Synthese* 26, pp. 218–58.

Tennant, Neil (2017). Logicism and Neologicism. In: *The Stanford Encyclopedia of Philosophy.* Ed. by Edward N. Zalta. Winter 2017. Metaphysics Research Lab, Stanford University. URL: https://plato.stanford.edu/archives/win2017/entries/logicism/.

Tentori, Katya, Vincenzo Crupi, and Selena Russo (2013). On the Determinants of the Conjunction Fallacy: Probability versus Inductive Confirmation. *Journal of Experimental Psychology: General* 142, pp. 235–55.

Titelbaum, Michael G. (2008). The Relevance of Self-locating Beliefs. *Philosophical Review* 117, pp. 555–605.

Titelbaum, Michael G. (2010). Not Enough There There: Evidence, Reasons, and Language Independence. *Philosophical Perspectives* 24, pp. 477–528.

Titelbaum, Michael G. (2013a). *Quitting Certainties: A Bayesian Framework Modeling Degrees of Belief.* Oxford: Oxford University Press.

Titelbaum, Michael G. (2013b). Ten Reasons to Care about the Sleeping Beauty Problem. *Philosophy Compass* 8, pp. 1003–17.

Titelbaum, Michael G. (2015a). Continuing On. *Canadian Journal of Philosophy* 45, pp. 670–91.

Titelbaum, Michael G. (2015b). Rationality's Fixed Point (Or: In Defense of Right Reason). In: *Oxford Studies in Epistemology.* Ed. by Tamar Szabó Gendler and John Hawthorne. Vol. 5. Oxford University Press, pp. 253–94.

Titelbaum, Michael G. (2016). Self-locating Credences. In: *The Oxford Handbook of Probability and Philosophy*. Ed. by Alan Hájek and Christopher R. Hitchcock. Oxford: Oxford University Press, pp. 666–80.

Titelbaum, Michael G. (2019). Precise Credences. In: *The Open Handbook of Formal Epistemology*. Ed. by Richard Pettigrew and Jonathan Weisberg. The PhilPapers Foundation, pp. 1–56.

Tversky, Amos and Daniel Kahneman (1974). Judgment under Uncertainty: Heuristics and Biases. *Science* 185, pp. 1124–31.

Tversky, Amos and Daniel Kahneman (1983). Extensional Versus Intuitive Reasoning: The Conjunction Fallacy in Probability Judgment. *Psychological Review* 90, pp. 293–315.

Tversky, Amos and Daniel Kahneman (1992). Advances in Prospect Theory: Cumulative Representation of Uncertainty. *Journal of Risk and Uncertainty* 5, pp. 297–323.

Tversky, Amos and Derek J. Koehler (1994). Support Theory: A Nonextensional Representation of Subjective Probability. *Psychological Review* 101, pp. 547–67.

van Enk, Steven J. (2015). Betting, Risk, and the Law of Likelihood. *Ergo* 2, pp. 105–21.

van Fraassen, Bas C. (1980). Rational Belief and Probability Kinematics. *Philosophy of Science* 47, pp. 165–87.

van Fraassen, Bas C. (1981). A Problem for Relative Information Minimizers. *British Journal for the Philosophy of Science* 32, pp. 375–79.

van Fraassen, Bas C. (1982). Rational Belief and the Common Cause Principle. In: *What? Where? When? Why?* Ed. by Robert McLaughlin. Dordrecht: Reidel, pp. 193–209.

van Fraassen, Bas C. (1983). Calibration: A Frequency Justification for Personal Probability. In: *Physics Philosophy and Psychoanalysis*. Ed. by R. Cohen and L. Laudan. Dordrecht: Reidel, pp. 295–319.

van Fraassen, Bas C. (1984). Belief and the Will. *The Journal of Philosophy* 81, pp. 235–56.

van Fraassen, Bas C. (1989). *Laws and Symmetry*. Oxford: Clarendon Press.

van Fraassen, Bas C. (1995). Belief and the Problem of Ulysses and the Sirens. *Philosophical Studies* 77, pp. 7–37.

van Fraassen, Bas C. (1999). Conditionalization: A New Argument For. *Topoi* 18, pp. 93–6.

van Fraassen, Bas C. (2005). Conditionalizing on Violated Bell's Inequalities. *Analysis* 65, pp. 27–32.

Van Horn, Kevin S. (2003). Constructing a Logic of Plausible Inference: A Guide to Cox's Theorem. *International Journal of Approximate Reasoning* 34, pp. 3–24.

Venn, John (1866). *The Logic of Chance*. London, Cambridge: Macmillan.

Vineberg, Susan (2011). Dutch Book Arguments. In: *The Stanford Encyclopedia of Philosophy*. Ed. by Edward N. Zalta. Summer 2011. URL: https://plato.stanford.edu/archives/su2011/entries/dutch-book/.

von Mises, Richard (1928/1957). *Probability, Statistics and Truth*. (English edition of the original German *Wahrscheinlichkeit, Statistik und Wahrheit*.) New York: Dover.

von Neumann, J. and O. Morgenstern (1947). *Theory of Games and Economic Behavior*. 2nd edition. Princeton, NJ: Princeton University Press.

Vranas, Peter B.M. (2004). Hempel's Raven Paradox: A Lacuna in the Standard Bayesian Solution. *British Journal for the Philosophy of Science* 55, pp. 545–60.

Wainer, Howard (2011). *Uneducated Guesses: Using Evidence to Uncover Misguided Education Policies*. Princeton, NJ: Princeton University Press.

Walley, Peter (1991). *Statistical Reasoning with Imprecise Probabilities*. London: Chapman and Hall.

Wasserstein, Ronald L. and Nicole A. Lazar (2016). The ASA's Statement on p-values: Context, Process, and Purpose. *The American Statistician*. DOI: 10.1080/00031305.2016.1154108.

Weatherson, Brian (2002). Keynes, Uncertainty and Interest Rates. *Cambridge Journal of Economics* 26, pp. 47–62.

Weatherson, Brian (2011). Stalnaker on Sleeping Beauty. *Philosophical Studies* 155, pp. 445–56.

Weatherson, Brian (2015). For Bayesians, Rational Modesty Requires Imprecision. *Ergo* 2, pp. 529–45.

Weatherson, Brian and Andy Egan (2011). Epistemic Modals and Epistemic Modality. In: *Epistemic Modality*. Ed. by Andy Egan and Brian Weatherson. Oxford: Oxford University Press, pp. 1–18.

Weintraub, Ruth (2001). The Lottery: A Paradox Regained and Resolved. *Synthese* 129, pp. 439–49.

Weirich, Paul (2012). Causal Decision Theory. In: *The Stanford Encyclopedia of Philosophy*. Ed. by Edward N. Zalta. Winter 2012. URL: https://plato.stanford.edu/archives/win2012/entries/decision-causal/.

Weisberg, Jonathan (2007). Conditionalization, Reflection, and Self-Knowledge. *Philosophical Studies* 135, pp. 179–97.

Weisberg, Jonathan (2009). Varieties of Bayesianism. In: *Handbook of the History of Logic*. Ed. by Dov. M Gabbya, Stephan Hartmann, and John Woods. Vol. 10: Inductive Logic. Oxford: Elsevier.

White, Roger (2005). Epistemic Permissiveness. *Philosophical Perspectives* 19, pp. 445–59.

White, Roger (2006). Problems for Dogmatism. *Philosophical Studies* 131, pp. 525–57.

White, Roger (2010). Evidential Symmetry and Mushy Credence. In: *Oxford Studies in Epistemology*. Ed. by Tamar Szabó Gendler and John Hawthorne. Vol. 3. Oxford: Oxford University Press, pp. 161–86.

Williams, J. Robert G. (ms). A Non-Pragmatic Dominance Argument for Conditionalization. Unpublished manuscript.

Williams, J. Robert G. (2016). Probability and Non-Classical Logic. In: *Oxford Handbook of Probability and Philosophy*. Ed. by Alan Hájek and Christopher R. Hitchcock. Oxford: Oxford University Press.

Williamson, Timothy (2000). *Knowledge and its Limits*. Oxford: Oxford University Press.

Williamson, Timothy (2007). How Probable Is an Infinite Sequence of Heads? *Analysis* 67, pp. 173–80.

Wittgenstein, Ludwig (1921/1961). *Tractatus Logico-Philosophicus*. Translated by D.F. Pears and B.F. McGuinness. London: Routledge.

Wong, S.M. et al. (1991). Axiomatization of Qualitative Belief Structure. *IEEE Transactions on Systems, Man and Cybernetics* 21, pp. 726–34.

Woolston, Chris (2015). Psychology Journal Bans *P* Values. *Nature* 519.9. URL: https://doi.org/10.1038/519009f.

Wroński, Leszek and Godziszewski, Michał Tomasz (2017). *The Stubborn Non-probabilist—'Negation Incoherence' and a New Way to Block the Dutch Book Argument*. International Workshop on Logic, Rationality and Interaction. Berlin, Heidelberg: Springer, pp. 256–67.

Yalcin, Seth (2012). A Counterexample to Modus Tollens. *Journal of Philosophical Logic* 41, pp. 1001–24.

Yule, G.U. (1903). Notes on the Theory of Association of Attributes in Statistics. *Biometrika* 2, pp. 121–34.

Zynda, Lyle (1995). Old Evidence and New Theories. *Philosophical Studies* 77, pp. 67–95.

Zynda, Lyle (2000). Representation Theorems and Realism about Degrees of Belief. *Philosophy of Science* 67, pp. 45–69.

Index of Names in Volumes 1 & 2